Cerebral Blood Flow

Mathematical Models,
Instrumentation, and
Imaging Techniques

NATO ASI Series

Advanced Science Institutes Series

A series presenting the results of activities sponsored by the NATO Science Committee, which aims at the dissemination of advanced scientific and technological knowledge, with a view to strengthening links between scientific communities.

The series is published by an international board of publishers in conjunction with the NATO Scientific Affairs Division

A	**Life Sciences**	Plenum Publishing Corporation
B	**Physics**	New York and London
C	**Mathematical and Physical Sciences**	Kluwer Academic Publishers Dordrecht, Boston, and London
D	**Behavioral and Social Sciences**	
E	**Applied Sciences**	
F	**Computer and Systems Sciences**	Springer-Verlag
G	**Ecological Sciences**	Berlin, Heidelberg, New York, London,
H	**Cell Biology**	Paris, and Tokyo

Recent Volumes in this Series

Series A: Life Sciences

Cerebral Blood Flow

Mathematical Models, Instrumentation, and Imaging Techniques

Edited by

Aldo Rescigno

University of Ancona
Ancona, Italy

and

Andrea Boicelli

San Raffaele Institute
Milan, Italy

Plenum Press
New York and London
Published in cooperation with NATO Scientific Affairs Division

Proceedings of a NATO Advanced Study Institute on
Ceerebral Blood Flow: Mathematical Models, Instrumentation,
and Imaging Techniques for the Study of CBF,
held June 2–13, 1986,
in L'Aquila, Italy

Library of Congress Cataloging in Publication Data

NATO Advanced Study Institute on Cerebral Blood Flow: Mathematical Models,
    Instrumentation, and Imaging Techniques for the Study of CBF (1986: L'Aquila,
    Italy)
    Cerebral blood flow: mathematical models, instrumentation, and imaging tech-
niques / edited by Aldo Rescigno and Andrea Boicelli.
        p.   cm.—(NATO ASI series. Series A, Life sciences; v. 153)
    "Proceedings of a NATO Advanced Study Institute on Cerebral Blood Flow,
Mathematical Models, Instrumentation, and Imaging Techniques for the Study of
CBF, held June 2–13, 1986, in L'Aquila, Italy"—T.p. verso.
    "Published in cooperation with NATO Scientific Affairs Division."
    Includes bibliographies and index.
    ISBN 978-1-4684-5567-0   ISBN 978-1-4684-5565-6 (eBook)
    DOI 10.1007/978-1-4684-5565-6

    1. Cerebral circulation—Mathematical models—Congresses. 2. Cerebral cir-
culation—Measurement—Mathematics—Congresses. 3. Cerebral circulation—
Imaging—Congresses. I. Rescigno, Aldo. II. Boicelli, Andrea. III. North Atlantic
Treaty Organization. Scientific Affairs Division. IV. Title. V. Series. [DNLM: 1.
Autoradiography—congresses. 2. Cerebrovascular Circulation—congresses. 3.
Mathematics—congresses. 4. Models, Cardiovascular—congresses. 5. Nuclear
Magnetic Resonance—congresses. 6. Tomography, Emission Computed—con-
gresses. WL 302 N279c 1986]
QP108.5.C4N37   1986
612'.825—dc19
DNLM/DLC   88-25519
for Library of Congress   CIP

© 1988 Plenum Press, New York
Softcover reprint of the hardcover 1st edition 1988

A Division of Plenum Publishing Corporation
233 Spring Street, New York, N.Y. 10013

PREFACE

The NATO Advanced Study Institute on "Cerebral Blood Flow: Mathematical Models, Instrumentation, and Imaging Techniques" was held in L'Aquila, Italy, June 2-13, 1986. Contributions to this program were received from the University of L'Aquila, Consiglio Nazionale delle Ricerche, Siemens Elettra S.p.A., and Bracco S.p.A.

Recent studies of the cerebral blood circulation have lagged behind analysis of other parameters such as glucose utilization, transmitter distribution, and precursors. This Advanced Study Institute tried to fill this gap by analyzing in detail different physical techniques such as Autoradiography (including Double-Tracer Autoradiography and highly specific tracers as Iodoantipyrine, Microspheres), Single Photon Emission Computed Tomography, Nuclear Magnetic Resonance. Each method was analyzed in regards to its precision, resolution, response time.

A considerable part of this Institute was devoted to the mathematics of CBF measurement, in its two aspects, i.e. the modeling of the underlying kinetic system and the statistical analysis of the data. The modeling methods proposed included the development of a differential algebra whereby the differential and integral equations involved could be solved by simple algebraic methods, including graph-theoretical ones; the statistical methods proposed included the illustration of different parametrizations of possible use in the interpretation of experimental results.

Aldo Rescigno
Andrea Boicelli

CONTENTS

NUCLEAR MAGNETIC RESONANCE

MATHEMATICAL MODELS

STOCHASTIC MODELS AND LINEAR TRACER KINETICS

Aldo Rescigno

Section of Neurosurgery, Yale University School of
Medicine, New Haven, CT, U.S.A.

Institute of Experimental and Clinical Medicine
University of Ancona, Ancona, Italy

1. MEANING OF A MODEL

The purpose of a model is to verify whether some hypotheses made on a real system are valid (Rescigno and Beck,
1987), and subordinately to determine the values of some of
the parameters that represent specific properties of the
system under study (Zierler, 1981). For instance the model
commonly used when dilution phenomena are expected, is a set
of ordinary differential equations with constant coefficients,
while when diffusion phenomena are the case, partial differential equations are to be used, finite difference equations for
delay phenomena, etc.

In Nuclear Medicine and in Pharmacokinetics the compartment model is often used (Rescigno and Segre, 1964); it
consists of a set of linear differential equations of order
one with constant coefficients, and it implies the hypotheses
that the system described contains a finite number of components, and that each component is homogeneous. These hypotheses exclude the presence of diffusion and of age-dependent
processes, or in general of transport of a non-Markovian
nature. The parameters computed from the experimental data
using this model are the transfer rates between compartments
and the turnover rates of compartments; they have a defined
physical meaning only if the model is appropriate (Beck and
Rescigno, 1970).

With n compartments the number of parameters necessary to
characterize the equations is n^2, therefore, ignoring all
experimental errors, only n^2 measurements are needed completely to describe the system under observation. With the
presence of experimental errors this number should be
increased to improve the reliability of the computed parameters. The cost of the measurements may be a factor of
importance. The strategy of the measurements may also be
important, i.e. measurements made at certain times may carry
more information than if made at other times. Some of the
parameters may be more "important" than others, i.e. their

3

information may be more "valuable". Above all, other parametrizations besides the classical one of the transfer rates
and turnover rates, may be more valuable in the sense that
they may be more directly connected to some externally measurable or physically well-defined quantities, or they may be
more sensitive to specific treatments (Matis, Wehrly and
Gerald, 1983).

Of course if the hypotheses incorporated into the model
are not appropriate, the computed parameters have no gnoseo-
logical value. If we make some weaker hypotheses there is a
lesser chance of rejection of the model, but the real system
will not be described as fully as with a stronger model, i.e.
we will be able to determine fewer parameters; those parameters though are more reliable and may even be more
"valuable" in the sense hinted above.

The ideal situation would be to define parameters that
depend on the smallest possible number of hypotheses, but that
still have a physically meaningful interpretation.

One such example is given by Rescigno and Gurpide (1973).
While that approach is not completely "model-free", it is certainly very "robust", i.e. it allows to compute parameters
that are not very dependent of the assumptions of a specific
model. All this considered, we think that the experimental
data available to the investigator should be examined in terms
of a model implying a minimum number of assumptions and giving
the best physical interpretation to the parameters involved.

2. LINEAR STOCHASTIC KINETICS

Consider a particle in a living system and suppose that
that particle can be recognized in two different states of the
system, where by state we mean a particular location or a particular chemical form, or both. If one state is the precursor
of the other (not necessarily the immediate precursor), then
we can study the relationship among event A (presence of the
particle in the precursor state), event B (transition from
precursor to successor state), and event C (presence of the
particle in the successor state).

For any t and τ such that $0 < \tau < t$, call $f(\tau)$ the
probability of A at time τ and $h(t)$ the probability of C at
time t; the range of both functions is 0, 1. Suppose now that
B depends only on the interval of time separating A and C, so
that we can call now $g(t-\tau)d\tau$ the conditional probability that
the particle is in C at time t if it left A in the interval
from τ to $\tau+d\tau$.

The product

$$f(\tau).g(t-\tau)d\tau$$

therefore is the absolute probability of A at time τ and of C
at time t.

By integration of the above product we must obtain the
probability of C at time t irrespective of A, i.e.

(1) $$\int_{o}^{t} f(\tau)g(t-\tau)d\tau = h(t).$$

This is the well known convolution integral representing
the relationship among the variables of a linear, invariant
system, without invoking the properties of homogeneous, well-
mixed compartments.

By linear system is meant that two different solutions of equation (1) can be added to give a new solution; in fact if a solution of equation (1) is given by $f_1(t)$, $h_1(t)$, and another one by $f_2(t)$, $h_2(t)$, then a third solution of equation (1) is $f_1(t)+f_2(t)$, $h_1(t)+h_2(t)$, as can be easily verified.

By invariant system I mean that a solution does not change if the time origin is changed. In fact suppose that $f(t)$, $h(t)$ is a solution of equation (1), and consider the new function

$$f_1(t) = 0 \qquad \text{for } 0 \leq t < t_o$$
$$= f(t-t_o) \qquad \text{for } t \geq t_o.$$

For this function

$$\int_0^t f_1(\tau) g(t-\tau) d\tau = \int_{t_o}^t f(\tau-t_o) g(t-\tau) d\tau$$
$$= \int_0^{t-t_o} f(\tau) g(t-t_o-\tau) d\tau;$$

using now equation (1),

$$\int_0^t f_1(\tau) g(t-\tau) d\tau = 0 \qquad \text{for } 0 \leq t < t_o$$
$$= h(t-t_o) \qquad \text{for } t \leq t_o,$$

i.e. $f(t)$ and $h(t)$ are shifted along the time axis by the same quantity.

If we think of an experiment where a very large number of identical particles is used, then the number of particles in A and in C are good estimators of functions $f(t)$ and $h(t)$ respectively. Function $g(t)$ represents the probability that a particle that left A at time zero will still be in C at time t; therefore in a hypothetical experiment where all identical particles left the precursor at time zero, the number of particles found in the successor will be given by $g(t)$.

In the following pages I shall try to show a number of properties of equation (1) and how to use those properties to interpret the results of some experiments.

3. DEFINITION OF MOMENTS

Given a generic function $f(t)$ defined for all values of t from 0 to $+\infty$, define the <u>moments</u>,

(2) $F_i = \int_0^\infty t^i/i! \, f(t) dt, \quad i=0,1,2,\dots$

and the <u>relative moments</u>,

(3) $f_i = \int_0^\infty t^i/i! \, f(t) dt/F_o, \quad i=1,2,3,\dots$

provided that the integrals above converge.

I shall show in section 6 what can be done when one of those integrals does not converge; for the time being we suppose that all those moments do exist.

Definition (3) applies only to values of <u>i</u> larger than zero; for convenience we complete that definition with

$$f_o = 1.$$

Frequently the moments of a function are defined without the factor $1/i!$ shown in (2); I prefer to use this factor because the moment generating functions defined in section 8 actually generate the moments as in (2) and (3) rather than the moments defined without the factor $1/i!$, and because the expressions we shall find later on will be considerably simpler. Observe also that these moments are just integral transforms of function $f(t)$ with kernel $t^i/i!$, very similar to the Mellin transform, whose kernel is t^{i-1}; even though in the Mellin transform i is a complex variable while in these moments it is a non-negative integer, they have many interesting properties in common. For more details see for instance Bateman (1954).

4. PROPERTIES OF THE CONVOLUTION

Multiply both sides of equation (1) by $t^i/i!$ and integrate from 0 to $+\infty$,

$$\int_0^\infty t^i/i! \int_0^t f(\tau)g(t-\tau)\,d\tau\,dt = \int_0^\infty t^i/i!\,h(t)dt,$$

where i is any non-negative integer; change the order of integration,

$$\int_0^\infty f(\tau)\int_\tau^\infty t^i/i!\,g(t-\tau)dt.d\tau = h_i;$$

change the variable of the inner integral,

$$\int_0^\infty f(\tau)\int_0^\infty (t+\tau)^i/i!\,g(t)dt.d\tau = h_i;$$

after expanding the binomial we obtain finally,

$$(4) \qquad \sum_{j=0}^{i} F_{i-j}.G_j = H_i; \quad i=0,1,2,\ldots$$

in particular,

$$F_0G_0 = H_0$$

$$F_1G_0 + F_0G_1 = H_1$$

$$F_2G_0 + F_1G_1 + F_0G_2 = H_2$$

If we divide both sides of equation (4) by F_0G_0 we get

$$(5) \qquad \sum_{j=0}^{i} f_{i-j}g_j = h_i. \quad i=1,2,3,\ldots$$

The number of particles in a given state is in general very large; if it can be observed as a function of time it represents a very good approximation of the probability density function defined in section 2. This means that if the functions $f(t)$, $h(t)$ corresponding to two given states can be measured, then their moments can be computed and the moments of the unknown function $g(t)$ calculated using equations (4) or (5).

These last moments can be given a clear physical meaning; for instance G_0 is the fraction of particles leaving the first state that actually reach the second state (a quantity analogous to the Bioavailability as defined in Pharmacokinetics), g_1 is the expected interval of time for a particle to move from the first to the second state, $g_2-f_1/2$ is the variance of this time divided by two, etc. (Rescigno and Michels, 1973).

5. MOMENTS OF A COMPARTMENT

As an example we can evaluate the moments of a specific system. Take a single compartment, i.e. a well mixed pool of homogeneous particles, all with the same probability m.dt of leaving it in the interval from t to t+dt, where m is a constant; in other words the probability of leaving the compartment does not depend on the absolute time or on the time when a particle entered it. If x(t) is the characteristic function of that compartment, i.e. the probability that a particle is in the compartment at time t, then

$$x(t+dt) = x(t).(1-m.dt)$$

is the probability that a given particle present in that compartment at time t is still there at time t+dt; rearranging this equation,

$$dx/dt = -m.x(t),$$

and integrating,

$$x(t) = x(0).e^{-mt},$$

where x(0), the constant of integration, is the probability that a given particle is present in the compartment at the initial time. Using definitions (2) and (3) we get for a single compartment,

$$X_i = X(0)/m^{i+1}, \quad i=0,1,2,\ldots$$

$$x_i = 1/m^i. \quad i=1,2,3,\ldots$$

6. NON-CONVERGING MOMENTS

In section 3 the moments and relative moments were defined subject to the condition that the integral

(6) $$\int_0^\infty t^i/i! \; f(t)dt$$

converges; this requires that function f(t) decreases fast enough when t increases. It is well known that most functions used to describe biological systems are of exponential order, i.e. they have the property that a constant c>0 exists such that the product $e^{-ct}f(t)$ is bounded for all values of t larger than some finite value; for a function of exponential order the integral (6) always converges, no matter how large i is. An exception is given by the functions describing a closed system, i.e. a system from where not all particles are eventually lost; in this case

$$\underset{t \to \infty}{\text{limit}} \; f(t) \neq 0,$$

and the integral (6) does not converge for any non-negative value of i. This is an obvious consequence of the fact that the average time spent by a particle in such a system is infinite. We shall not consider this case, but the more interesting case when function g(t), as defined in section 2, is of exponential order, while function f(t) is bounded but does not approach zero as $\underline{t}$ approaches infinity. This corresponds to feeding a "regular" system with an endless stream of particles. Equation (1) shows that if function $f(t)$ does not approach zero when $\underline{t}$ goes to infinity, neither function $h(t)$ will; therefore both $f(t)$ and $h(t)$ have undefined moments, while the moments of $g(t)$ are defined, but unknown.

From the hypothesis that $f(t)$ is bounded, it follows that $e^{-ct}f(t)$ is of exponential order for any $c>0$; equation (1) can be rewritten

$$\int_0^t e^{-c\tau}f(\tau) \cdot e^{-c(t-\tau)}g(t-\tau)d\tau = e^{-ct}h(t),$$

showing that multiplying both $f(t)$ and $h(t)$ by e^{-ct} is equivalent to multiplying $g(t)$ by the same exponential. The new functions $e^{-ct}f(t)$ and $e^{-ct}h(t)$ have finite moments; they can be used to compute the moments of the modified function $e^{-ct}g(t)$; calling $G_i{}^*$ the moments of this last function, then

$$G_i{}^* = \int_0^\infty t^i/i! \cdot e^{-ct}g(t)dt$$

$$= \int_0^\infty \sum_{j=0}^\infty (-ct)^j/j! \cdot t^i/i! \cdot g(t)dt$$

$$G_i{}^* = \sum_{j=0}^\infty (-1)^j c^j (i+j)!/i!j! \cdot G_{i+j},$$

$$i=0,1,2,\ldots$$

and by inversion

(7) $$G_i = \sum_{j=0}^\infty c^j (i+j)!/i!j! \cdot G_{i+j}{}^*, \quad i=0,1,2,\ldots$$

In the frequent case when the functions $f(t)$ and $h(t)$ are evaluated by measuring the activity of a radioactive tracer in two states of the system, then the "true" probability density functions are multiplied by the exponential function e^{-ct}, where $\underline{c}$ is the decay constant of the nuclide used. If the "measured" functions $f(t)$ and $h(t)$ are not corrected for the radioactive decay, then the moments $G_i{}^*$ should be corrected using equations (7); these last corrections are frequently easy because in general the infinite series in (7) converges very rapidly. More important is the fact that the direct correction of $f(t)$ and $h(t)$ for the disintegration rate of the nuclide involves a non-negligible error when that rate is very high (Rescigno and Lambrecht, 1985), as shown in section 7.

7. RECORDING THE MOMENTS

If the functions $f(t)$ and $h(t)$ are known, their moments can be computed using the definitions given in section 3. However, the integrations required for the computation of the moments of any function introduce some errors that are added to the errors intrinsic to the measurement of the original function.

Furthermore, any particle counter has a finite integration time, therefore it does not measure the exact number of particles present at time t, but the average number of particles present in a certain interval of time; this implies a non negligible error when the rate of change of the number of particles to be counted is large compared to the integration time of the counter (Duncan et al., 1983). To have an idea of the errors involved, consider the simple function

$$f(t) = e^{-mt} \quad ;$$

its average value over the interval of time t_1, t_2 is

$$\int_{t_1}^{t_2} e^{-mt}dt/(t_2-t_1),$$

while its value at the center of the interval is

$$\exp[-m(t2-tl)/2];$$

a simple computation shows that the error committed when taking the former expression for the latter is larger than 10% if $t_2-t_1= 1.6/m$, and is larger than 100% if $t_2-t_1= 4.4/m$.
With tracers having a physical half-life of less than a minute, the acquisition time of a typical tomograph will cause an excessive error in the determination of the true values of the activity.

Both errors can be eliminated if the transformation function ---> moment is bypassed and the moments are recorded directly at the source, i.e. if the detector itself acts as an encoder.

Call $X(t)$ the number of radioactive particles present in a voxel; the probability that an event will be recorded by an appropriate detector in the interval of time from t to t+dt is $kX(t)$, where k is a constant depending upon the efficiency of the detector and upon the disintegration rate of the radiotracer used; the probability of a double event in the infinitesimal time interval dt is an infinitesimal of order higher than dt and can be neglected. The constant k can be determined by measuring for a sufficiently long interval of time a calibrated source.

I shall ignore here the effect of the finite resolving time of the detector, negligible if the counting rate is not too fast.

Define the random variable $N(t)$ equal to the number of events recorded in the interval of time from 0 to t, with distribution

$$p(r,t) = \text{Prob}\{N(t)=r\}.$$

Note that if $X(t)$ is the number of radioactive particles present in a voxel, then $X(t)$ itself is a random variable, because that number depends not only on the macro-processes taking place in that voxel, but also on the number of disintegrations having taken place in the preceding time interval 0, t.

The counter reads zero at the beginning of an experiment; then, each time an event is recorded, the value of the random variable $N(t)$ increases by one unit; it does not change if no events are recorded; therefore

$$p(0,0) = 1$$

$$p(0,t+dt) = [1-kX(t)dt].p(0,t)$$

$$p(r,t+dt) = kX(t)dt.p(r-1,t) + [1-kX(t)dt].p(r,t), \quad r>0$$

Divide by dt,

$$\partial p(0,t)/\partial t = -kX(t).p(0,t)$$

$$\partial p(r,t)/\partial t = -kX(t).[p(r,t)-p(r-1,t)], \quad r>0$$

and integrate,

$$p(r,t) = 1/r!.[\int_0^t kX(\tau)d\tau]^r.\exp[-\int_0^t kX(\tau)d\tau]; \quad r=0,1,2,\ldots$$

this **expression shows** that $N(t)$ is a Poisson random variable, with **intensity** $\varphi(t)$ given by the integral

$$\varphi(t) = \int_0^t kX(\tau)d\tau;$$

its **expected value**, variance, and third central moment are

$$E[N(t)] = \text{Var}[N(t)] = M_3[N(t)] = \varphi(t).$$

8. RECORDING HIGHER MOMENTS

Define now the random variable $N_i(t)$ equal to the sum of the i-th **power** of the times of recording an event in the interval of **time** from 0 to t, divided by i!, where $i=0,1,2,3,\ldots$; in other **words**, if the first, second, ..., n-th events are recorded at times t_1, t_2, ..., t_n respectively, with

$$0 < t_1 < t_2 < \ldots < t_n < t,$$

then

$$N_i(t) = (t_1{}^i+t_2{}^i+\ldots+t_n{}^i)/i!. \quad i=0,1,2,\ldots$$

in particular

$$N_0(t) = n$$

is a **random** variable identical to $N(t)$ as defined in the previous section.

Each time an event is detected, its instant of occurrence is **read** on a clock, this value is raised to the power i, divided by i!, and this number added to a corresponding register. When **time** t is reached, the different registers show a particular **realization** of the random variables $N_0(t)$, $N_1(t)$, $N_2(t)$, ...

If **in** the interval of time from t to t+dt an event is recorded, **the** value of the random variable $N_i(t)$ increases by an **amount** $t^i/i!$; the random variable does not change when no pulses **are** recorded.

For **any** positive integer i,

$$p_i(r,t+dt)dr = [1-kX(t)dt].p_i(r,t)dr, \quad r<t^i/i!$$

$$= kX(t)dt.p_i(r-t^i/i!,t)dr+[1-kX(t)dt].p_i(r,t)dr.$$
$$r\geqq t^i/i!$$

Divide by dt,

(8) $\quad \partial p_i/\partial t.dr = -kX(t).p_i(r,t)dr, \qquad r<t^i/i!$

(9) $\quad \partial p_i/\partial t.dr = -kX(t).[p_i(r,t)dr - p_i(r-t^i/i!,t)dr].$
$$r\geqq t^i/i!$$

Define the moment generating function

$$L_i(s,t) = \int_0^\infty e^{rs} p_i(r,t)\,dr;$$

multiply both equations (8) and (9) by e^{rs}, then integrate (8) from 0 to $t^i/i!$ and (9) from $t^i/i!$ to $+\infty$; add the two results together,

$$\partial L_i/\partial t = -kX(t)[1-\exp(st^i/i!)]L_i(s,t);$$

at time 0 all registers read zero, therefore

$$L_i(s,0) = 1;$$

by integration we get

$$L_i(s,t) = \exp\left\{-\int_0^t (1-\exp[s\tau^i/i!])kX(\tau)\,d\tau\right\}. \quad i=1,2,\ldots$$

Take the derivatives with respect to s of this last equation,

$$\partial L_i/\partial s = \int_0^t \tau^i/i!.\exp(s\tau^i/i!).kX(\tau)\,d\tau.L_i$$

$$\partial^2 L/\partial s^2 = \int_0^t (\tau^i/i!)^2.\exp(s\tau^i/i!).kX(\tau)\,d\tau.L_i +$$
$$+ \int_0^t \tau^i/i!.\exp(s\tau^i/i!).kX(\tau)\,d\tau.\partial L_i/\partial s$$

$$\partial^3 L/\partial s^3 = \int_0^t (\tau^i/i!)^3.\exp(s\tau^i/i!).kX(\tau)\,d\tau.L_i +$$
$$+ 2\int_0^t (\tau^i/i!)^2.\exp(s\tau^i/i!).kX(\tau)\,d\tau.\partial L_i/\partial s +$$
$$+ \int_0^t \tau^i/i!.\exp(s\tau^i/i!).kX(\tau)\,d\tau.\partial^2 L_i/\partial s^2;$$

make s=0,

$$E[N_i(t)] = \int_0^t \tau^i/i!.kX(\tau)\,d\tau$$

$$E[N_i(t)^2] = \int_0^t (\tau^i/i!)^2.kX(\tau)\,d\tau + [\int_0^t \tau^i/i!.kX(\tau)\,d\tau]^2$$

$$E[N_i(t)^3] = \int_0^t (\tau^i/i!)^3.kX(\tau)\,d\tau +$$
$$+ 3\int_0^t (\tau^i/i!)^2.kX(\tau)\,d\tau.\int_0^t \tau^i/i!.kX(\tau)\,d\tau +$$
$$+ [\int_0^t \tau^i/i!.kX(\tau)\,d\tau]^3$$

and finally

$$\mathrm{Var}[N_i(t)] = \int_0^t (\tau^i/i!)^2.kX(\tau)\,d\tau$$
$$= (2i)!/(i!)^2.E[N_{2i}(t)]$$

$$M_3[N_i(t)] = \int_0^t (\tau^i/i!)^3.kX(\tau)\,d\tau$$
$$= (3i)!/(i!)^3.E[N_{3i}(t)]$$

Equations (8) and (9) were written only for positive integer values of i, but the moments computed here are valid for any non-negative integer i, because for i=0 they coincide with the results of section 7.

In summary, for any particular experimental set-up the value of k can be determined by measuring $N_0(t)$ with a

11

calibrated phantom; then the successive moments of the function $kX(t)$ are accumulated by separate registers; the value read on register i is the best estimator of the i-moment of $kX(t)$, while from the value read on register 2i we estimate its variance, and so forth. While ordinary ratemeters operate by analog transformations, with this method the measured function is digitized, therefore the only errors introduced are due to the measure of t, which can be extremely precise, and to the random fluctuations of the photon production and transmission; these last errors of course are estimated by the successive moments.

9. SCALING THE COUNTING RATE

With high counting rates it may be necessary to scale the events before recording their time of occurrence to cope with the finite resolution time of the registers. Suppose that only every m-th event is recorded; define, for i=1,2,...,

$N_i^{(m)}$ = sum of the i-th power of the time of recording every m-th pulse in the interval 0, t;

$p_i^{(m)}(r,a,t)dr = \text{Prob}\{r<N_i^{(m)}(t)<r+dr . N(t)\equiv a \pmod m\};$
$$0 \leq a < m$$

proceeding as before,

$$p_i^{(m)}(r,0,t+dt)dr = [1-kX(t)dt].p_i^{(m)}(r,0,t)dr, \qquad r<t^i$$

$$= kX(t)dt.p_i^{(m)}(r-t^i/i!,m-1,t)dr +$$

$$+ [1-kX(t)dt].p_i^{(m)}(r,0,t)dr. \quad r\geq t^i$$

$$p_i^{(m)}(r,a,t+dt)dr = kX(t)dt.p_i^{(m)}(r,a-1,t)dr +$$

$$+ [1-kX(t)dt].p_i^{(m)}(r,a,t)dr,$$
$$a=1,2,...,m-1$$

Divide by dt,

$$\partial p_i^{(m)}(r,0,t)/\partial t.dr = -kX(t).p_i^{(m)}(r,0,t)dr, \qquad r<t^i$$

$$= -kX(t)[p_i^{(m)}(r,0,t)dr - p_i^{(m)}(r-t^i/i!,m-1,t)dr],$$
$$r\geq t^i$$

$$\partial p_i^{(m)}(r,a,t)/\partial t.dr =$$

$$= -kX(t)[p_i^{(m)}(r,a,t)dr-p_i^{(m)}(r,a-1,t)dr];$$
$$a=1,2,...,m-1$$

define

$$P_i^{(m)}(r,t)dr = \sum_{a=0}^{m-1} p_i^{(m)}(r,a,t)dr$$

$$= \text{Prob}\{r<N_i^{(m)}(t)<r+dr\}$$

and add the equations above for all values of a,

$$(10) \qquad \partial P_i^{(m)}/\partial t.dr = -kX(t).p_i^{(m)}(r,m-1,t)dr,$$
$$r<t^i$$

12

(11) $\partial P_i^{(m)}/\partial t.dr = -kX(t).[p_i^{(m)}(r,m-1,t)dr -$

$$- p_i^{(m)}(r-t^i/i!,m-1,t)dr]. \quad r \geq t^i$$

Define the moment generating functions

$$L_i^{(m)}(s,t) = \int_o^\infty e^{rs}p_i^{(m)}(r,t)dr,$$

$$l_i^{(m)}(s,t) = \int_o^\infty e^{rs}p_i^{(m)}(r,m-1,t)dr;$$

multiply equations (10) and (11) by e^{rs}, then integrate the first from 0 to $t^i/i!$ and the second from $t^i/i!$ to $+\infty$; add the two results together,

$$\partial L_i^{(m)}/\partial t = -kX(t)[1-\exp(st^i/i!)]l_i^{(m)}(s,t).$$

Take the successive derivative with respect to s,

$$\partial^2 L_i^{(m)}/\partial s\partial t =$$

$$= kX(t)\{t^i/i!.\exp(st^i/i!)l_i^{(m)}(s,t) -$$

$$- [1-\exp(st^i/i!)]\partial l_i^{(m)}/\partial s\}$$

$$\partial^3 L_i^{(m)}/\partial s^2\partial t =$$

$$= kX(t)\{(t^i/i!)^2\exp(st^i/i!)l_i^{(m)}(s,t) +$$

$$+ 2t^i/i!.\exp(st^i/i!)\partial l_i^{(m)}/\partial s -$$

$$- [1-\exp(st^i/i!)]\partial^2 f_i^{(m)}/\partial s^2\};$$

thence, making s = 0,

$$dE[N_i^{(m)}(t)]/dt = kX(t).t^i/i!.l_i^{(m)}(0,t),$$

$$dE[N_i^{(m)}(t)^2]/dt =$$

$$= kX(t).t^i/i![t^i/i!.l_i^{(m)}(0,t) +2\partial l_i^{(m)}(0,t)/\partial s].$$

Observe that

$$l_i^{(m)}(0,t) = \int_o^\infty p_i^{(m)}(r,m-1,t)dr$$

$$= \text{Prob}\{N(t)\equiv m-1(\text{mod } m)\}$$

and

$$\partial l_i^{(m)}(0,t)/\partial s = \int_o^\infty rp_i^{(m)}(r,m-1,t)dr$$

$$= E[N_i^{(m)}(t)|N(t)\equiv m-1(\text{mod } m)].\text{Prob}\{N(t)\equiv m-1(\text{mod } m)\};$$

from section 7 we have

$$\text{Prob}\{N(t)\equiv m-1(\text{mod } m)\} = \sum_{j=1}^\infty p(jm-1,t)$$

$$= \sum_{j=1}^\infty 1/(jm-1)!.[\int_o^t kX(\tau)d\tau]^{jm-1}.\exp[-\int_o^t kX(\tau)d\tau];$$

therefore,

(12) $dE[N_i^{(m)}(t)]/dt =$

$$= kX(t).t^i/i!.\exp[-\int_0^t kX(\tau)d\tau].\sum_{j=1}^{\infty}[\int_0^t kX(\tau)d\tau]^{jm-1}/(jm-1)!$$

$dE[N_i^{(m)}(t)^2]/dt = dE[N_i^{(m)}(t)]/dt.$

$$.\left\{t^i/i!+2E[N_i^{(m)}(t)\mid N(t)\equiv m-1(\text{mod } m)]\right\},$$

thence

$$dVar[N_i^{(m)}(t)]/dt = dE[N_i^{(m)}(t)^2]/dt -$$

$$- 2E[N_i^{(m)}(t)].dE[N_i^{(m)}(t)]/dt$$

(13) $\quad dVar[N_i^{(m)}(t)]/dt = dE[N_i^{(m)}(t)]/dt.$

$$.\left\{t^i/i!+2E[N_i^{(m)}(t)\mid N(t)\equiv m-1(\text{mod } m)]-2E[N_i^{(m)}(t)]\right\}.$$

Some of the properties of equations (12) and (13) are described in a separate paper (Rescigno, 1987). For all practical purposes we can write,

$$E[N_i^{(m)}(t)] = 1/m.E[N_i(t)]$$

and

$$Var[N_i^{(m)}(t)] = 1/m.Var[N_i(t)]$$

provided that m is not very large. This approximation depends of course on m and on the particular function kX(t) one is trying to evaluate.

10. ANALYSIS OF THE MOMENTS

Equation (1) describes the fate of a particle in its transitions from a precursor to a successor. A complex system may involve many such transitions, i.e. it may be formed by a number of sub-systems, each one of them described by an equation of that form.

Consider two systems in series, i.e. two systems such that all particles entering the second of them are originating from the first; the first system is called the unique precursor of the second (Rescigno and Segre, 1961). Call

A(t)dt = probability that a particle present in the first system at time 0 leaves it in the time interval t, t+dt;

B(t)dt = probability that a particle present in the second system at time 0 leaves it in the time interval t, t+dt;

then, ignoring infinitesimals of order higher than one,

$$A(\tau)d\tau.B(t-\tau)dt$$

is the probability that a particle present in the first system at time 0 moves to the second in the interval τ, $\tau+d\tau$ and leaves this one in the interval t, t+dt. The integral of the above product for all values of τ from 0 to t, is equal to the probability that a particle present in the first system at time 0 leaves the second at a time between t and t+dt, no matter when it moved from the first to the second, therefore the convolution

$$\int_o^t A(\tau)B(t-\tau)d\tau = C(t)$$

is the probability density function characteristic of the system formed by two sub-systems in series.

Obviously the properties of the convolution described in section 4 apply in this case also; in particular we can write

$$A_0 B_0 = C_0,$$

$$A_1 B_0 + A_0 B_1 = C_1,$$

$$A_2 B_0 + A_1 B_1 + A_0 B_2 = C_2,$$

$$a_1 + b_1 = c_1,$$

$$a_2 + a_1 b_1 + b_2 = c_2,$$

and so forth.

Consider two systems in parallel, i.e. such that a particle can enter them by two different mechanisms, but that, once in, behaves in a unique way; calling $A(t)$ and $B(t)$ the probability density functions of the two separate sub-systems, the probability density function of the whole system is $A(t)+B(t)$. Obviously in this case the moments of the resulting system are

$$A_i + B_i, \quad i=0,1,2,\dots$$

and the relative moments

$$\frac{A_i + B_i}{A_0 + B_0} = \frac{A_i}{A_0} \cdot \frac{A_0}{A_0+B_0} + \frac{B_i}{B_0} \cdot \frac{B_0}{A_0+B_0}$$

$$= a_i w_a + b_i w_b, \quad i=1,2,\dots$$

where

$$w_a = A_0/(A_0+B_0), \quad w_b = B_0/(A_0+B_0)$$

are called the <u>weights</u> of the two systems (Rescigno, 1973), because they represent the probability that a particle goes through one rather than the other sub-system.

Consider a system followed by a loop, i.e. such that a particle can reenter it after leaving it, and define

$A(t)dt$ = probability that a particle present in the system at time 0 leaves it for the first time in the time interval t, t+dt;

$M(t)dt$ = probability that a particle present in the system at time 0 leaves it in the time interval t, t+dt, irrespective of the number of passages through it;

then the convolution of $M(t)$ and $A(t)$ is the probability density function of a particle going through that system two or more times; by adding to it the probability density function of a particle going through it only once, we should get again the function $M(t)$, i.e.,

$$M(t) = \int_o^t M(\tau)A(t-\tau)d\tau + A(t);$$

using now equation (4),

$$M_0 = M_0 A_0 + A_0$$

$$M_1 = M_0 A_1 + M_1 A_0 + A_1$$

$$M_2 = M_0 A_2 + M_1 A_1 + M_2 A_0 + A_2$$

thence

$$M_0 = A_0/(1-A_0)$$

$$M_1 = A_1(M_0+1)/(1-A_0),$$

$$M_2 = [A_2(M_0+1)+A_1 M_1]/(1-A_0),$$

and for the relative moments,

$$m_1 = a_1(M_0+1)$$

$$m_2 = a_2(M_0+1) + a_1 M_i$$

. . . .

$$m_i = a_i(M_0+1) + \sum_{j=1}^{i-1} a_j M_{i-j} .$$

If enough moments of a system have been measured, all equations of this section can be inverted and the moments of the sub-systems computed; they describe the behaviour of a particle during transfers not directly observable (Rescigno and Duncan, 1985).

11. MULTIDIMENSIONAL MOMENTS

Given the function $f(t;x,y,z)$ defined for $t>0$ and for the values of x, y, z within a given volume D, we can define the (multidimensional) moments

$$F_{ijkl} = \int\int\int\int t^i/i! . x^j/j! . y^k/k! . z^l/l! . f(t;x,y,z) dt . dx . dy . dz,$$

$$i,j,k,l = 0,1,2,...$$

where the integral is computed for all values of t from 0 to $+\infty$ and for all values of x, y, z within D. In a similar way we can define the analogous relative moments.

These moments can be recorded by accumulating in appropriate registers the time of recording raised to the power i, multiplied by the coordinates of the source raised to the power j, k, l respectively, and divided by the factorials $i!j!k!l!$. They have all the properties of the moments defined before, plus some specific properties of their own. A list of all those properties is beyond the scope of this chapter, but we shall present just a few examples.

Consider first the simplest case where the function $f(t;x,y,z)$ is a single "box", i.e.

$$f(t;x,y,z) = g(t) \text{ for } a_1<x<a_2, b_1<y<b_2, c_1<z<c_2,$$

$$= 0 \quad \text{for all other values of } x, y, z,$$

and $g(t)$ is a function with all the moments as defined in section 3. Then,

$$F_{0000} = (a_2-a_1)(b_2-b_1)(c_2-c_1).G_0 = V.G_0,$$

where V is the volume of the "box"; and

$$F_{ijkl} = (a_2^{j+1}-a_1^{j+1})/(j+1)!.$$
$$.(b_2^{k+1}-b_1^{k+1})/(k+1)!.$$
$$.(c_2^{l+1}-c_1^{l+1})/(l+1)!.G_i.$$

It follows that from the relative moments

$$f_{i100} = (a_1+a_2)/2.g_i,$$
$$f_{i010} = (b_1+b_2)/2.g_i,$$
$$f_{i001} = (c_1+c_2)/2.g_i,$$

we can determine the coordinates of the center of the "box". Furthermore from

$$f_{i200} = (a_1^2+a_1a_2++a_2^2)/3!.g_i,$$
$$f_{i020} = (b_1^2+b_1b_2++b_2^2)/3!.g_i,$$
$$f_{i002} = (c_1^2+c_1c_2++c_2^2)/3!.g_i,$$

and higher moments we can check the shape of the "box".

Consider now the case where a tracer diffuses along the X axis; its concentration $f(t;x,y,z)$ is described by the partial differential equation

(14) $$\partial f/\partial t = -D.\partial^2 f/\partial x^2,$$

where D is the diffusion coefficient. From the hypotheses that

$$\lim_{t \to \infty} f(t;x,y,z) = \lim_{t \to \infty} \partial f/\partial t = 0,$$

it follows that, for $i>0$, the i,j,k,l moment of $\partial f/\partial t$ is the $i-1,j,k,l$ moment of $-f(t;x,y,z)$, and for $j>1$ the i,j,k,l moment of $\partial^2 f/\partial x^2$ is the $i,j-2,k,l$ moment of $f(t;x,k,l)$. From equation (14) we have

$$F_{i-1,j,k,l} = D.F_{i,j-2,k,l}; \quad i>0, \; j>1$$

from a single ratio, say $-F_{0200}/F_{1000}$, we can compute the diffusion coefficient D; other ratios can be used to verify the hypothesis of the one-dimensional diffusion.

In general, a function is uniquely determined by its moments (Wald, 1939); if only few of its moments are known, a function can be reconstructed with a certain degree of accuracy using Chebyshev's inequality and its generalizations. Therefore if we observe the activity of a tracer by recording directly a number of its moments, thus reducing the errors due to the operations of integration, we can compute some parameters that express in mathematical form the underlying physical phenomenon. For instance from sequential autoradiographies of the brain we can determine the local rate of utilization of a particular drug.

REFERENCES

H.Bateman, 1954. _Tables of Integral Transforms_, Vol.I.
McGraw-Hill, New York, pages 303-366.

J.S.Beck and A.Rescigno, 1970. Calcium kinetics: the
philosophy and practice of science. _Phys. Med. Biol._ 15:
566-567.

C.C.Duncan, R.M.Lambrecht, A.Rescigno, C.Y.Shiue, C.W.Bennett,
L.R.Ment, 1983. The ramp injection of radiotracers for
blood flow measurement by emission tomograpgy. _Phys. Med.
Biol._ 8: 963-972.

J.H.Matis, T.E.Wehrly and K.B.Gerald, 1983. The statistical
analysis of pharmacokinetic data. _Tracer Kinetics and
Physiologic Modeling_ (R.M.Lambrecht and A.Rescigno
editors). Springer-Verlag, Berlin, pages 1-58.

A.Rescigno, 1973. On transfer times in tracer experiments. _J.
Theor. Biol._ 39: 9-27.

A.Rescigno, 1987. Image analysis and the method of moments.
Physics and Engineering of Medical Imaging (R. Guzzardi
editor). Martinus Nijhoff, The Hague, in press.

A.Rescigno and J.S.Beck, 1987. The use and abuse of models. _J.
Pharmacokin. Biopharm._ in press.

A.Rescigno and C. C. Duncan, 1985. Tracer kinetics and
stochastic models: from compartment analysis to image
reconstruction. _Mathematics and Computers in Biomedical
Applications_ (J. Eisenfeld and C. DeLisi editors).
Elsevier, Amstermam, pages 205-218.

A.Rescigno and E.Gurpide, 1973. Estimation of average times of
residence, recycle and interconversion of blood-borne
compounds using tracer methods. _J. Clin. Endocrinol.
Metab._ 36: 263-276.

A.Rescigno and R.M.Lambrecht, 1985. An algorithm for
reconstruction of count rate curves from total counts.
Mathematics in Biology and Medicine (V.Capasso, E. Grosso
and S.L.Paveri-Fontana editors). Berlin, Springer-Verlag,
pages 399-408.

A.Rescigno and L.D.Michels, 1973. Compartment modeling from
tracer experiments. _Bull. Math. Biol._ 35: 245-257.

A.Rescigno and G.Segre, 1961. The precursor-product
relationship. _J. Theor. Biol._ 1: 498-513.

A.Rescigno and G.Segre, 1964. _Drug and Tracer Kinetics_.
Blaisdell, Waltham, Mass.

A.Wald, 1939. Limits of a distribution function determined by
absolute moments and inequalities satisfied by absolute
moments. _Trans. Amer. Math. Soc._ 46: 280-306.

K.Zierler, 1981. A critique of compartmental analysis. _Ann.
Rev. Biophys. Bioeng._ 10: 531-562.

ON MODELING FLOW DATA USING GENERALIZED

STOCHASTIC COMPARTMENTAL MODELS

J. H. Matis[1] and K. B. Gerald[2]

[1]Department of Statistics
Texas A&M University
College Station, T.X.

[2]Department of Statistics
University of Kansas Medical Center
39th & Rainbow
Kansas City, KS   66103

1. INTRODUCTION

Compartmental models have played a central role in the modeling of drug, tracer and particle kinetics for many decades. The historical origins of the theory and applications of compartmental models are reviewed in Rescigno and Segre (1966) and many recent developments are presented in Jacquez (1985) and Godfrey (1983). Most of the theory and virtually all of the application of compartmental analysis have been based on a deterministic model formulation. Indeed, the term "compartment" has been defined through a deterministic model as "a variable X(t)... governed by the differential equation

$$dX(t)/dt = - k\ X(t) + f(t) \qquad (1)$$

with k constant" (Rescigno et al., 1983, p. 63). The mathematical model describing the kinetics of a compartmental system by a set of linear, first-order differential equations leads to the familiar sums of exponentials solution of the model. The sum of exponentials solution may then be fitted to observed data, usually using nonlinear least squares estimation procedures.

This paper presents two major departures from the classical approach. The first part of the paper presents a stochastic formulation of the model where the retention time random variables are assumed to be exponentially distributed. The stochastic formulation broadens the foundations of compartmental theory to incorporate, explicitly, particle stochasticity. Consequently, the conceptual focus of compartmental analysis shifts toward a "survivorship model" context with its discrete state space and assumed retention time distributions. It is shown that the assumption of exponential retention time distributions is the stochastic analog of the assumed constant k in Eq. (1) of the deterministic formulation. The second part of the paper generalizes the stochastic formulation to include retention times which are assumed to follow a gamma distribution. Recent research on such models is outlined and a case is made for the utility of these generalized models in practical data analysis.

2. A STOCHASTIC COMPARTMENT MODEL

2.1 Preliminaries

Most recent books on compartmental modeling, including Anderson (1983), Godfrey (1983), and Jacquez (1985), give general reviews of the stochastic modeling of compartmental systems. The general consensus seems to be that "every real system must be considered to be subject to uncertainties of one type or another, all of which are ignored in the formulation of a deterministic model" Gold (1977, p. 96). Several different types of stochasticity are identified in the above references. Most stochastic compartment models are based on only one type of stochasticity, although some recent papers have combined multiple sources of stochasticity [see e.g. Matis and Wehrly (1979)]. The present paper considers the stochasticity, or uncertainty, only in the location of individual particles. The uncertainty is modeled using retention time distributions to describe the random time spent in various compartments by individual particles during a single visit. The assumed retention time distributions give rise to residence time distributions and moments to describe the cumulative time which particles spend in the various compartments prior to leaving the system. A detailed case for including stochasticity in one particular area of flow kinetics, namely the modeling of digesta flow, is developed in Matis (1987).

The following notation, which is related to Figure 1, will be useful subsequently:

1) Let $p_{ij}(t)$; $i,j = 1,2$; denote the probability that a particle originating in i at time 0 is in j at time t.

2) Let $X_{ij}(t)$; $i,j = 1,2$; denote the number of particles that originated in i at time 0 and are in j at time t.

3) Let $\dot{p}_{ij}(t)$ and $\dot{X}_{ij}(t)$ be the (time-) derivatives of $p_{ij}(t)$ and $X_{ij}(t)$, respectively.

4) Let R_{ij}; $i = 1,2$, $j = 0, 1, 2$; $i \neq j$; denote the random retention time (or "lifetime") of a random particle in i whose next transfer is to j; where $j = 0$ denotes the exterior.

5) Let S_{ij}, $i,j = 1, 2$, denote the total residence time (or "sojourn time") that a particle originating in i will accumulate in j prior to exiting the system. Thus S_{ij} is a compound distribution, the sum of independent retention times.

This section develops first the one-compartment model with exponential retention times, and then extends the model to multiple compartments.

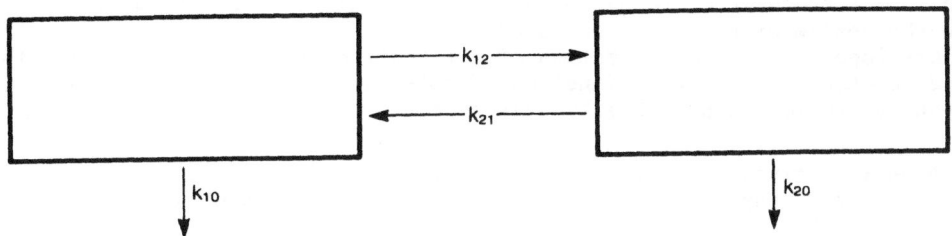

Figure 1. Two-compartment model.

2.2 One-Compartment Model with Exponential Retention Times

We consider first the stochastic one-compartment model with the following assumption:

Assumption 1. The retention time, R_{10}, of any particle in the single compartment model is an exponentially distributed random variable with parameter k_1.

The density function for R_{10} is

$$f(t) = k_1 \exp\{-k_1 t\} \text{ , for } t > 0 \qquad , \qquad (2)$$

which yields the following distribution and survivorship functions:

$$F(t) = \text{Prob } [R_{10} \leq t] = 1 - \exp\{-k_1 t\} \text{ , and} \qquad (3)$$

$$S(t) = \text{Prob}[R_{10} > t] = \exp\{-k_1 t\} \text{ .} \qquad (4)$$

For this model, the $p_{11}(t)$ probability function is equivalent to the survival probability, hence one has

$$p_{11}(t) = \exp\{-k_1 t\} \qquad (5)$$

which provides a regression model to which one may fit observed data and thereby estimate k_1. It is also clear from Eq. (5) that

$$\dot{p}_{11}(t) = k_1 p_{11}(t), \qquad (6)$$

which is the stochastic analog to the differential equation in Eq. (1) on which the deterministic model is based.

One important characterization of a retention time distribution is its hazard function. The hazard function, denoted as $k(t)$, may be defined through the following probability statement

Prob {any given particle transfers out of the compartment in the time interval $(t, t + \Delta t)$ given that it was present at t}

$$= k(t)\Delta t , \qquad (7)$$

as Δt approaches 0. Conceptually, the hazard function gives the instantaneous transfer rate at time t and it follows from Eq. (7) that it may be obtained analytically, [see e.g. Gross and Clark (1975)], as

$$k(t) = f(t)/S(t) \cdot \qquad (8)$$

The hazard rate for the assumed exponential retention time distribution is

$$k(t) = k_1 , \qquad (9)$$

which is the stochastic counterpart to the assumed constant turnover rate of the deterministic model. The time-invariance of the present hazard function is often called the "lack of memory" property and it is unique to the exponential distribution. Thus the assumption of a constant hazard function would be equivalent to Assumption 1, however, the present focus on retention time distributions provides a more convenient framework for subsequent generalization.

The statistical moments of the retention time distributions are also of subsequent importance. The mean and variance of the exponential distribution are

$$E[R_1] = k_1^{-1} \quad \text{and} \qquad (10)$$

$$V[R_1] = k_1^{-2} . \qquad (11)$$

2.3 Multicompartment Models with Exponential Retention Times

We consider now the stochastic two-compartment model in Figure 1 with the following assumption:

Assumption 2. The retention times R_{10}, R_{12}, R_{20} and R_{21} for the particles in the model are exponentially distributed random variables with parameters k_{10}, k_{12}, k_{20} and k_{21}, respectively.

Using Eq. (9), it follows that the k_{ij} parameters are again the respective hazard rates defining the instantaneous transfer probabilities. Through relationships such as

22

$$P_{11}(t+\Delta t) = P_{11}(t)[1-(k_{10}+k_{12})\Delta t] + P_{12}(t) [k_{21}\Delta t] + o(\Delta t) \, ,$$

also called Chapman-Kolmogorov equations, one may obtain equations such as

$$\dot{P}_{11}(t) = -(k_{10}+k_{12}) P_{11}(t) + k_{21} P_{12} (t) \, ,$$

which are called Kolmogorov forward equations (see e.g. Chiang (1980) p. 417). The full set of Kolmogorov equations for the present two-compartment model is

$$\dot{P}_{i1}(t) = -(k_{10} + k_{12}) P_{i1}(t) + k_{21} P_{i2}(t) \tag{12}$$

$$\dot{P}_{i2}(t) = k_{12} P_{i1}(t) - (k_{20} + k_{21}) P_{i2}(t) \, , \text{ for } i = 1,2 \, . $$

Letting $P(t) = [p_{ij}(t)]$ denote the matrix of probabilities, $\dot{P}(t) = [\dot{p}_{ij}(t)]$ the matrix of derivatives, and $K = [k_{ij}]$ the matrix of coefficients where $k_{ii} = -\sum_{j\neq i} k_{ij}$, the Kolmogorov equations may be written in matrix form as

$$\dot{P}(t) = P(t)K \, . \tag{13}$$

The general matrix solution to Eq. (13) is

$$P(t) = P(0) \, e^{Kt} \, . \tag{14}$$

Substituting $P(0) = I$, and assuming unequal eigenvalues, Eq. (14) may be written as

$$P(t) = T \, e^{\Lambda t} \, T^{-1} \tag{15}$$

where T is the matrix of eigenvectors, with inverse T^{-1}, and $e^{\Lambda t}$ is a diagonal matrix of $e^{\lambda_i t}$ elements with λ_i denoting the ith eigenvalue. Letting M denote the matrix of mean residence times (MRT) with elements $M_{ij} = E[R_{ij}]$, and V the corresponding matrix of variances of residence times (VRT), one can show that

$$M = -K^{-1} \text{ and} \tag{16}$$

$$V = 2MM_D - M_{(2)} \tag{17}$$

where M_D and $M_{(2)}$ are matrices of diagonal elements and of squared elements, respectively, of M [see e.g. Matis et al. (1983)].

Although the formulas (13) - (17) are illustrated here only in the context of a two-compartment model, it should be noted that they generalize to any n-compartment model. The special case of n = 1 yields the previous solutions in Eq. (5), (9), (10) and (11). Equation (15) implies that all $p_{ij}(t)$ solutions have the sums of exponentials form. The explicit solutions for the general two-compartment model and many related models are catalogued in Jacquez (1985).

3. A GENERALIZED STOCHASTIC COMPARTMENTAL MODEL

3.1 Introduction

The previous stochastic models were based on the exponential retention time distribution, which leads to the constant hazard rate, k_1, as illustrated in Eq. (7). The constant hazard rate property of the exponential distribution is a stringent one in biomedical applications as, by definition, it rules out any dependency of the transfer probability on the elapsed time that the particle has spend in the compartment(s). Many natural biological processes have such general phenomena as noninstantaneous "mixing" and particle "aging" which suggest nonexponential distributions.

We consider first the one-compartment model with certain parametric forms of retention time distributions. The models are equivalent to the survivorship models in biostatistics (see e.g. Gross and Clark (1975) and Lee (1980)). The principal objective of survival analysis is to give a robust and tractable model which may be fitted to data. The stochastic compartmental model has a different emphasis in that it aims not only to describe the data but also to provide a mechanistic description of the process itself. This objective of a mechanistic model suggests more model structure, and leads to multicompartment extensions of the simple one-compartment, survivorship models.

3.2 One-Compartment Model With Nonexponential Retention Times

We consider now the stochastic model with the following assumption:

Assumption 3. The retention time, R_1, of any particle in the single compartment model is a nonexponential random variable with density function f(t), survivorship function S(t) and time-varying hazard rate k(t).

Three common nonexponential retention time distributions are outlined in this section. The first is the gamma distribution with density function

$$f(t) = \lambda^{\alpha} \, t^{\alpha-1} \, \exp(-\lambda t)/\Gamma(\alpha) \ . \tag{18}$$

The survivorship function and the hazard rates are not available in a simple analytical form, but are illustrated in Figure 2A. Note that the rate is monotonically increasing, constant, or monotonically decreasing for $\alpha > 1$, $\alpha = 1$, or $\alpha < 1$ respectively.

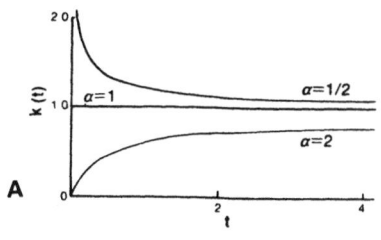 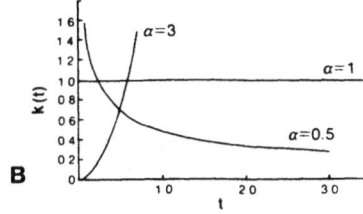

Figure 2.   Hazard rate functions for some nonexponential
retention time distributions.

    A.   Gamma $(\alpha, \lambda = 1)$ for $\alpha = 1/2,\ 1,\ 2$
    B.   Weibull $(\alpha, \lambda = 1)$ for $\alpha = 1/2,\ 1,\ 2$

A second common retention time distribution is the Weibull, with
density and survivorship functions as follows:

$$f(t) = \alpha\lambda^{\alpha}t^{\alpha-1}\exp[-(\lambda t)^{\alpha}]$$

$$S(t) = \exp[-(\lambda t)^{\alpha}] .$$

The hazard rate function,

$$k(t) = \alpha\lambda^{\alpha}\ t^{\alpha-1} , \tag{19}$$

is illustrated for a few sample parameter values in Figure 2B. Again the
rate is increasing, constant, or decreasing as $\alpha>1$, $\alpha=1$, or $\alpha<1$. The
analytical form of the $S(t)$ function enables it to be readily fitted to
data.

The third distribution is a special case of the gamma, called the
Erlang, distribution where α is a positive integer, denoted n. The density
and survivorship functions using parameters k_1 and n are then obtained from
Eq. (18) as:

$$f(t) = k_1^{\,n}\ t^{n-1}\ \exp(-k_1 t)/(n-1)! \quad \text{and} \tag{20}$$

$$S(t) = \exp(-k_1 t)\ \sum_{i=0}^{n-1}\ (k_1 t)^i/i! . \tag{21}$$

The hazard rate function is immediately

$$k(t) = \frac{k_1^n t^{n-1}/(n-1)!}{\sum_{i=0}^{n-1} (k_1 t)^i/i!} \qquad (22)$$

which is sketched for a few values of n in Figure 3.

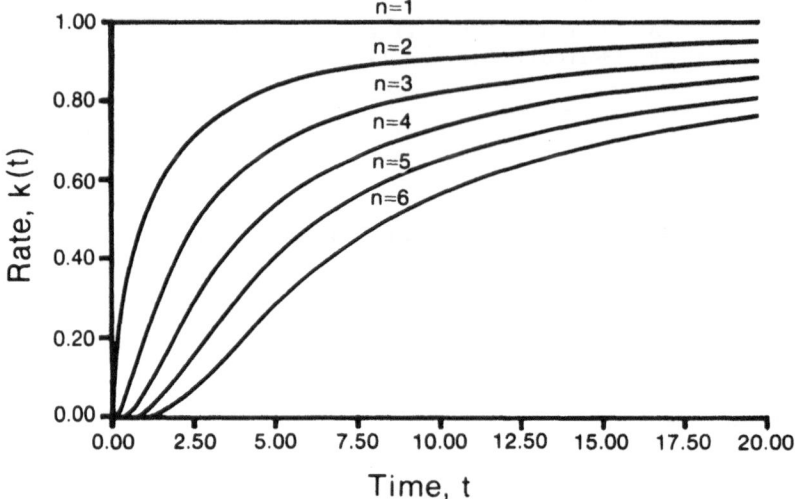

Figure 3. Hazard rate functions for Erlang (n, $\lambda = 1$)
distributions with n = 1, ... , 6.

    The third model, the Erlang, has been widely fitted to animal nutrition data. As noted in Matis and Wehrly (1985), the hazard rate has a useful theoretical form of starting at 0 and then asymptoting to a constant, k_1, after the initial "mixing" and/or "aging" is complete. The model has only two parameters, n and k_1, and yet it is very rich in form. This is immediately apparent in the survivorship function, S(t), in Eq. (21). This function, which for a one compartment model is also the probability function $p_{11}(t)$, is not restricted to the sums of exponentials form. Thus, one has available a new rich family of models which contain only two parameters.

3.3 General Solution of Models with Erlang Retention Times

    In most applications of compartmental modeling, one has some understanding of the biological or ecological system (or mechanism), and typically this leads to a conceptual model with two or more compartments.

Often the assumption of retention times with constant hazard functions in these compartments is not satisfactory, as previously noted, due to such phenomena as noninstantaneous mixing, physical particle transport, particle aging and delayed biological response. These considerations lead to multi-compartmental models with nonexponential retention times, which have been called generalized compartment models.

A number of recent papers have developed generalized multicompartment models based on Erlang-distributed retention times. The assumption for these models is:

> <u>Assumption 4</u>. The retention times R_{10}, R_{12}, R_{20} and R_{21} are independent gamma (Erlang) distributed random variables with parameter sets (n_1, k_{10}), (n_1, k_{12}), (n_2, k_{20}) and (n_2, k_{21}) respectively.

An equivalent form of the assumption may be based on random variables, R_1 and R_2, which represent the retention times in compartments 1 and 2, respectively, without regard to the subsequent destination. The hazard functions from a common origin are additive, hence, Assumption 4 could be reformulated to state that 1) R_1 and R_2 are Erlang distributed random variables with parameter sets $(n_1, k_{10} + k_{12})$ and $(n_2, k_{20} + k_{21})$ and 2) the conditional probabilities of transfering to the opposite compartment given departure from the present compartment are $\alpha_1 = k_{12}/(k_{10} + k_{12})$ and $\alpha_2 = k_{21}/(k_{20} + k_{21})$. This reformulation describes the process in a so-called semi-Markov process context, and is illustrated in Figure 4.

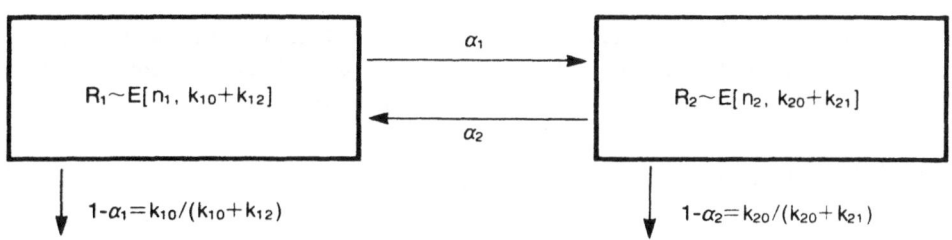

Figure 4. Generalized two-compartment model with Erlang retention time distribution.

Moreover, capitalizing on the fact that the sum of m independent and identical exponential random variables is an Erlang random variable with parameter m, one can <u>solve</u> the model with Erlang retention times using an equivalent model of compartments with exponential retention times. The equivalent model is represented in Figure 5. This

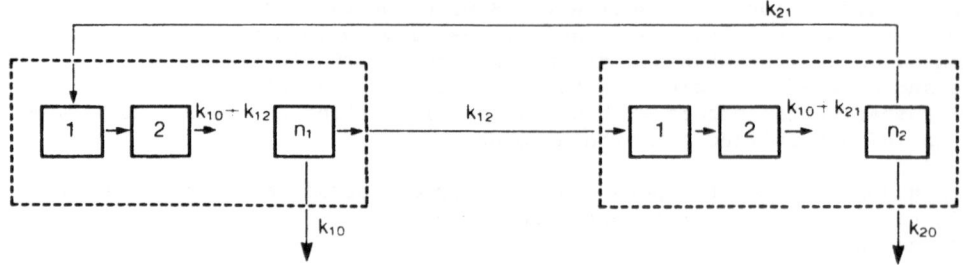

Figure 5.   Standard (exponential) representation of a generalized two-compartment model with Erlang retention times.

approach utilizes (n_1+n_2) so-called "pseudo-compartments" which are not envisioned to correspond to physical subdivisons of the compartment but which merely constitute a mathematical artifice to solve the system. Letting $\pi_{ij}(t)$ denote the probability that a particle originating in pseudo-compartment i at t = 0 is in j at time t, one has the relationships

$$p_{11}(t) = \sum_{j=1}^{n_1} \pi_{1j}(t) \tag{23}$$

$$p_{12}(t) = \sum_{j=n+1}^{n_1+n_2} \pi_{ij}(t)$$

Clearly, the usual initial condition of $p_{11}(0) = 1$ is transformed into $\pi_{11}(0) = 1$ and $\pi_{1j}(0) = 0$ for all j > 1. A set of Kolmogorov equations which is analogous to the previous Eq. (12) is now:

$$\dot{\pi}_{i1}(t) = -(k_{10}+k_{12})\,\pi_{i1}(t) + k_{21}\,\pi_{i,n_1+n_2}(t) \tag{24}$$

$$\dot{\pi}_{ij}(t) = -(k_{10}+k_{12})\,\pi_{ij}(t) + (k_{10}+k_{12})\,\pi_{i,j-1}(t) \quad \text{for } j = 2, \cdots, n_1$$

$$\dot{\pi}_{i,n_1+1}(t) = -(k_{20}+k_{21})\,\pi_{i,n_1+1}(t) + k_{12}\,\pi_{i,n_1}(t)$$

$$\dot{\pi}_{i,n_1+j}(t) = -(k_{20}+k_{21})\,\pi_{i,n_1+j}(t) + (k_{20}+k_{21})\,\pi_{i,n_1+j-1}(t)$$

$$\text{for } j = 2, \cdots, n_2$$

Letting $\Pi(t) = [\pi_{ij}(t)]$ denote the new matrix of probabilities and K the new coefficient matrix, it follows that the system of equations in Eq. (24) may be represented as

$$\dot{\Pi}(t) = \Pi(t)K \tag{25}$$

where K =

$$
\begin{bmatrix}
-(k_{10}+k_{12}) & (k_{10}+k_{12}) & \cdots & 0 & 0 & \cdots & 0 \\
0 & -(k_{10}+k_{12}) & \cdots & 0 & 0 & \cdots & 0 \\
\vdots & \vdots & & \vdots & \vdots & & \\
0 & 0 & \cdots & (k_{10}+k_{12}) & 0 & \cdots & 0 \\
0 & 0 & \cdots & -(k_{10}+k_{12}) & k_{12} & \cdots & 0 \\
0 & 0 & & 0 & -(k_{20}+k_{12}) & & 0 \\
\vdots & \vdots & & \vdots & \vdots & & \vdots \\
0 & 0 & \cdots & 0 & 0 & \cdots & (k_{20}+k_{21}) \\
k_{21} & 0 & & 0 & 0 & \cdots & -(k_{21}+k_{21})
\end{bmatrix}
$$

$\underbrace{\qquad\qquad\qquad}_{n_1 \text{ columns}} \qquad \underbrace{\qquad\qquad\qquad}_{n_2 \text{ columns}}$

The general solution to Eq. (25) is

$$\Pi(t) = e^{Kt}, \tag{26}$$

as before in Eq. (14), however, the eigenvalues in the nontrivial case $n_1 + n_2 > 2$ are not distinct and not necessarily real; hence the $\pi_{ij}(t)$, and subsequent $p_{ij}(t)$, solutions for this model do not have the sums of exponentials form given in Eq. (15). Some special cases of the above model which have been investigated in the literature are outlined below. A new model which is potentially useful for NMR data is then introduced.

3.4 Some Particular Models with Erlang Retention Times

One of the early papers to consider generalized compartment models was Matis (1972), which considers a two-compartment irreversible system where the first compartment is Erlang. The two variables used to define the model are $R_{12} \sim$ Erlang (n_1, λ_1) and $R_{20} \sim$ Erlang $(n_2 = 1, k_2)$, which corresponds to Figure 5 and Eq. (24) and (25) with $k_{10} = k_{21} = 0$. The solution to the model is

$$p_{11}(t) = \delta^n \exp(-k_2 t) + \exp(-\lambda_1 t) \sum_{i=0}^{n-1} (1-\delta^{n-i})(\lambda_1 t)^i/i! \qquad (27)$$

with $\delta = \lambda_1/(\lambda_1-k_2)$.

The model is relatively easy to fit to data for fixed n in order to estimate the parameters λ_1 and k_2. The model has been widely used in ruminant nutrition studies, and is discussed further by France et al (1985) and Matis (1987).

A subsequent paper, Hughes and Matis (1984), extends the previous irreversible model to the case of $n_2 > 1$. The $p_{11}(t)$ solution is given in the paper, however, it is unwieldy for large n_1 and/or n_2. The use of the model for small n_1 and n_2 is illustrated on a sample data set.

The generalized form of the common two-compartment model used in pharmacokinetics, which includes reversible flow ($k_{21} \neq 0$) but sets $k_{20} = 0$ to correspond to central elimination, was first studied in Matis and Wehrly (1984). The paper presents a theorem stating that with $k_{20} = 0$ the eigenvalues of the matrix K in Eq. (25) were complex in all models with $n_1 + n_2 > 5$ and in many models with $n_1 + n_2 \leq 4$. Assuming that one has 2s distinct complex roots and r real roots, with $2s + r = n_1 + n_2$, the $p_{ij}(t)$ solutions would be of the form:

$$p_{ij}(t) = \sum_{\ell=1}^{r} a_{ij\ell} e^{\lambda_\ell t} + \sum_{k=1}^{s} e^{v_k t} [b_{ijk}\cos(\zeta_k t) - c_{ijk}\sin(\zeta_k t)] \qquad (28)$$

where λ_ℓ, $\ell=1,\ldots r$, denotes a real root and the complex roots have form $v_k \pm i\zeta_k$ for $k = 1,\ldots,s$. It can be shown that the roots have negative real parts, i.e. $\lambda_\ell < 0$ and $v_k < 0$ for all ℓ and k, hence the $p_{ij}(t)$ solutions have damped oscillations. In practice, these oscillations are very subtle for common parameter values and generally they are not readily observable in plots of the functions. However, the analytical solutions of the roots as functions of the underlying k_{ij} parameters is very difficult. Hence, although one may use nonlinear least squares procedures to fit models of the form in Eq. (28) to data, the transformation of the parameter estimates in the model to estimates of the underlying k_{ij} parameters has proven very challenging.

An alternative parameter estimation procedure based on the method of moments is presented and illustrated in Matis and Wehrly (1985). The basic logic of the procedure is that moments for the system are obtainable from the inverse of the K matrix, and often such moments may be expressed as tractable analytical functions of the k_{ij}'s. The k_{ij}'s are therefore estimable using the method of moments, and from the k_{ij} parameters one can easily estimate the retention time distributions and the time-varying hazard rates of given generalized compartment models.

3.5 A Multicompartment Model for NMR data

Consider the model defined in Figure 6 and illustrated in pseudo-compartment form in Figure 7.

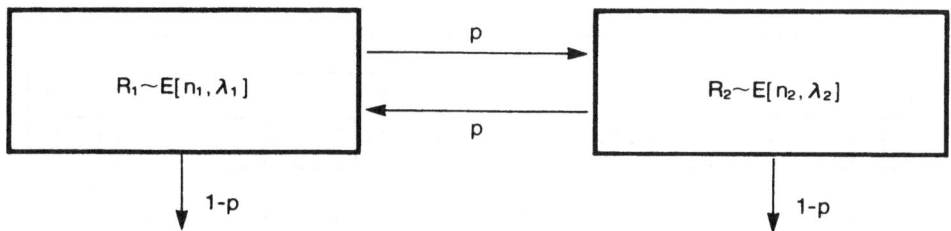

Figure 6. Proposed generalized compartment model for NMR.

The model involves only three parameters, λ_1, λ_2, and p, and has been suggested as being useful for modeling NMR data. The model generalization where the crossflow probabilities are p and $q \neq p$ is not estimable unless data are available for both $p_{11}(t)$ and $p_{12}(t)$, or for the aggregate.

The coefficient matrix of the model in the pseudo-compartment form in Figure 7 is

$$
K = \begin{bmatrix}
-\lambda_1 & \lambda_1 \cdots & 0 & 0 & \cdots & 0 \\
0 & -\lambda_1 \cdots & 0 & 0 & \cdots & 0 \\
\vdots & \vdots & \vdots & \vdots & & \vdots \\
0 & 0 \cdots & \lambda_1 & 0 & \cdots & 0 \\
0 & 0 \cdots & -\lambda_1 & p\lambda_1 & \cdots & 0 \\
0 & 0 \cdots & 0 & -\lambda_2 & \cdots & 0 \\
\vdots & & \vdots & \vdots & & \vdots \\
0 & 0 & 0 & 0 & & \lambda_2 \\
p\lambda_2 & 0 & 0 & 0 & \cdot & -\lambda_2
\end{bmatrix}
\qquad (29)
$$

$\underbrace{\qquad\qquad}_{n_1 \text{ columns}}$ $\underbrace{\qquad\qquad}_{n_2 \text{ columns}}$

This matrix also will lead to complex eigenvalues for large values of $n_1 + n_2$, and hence the analytical solution of the $p_{ij}(t)$ probability functions would be of the form in Eq. (28). As before, it would be challenging to fit such models to output data in order to estimate the three parameters.

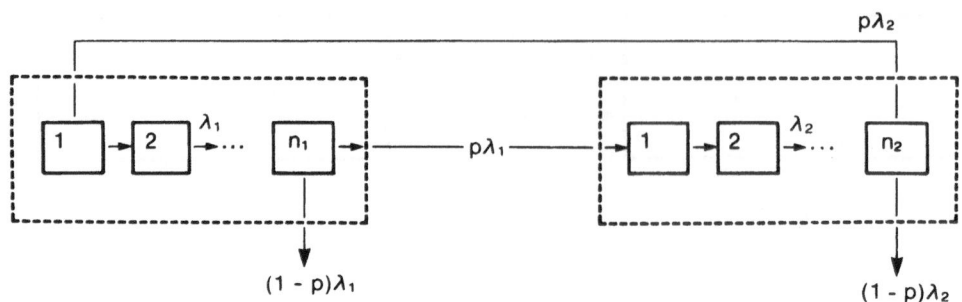

Figure 7. Standard (exponential) representation of the proposed generalized compartment model.

Two alternative estimation procedures are available. In principle, one could again use the method of moments procedures described previously. For example, the matrix of mean residence times (MRT) is relatively easy to obtain for the K matrix in Eq. (29). Assuming that the process starts in compartment 1, one could show that the required MRT's for the $n_1 + n_2$ pseudo-compartments are given by the first row of $-K^{-1}$ as follows:

$$
\begin{array}{ccc}
\overbrace{\qquad\qquad}^{n_1 \text{ columns}} & \overbrace{\qquad\qquad}^{n_2 \text{ columns}} \\
\end{array}
$$

$$
\left[\frac{1}{(1-p^2)\lambda_1} \quad \frac{1}{(1-p^2)\lambda_1} \ \ldots \ \frac{1}{(1-p^2)\lambda_1} \quad \frac{1}{(1-p^2)\lambda_2} \ \ldots \ \frac{1}{(1-p^2)\lambda_2} \right] \tag{30}
$$

With the proper aggregation, the MRT for generalized compartments 1 and 2 are $n_1/[(1-p^2)\lambda_1]$ and $n_2/[(1-p^2)\lambda_2]$ respectively. Higher order moments could be obtained in similar fashion, and the parameters estimated by equating the observed sample and the population moments.

Another alternative is the use of numerical analysis techniques to solve the differential equations of the form in Eq. (24) numerically. Computer software is now readily available to yield least squares solutions from such repeated numerical curve fitting. The present authors have utilized IMSL programs on mainframe computers and PCNONLIN, SCOPE, and SCOPEFIT programs on personal computers. The increasing availability of the necessary hardware and software for such estimation makes this last alternative very attractive.

In summary, the new generalized compartment model presented in Figure 6 can be fitted to data in several ways. The model is very rich in form despite its very small number of parameters, which makes it potentially very useful in modeling flow data.

References

Anderson, D. H., 1980, "Compartmental Modeling and Tracer Kinetics," New York: Springer Verlag.

Chiang, C. L., 1980, "An Introduction to Stochastic Processes and Their Applications," Huntington, NY: Krieger.

France, J., Thornly, J. H. M., Dhanoa, M. S. and Siddons, R. C., 1985, On the mathematics of digesta flow kinetics, J. Theo. Biol., 113:743.

Godfrey, K., 1983, "Compartmental Models and Their Application," New York: Academic.

Gold, H. J., 1977, "Mathematical Modeling of Biological Systems," New York: Wiley.

Gross, A. J. and Clark, V. A., 1975, "Survival Distributions: Reliability Applications in the Biomedical Sciences," New York: Wiley.

Hughes, T. H. and Matis, J. H., 1984, An irreversible two-compartment model with age-dependent turnover rates, Biometrics, 40:501.

Jacques, J. A., 1985, "Compartmental Analysis in Biology and Medicine", 2nd ed., Ann Arbor, MI: University of Michigan Press.

Lee, E. T., 1980, "Statistical Methods for Survival Data Analysis", Belmont, CA: Lifetime Learning Publ.

Matis, J. H., 1972, Gamma time-dependency in Blaxter's compartmental model, Biometrics, 28:597.

Matis, J. H., 1987, The case for stochastic models of digesta flow, J. Theo. Biol., in press.

Matis, J. H. and Wehrly, T. E., 1979, Stochastic models of compartmental systems, Biometrics, 35:199.

Matis, J. H. and Wehrly, T. E., 1984, On the use of residence time moments in the statistical analysis of age-dependent stochastic compartmental systems, In: "Mathematics in Biology and Medicine", S. L. Paveri-Fontana and V. Capasso (eds.), New York: Springer-Verlag.

Matis, J. H. and Wehrly, T. E., 1985, Stochastic compartmental models with gamma retention times: An application and estimation procedure, In: "Mathematics and Computers in Biomedical Applications", J. Eisenfeld and C. DeLisi (eds.), New York: Elsevier.

Matis, J. H., Wehrly, T. E., and Metzler, C. M., 1983, On some stochastic forumulations and related statistical moments in pharmacokinetic models, J. Pharmacokinet. Biopharm., 11:77-92.

Rescigno, A., Lambrecht, R. M. , and Duncan, C. C., 1983, Mathematical methods in the formulation of pharmacokinetic models, In: "Tracer Kenetics and Physiologic Modeling", R. M. Lambrecht and A. Rescigno (eds.), New York: Springer-Verlag.

Rescigno, A. and Segre, G., 1966, "Drug and Tracer Kinetics", Waltham, MA: Blaisdell.

AUTORADIOGRAPHY

BASIC PRINCIPLES IN IMAGING OF REGIONAL CEREBRAL METABOLIC RATES

WITH RADIOISOTOPES

Louis Sokoloff

National Institute of Mental Health
Laboratory of Cerebral Metabolism, Bldg. 36, Rm. 1A-05
Bethesda, MD   20892

INTRODUCTION

Radioisotopes are frequently used to facilitate the assay of rates of biochemical reactions. Usually they are used to study chemical reactions in vitro, and the procedure is to label one of the reactants and to measure the rate of accumulation of a labeled product. From assay or knowledge of the specific activity of the reactant molecule and the stoichiometry of the reaction the rate of the overall reaction can be calculated from the rate of radioactive product formation. The application of this approach generally necessitates, however, specialized biochemical procedures to identify and isolate the labeled product so as to limit the measurement of the radioactivity to a specific chemical product of the reaction.

Quantitative autoradiography makes it possible to measure local concentrations of isotopes in tissues of animals labeled in vivo. In a few cases the administration of a judiciously selected labeled chemical compound combined with an appropriately designed procedure has made it possible to use this capability to measure the local rates of a chemical process in tissues of animals in vivo. Emission tomography, particularly positron-emission tomography, extends this capability to man and makes it possible to assay the rates of biochemical processes in human tissues in vivo. Autoradiography or emission tomography do not, however, obviate the need to adhere to established principles of chemical and enzyme kinetics and tracer theory. Generally, all such methods, whether to be used in man with positron-emission tomography or in animals with autoradiography, must first be developed by research in animals with autoradiography because it is only in animals that the measurements needed to validate the model and basic assumptions of the method can be tested and evaluated.

It is a common misconception that the ability to image the distribution of the local concentrations of an isotope in the tissues is by itself sufficient to characterize a physiological or biochemical process in the tissues. It is almost certainly possible to administer radioactive shoe-polish to an animal and to observe differential concentrations of radioactivity in the tissues, but it is unlikely that the distribution of radioactivity would provide any useful information concerning any process in the tissues. The selection of the labeled compound to be used as the tracer and the timing and procedure are of critical importance to identify and assay the process of interest. The tissue uptake of label from a labeled compound circulating in the blood is dependent on at least three processes: 1) blood flow and delivery of the tracer to the tissues; 2) transport of the tracer compound between the blood and tissues; and 3) metabolism of the tracer compound in the tissue and the fate of the metabolic products. [^{14}C]Antipyrine is a diffusible, chemically inert tracer that is taken up by brain tissue according to the blood flow to the tissue and its relative solubility in tissue and blood (29). 3-0-[^{14}C]Methylglucose is distributed to brain tissue according to blood flow, tissue-blood distribution ratio, and carrier-mediated blood-brain barrier transport (2,3,25). 2-[^{14}C]Deoxyglucose is distributed to the brain tissue according to blood flow, tissue-blood distribution ratio, carrier mediated blood-brain barrier transport, and hexokinase-catalyzed phosphorylation (40). In Fig. 1 are presented the autoradiographs of brain sections obtained one minute and sixty minutes after each of these labeled compounds was administered as an intravenous pulse. The autoradiographs illustrate the distribution of the label in the cerebral tissues, the darker the density the higher the concentration of ^{14}C. At one minute the distribution of ^{14}C is essentially the same regardless of the labeled compound used, indicating that the influences of blood flow, solubility, transport, and metabolism cannot be discriminated. At sixty minutes after the pulse the autoradiographs obtaind with [^{14}C]antipyrine and [^{14}C]methylglucose are essentially homogeneous whereas that obtained with [^{14}C]deoxyglucose retains the same pattern as that obtained one

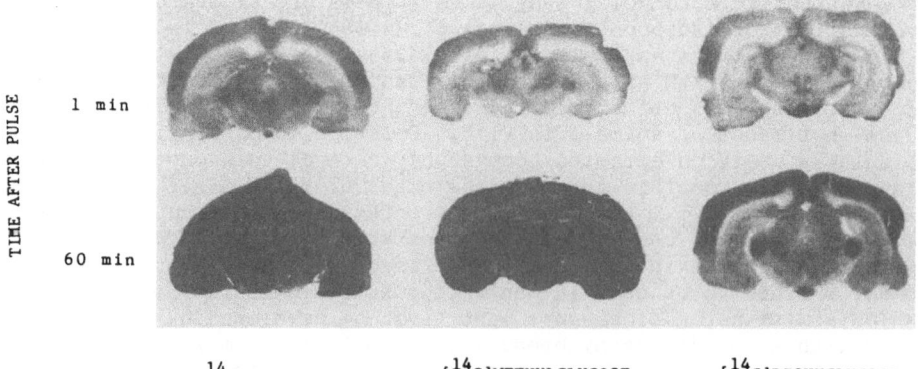

TIME AFTER PULSE

1 min

60 min

[^{14}C]ANTIPYRINE   [^{14}C]METHYLGLUCOSE   [^{14}C]DEOXYGLUCOSE

Fig. 1. Autoradiographs of brain sections from rats 1 and 60 min after a pulse of 51-95 µCi of [^{14}C]antipyrine, [^{14}C]methylglucose, or [^{14}C]deoxyglucose. Note that 60 min after the pulse only the autoradiograph from the rat receiving [^{14}C]deoxyglucose retains regional heterogeneity, indicating that local metabolism rather than blood flow and transport determines the local tissue concentrations of ^{14}C at long times after a pulse of [^{14}C]deoxyglucose. From Sokoloff et al. (40).

minute after the pulse. At sixty minutes the [14C]antipyrine and [14C]methylglucose autoradiographs reflect the tissue-blood distribution ratios whereas that of [14C]deoxyglucose reflects its metabolism. It is clear from these results that not only the chemical species of labeled compound but also the procedure and time are of critical importance in identifying and defining the relationship between the distribution of the label in the tissues and the nature of the physiological and/or biochemical processes responsible for the distribution.

GENERAL PRINCIPLES IN RADIOISOTOPIC ASSAY OF BIOCHEMICAL REACTIONS IN VIVO

A chemical reaction is the conversion of one species of molecule to another. The rate of this reaction can be measured by determining the rate of disappearance of one or more of the reactants or the formation of one or more of the products. The addition of a radioactive label to one of the reactants in molecular concentrations too negligible to alter the kinetics of the reaction facilitates the measurement of the radioactive reactants or products but does not solve all the problems. The rate of chemical transformation of the labeled species can relatively easily be measured, but this is not the rate of the reaction of interest. To derive the rate of the total overall reaction from measurement of the reaction rate of only the labeled species, it is necessary to know the integrated specific activity (i.e., the ratio of labeled to total molecules) of the precursor pool. Occasionally the labeled species exhibits a kinetic difference from the natural compound, the so-called isotope effect; this isotope effect can be evaluated and appropriate correction made for it. The general equation for the determination of the rate of a biochemical reaction with radioactive tracers is presented in Fig. 2.

In assays of biochemical reactions in vivo it is generally impossible to measure the integrated specific activity of the precursor pool directly. This would require the measurement of the complete time courses of the concentrations of the labeled and unlabeled precursor molecules in the tissue at the enzyme site. it is, therefore, necessary to determine the precursor pool specific activity indirectly from measurements in the blood supplied to the tissue. The specific activity in the arterial blood or plasma can be readily measured directly, and the precursor-specific activity can be calculated from it by correcting for the lag in the equilibration of the precursor pool in the tissue with that of the plasma. To apply this correction it is necessary to know the kinetics of the equilibration process between the precursor pools in the tissue and plasma.

GENERAL EQUATION FOR MEASUREMENT OF REACTION RATES WITH TRACERS:

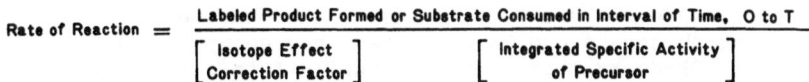

Fig. 2. General equation for the determination of rates of biochemical reactions with radioisotopes (see text).

The rate of a chemical reaction can be determined by measurement of either precursor disappearance or product accumulation; generally the errors are smaller with the latter measurement because the percent increase in amount of product is much greater than the percent loss in amount of precursor. Autoradiography and emission tomography, which measure only total concentration of all the radioactive molecules, cannot distinguish among the various chemical species which may be labeled, neither the precursor nor any of the possible labeled products. Strategies must, therefore, be developed that ensure that the radioactivity is contained exclusively in the precursor and/or in one or more of the products specific to the chemical reaction to be assayed. The labeled precursor should be so selected that its chemical transformations are limited as much as possible to the pathway under study.

These general principles have been more or less successfully applied in two currently operational methods for the measurement of energy metabolism in the nervous system of animals and man. One method is the steady-state O_2 consumption technique which is based on the measurement of substrate disappearance (7). Cerebral blood flow is measured by positron-emission tomography with $C^{15}O_2$ and, when multiplied by arterial O_2 content, provides the steady state delivery of O_2 to the tissues. Local cerebral O_2 extraction from the blood is measured by positron-emission tomography with $^{15}O_2$. The product of local oxygen extraction, blood flow, and arterial O_2 content provides the values for local O_2 consumption. The other method is the radioactive deoxyglucose technique for the measurement of local cerebral glucose utilization. It has been used extensively in animals with autoradiography (38,39,40), but it has been adapted to man by the use of $[^{18}F]$fluorodeoxyglucose and positron-emission tomography (27,30). The deoxyglucose technique is based on the measurement of product accumulation. Because it encompasses almost all the principles to be considered in the measurement of biochemical processes in vivo, it will serve as an informative example of their application.

THE DEOXYGLUCOSE METHOD

Theory

The method is derived from a model based on the biochemical properties of 2-deoxyglucose (Fig. 3A) (40). 2-Deoxyglucose (DG) is transported bi-directionally between blood and brain by the same carrier that transports glucose across the blood-brain barrier (2,3,24 . In the cerebral tissues it is phosphorylated by hexokinase to 2-deoxyglucose-6-phosphate (41). Deoxyglucose and glucose are, therefore, competitive substrates for both blood-brain barrier transport and hexokinase-catalyzed phosphorylation. Unlike glucose-6-phosphate, however, which is metabolized further eventually to CO_2 and water and to a lesser degree via the hexosemonophosphate shunt, deoxyglucose-6-phosphate (DG-6-P) cannot be converted to fructose-6-phosphate and is a poor substrate for glucose-6-phosphate dehydrogenase (41). There is very little glucose-6-phosphatase activity in brain (12) and even less deoxyglucose-6-phosphatase activity (40). Deoxyglucose-6-phosphate can be converted into deoxyglucose-1-phosphate and then into UDP-deoxyglucose and eventually into glycogen, glycolipids and glycoproteins, but thse reactions are slow, and in mammalian

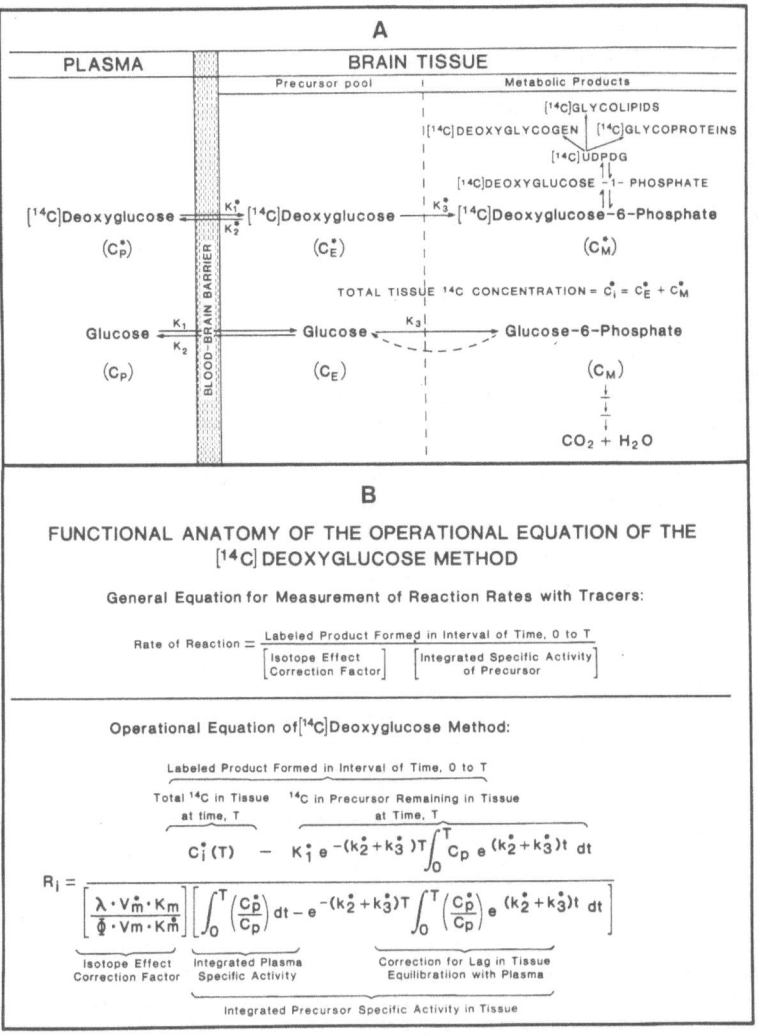

Fig. 3(A). Diagrammatic representation of the theoretical model of the deoxyglucose method. C_i* represents the total ^{14}C concentration in a single homogeneous tissue of the brain. C_P* and C_P represent the concentrations of [^{14}C]deoxyglucose and glucose in the arterial plasma, respectively; C_E* and C_E represent their respective concentrations in the tissue pools that serve as substrates for hexokinase. C_M* represents the combined concentration of [^{14}C]deoxyglucose-6-phosphate and its products in the tissue. The constants K_1*, k_2*, and k_3*, represent the rate constants for carrier-mediated transport of [^{14}C]deoxyglucose from plasma to tissue, for carrier-mediated transport back from tissue to plasma, and for phosphorylation by hexokinase, respectively. The constants K_1, k_2, and k_3 are the equivalent rate constants for glucose. [^{14}C]Deoxyglucose and glucose share and compete for the carrier that transports both between plasma and tissue and for hexokinase which phosphorylates them to their respective hexose-6-phosphates. The dashed arrow represents the possibility of glucose-6-phosphate hydrolysis by glucose-6-phosphatase activity, if any.

(continued)

tissues only a very small fraction of the deoxyglucose-6-phosphate formed proceeds to these products (22). In any case, these compounds are secondary, relatively stable products of deoxyglucose-6-phosphate, and all combined together represent the products of deoxyglucose phosphorylation. Deoxyglucose-6-phosphate and its derivatives, once formed, are, therefore, essentially trapped in the cerebral tissues, not forever but long enough to allow a reasonable period of measurement.

If the interval of time following intravenous administration of $[^{14}C]$deoxyglucose is kept short enough (e.g., less than one hour) to allow the assumption of negligible loss of $[^{14}C]$DG-6-P and/or its secondary products from the tissues, then the quantity of labeled products accumulated in any cerebral tissue at any given time following the introduction of $[^{14}C]$DG into the circulation is equal to the integral of the rate of $[^{14}C]$DG phosphorylation by hexokinase in that tissue during that interval of time. This integral is in turn related to the amount of glucose that has been phosphorylated over the same interval, depending on the time courses of the relative concentrations of $[^{14}C]$DG and glucose in the precursor pools and the Michaelis-Menten kinetic constants for hexokinase with respect to both $[^{14}C]$DG and glucose. With cerebral glucose consumption in a steady state, the amount of glucose phosphorylated during the interval of time equals the steady state flux of glucose through the hexokinase-catalyzed step times the duration of the interval, and the net rate of flux of glucose through this step equals the rate of glucose utilization.

These relationships can be mathematically defined and an operational equation derived if the following assumptions are made: 1) a steady state for glucose (i.e., constant plasma glucose concentration and constant rate of glucose consumption) throughout the period of the procedure; 2) homogeneous tissue compartment with respect to blood flow, transport, and concentrations of $[^{14}C]$DG and glucose; and 3) tracer concentrations of $[^{14}C]$DG (i.e., molecular concentrations of free $[^{14}C]$DG essentially equal to zero. The operational equation which defines R_i, the rate of glucose consumption per unit mass of tissue, i, in terms of measurable variables is presented in Figure 3B.

The rate constants, K_1^*, k_2^*, k_3^*, are determined in a separate group of animals by a non-linear, iterative process that provides the least squares best-fit of an equation which defines the time course of total tissue ^{14}C concentration in terms of the time, the history of the plasma concentration, and the rate constants to the experimentally determined time courses of tissue and plasma concentrations of ^{14}C (40). The rate constants have thus far been completely determined only in

---

Fig. 3(B). (continued)

Operational equation of radioactive deoxyglucose method and its functional anatomy. T represents the time at the termination of the experimental period; λ equals the ratio of the distribution space of deoxyglucose in the tissue to that of glucose; ϕ equals the fraction of glucose which, once phosphorylated, continues down the glycolytic pathway; and K_m^* and V_m^* and K_m and V_m represent the familiar Michaelis-Menten kinetic constants of hexokinase for deoxyglucose and glucose, respectively. The other symbols are the same as those defined in Figure 3A. Note the similarity in the structures of the operational equation and of the general equation in Fig. 2.

normal conscious albino rats (TABLE 1). Partial analyses indicate that the values are quite similar in the conscious monkey (17), dog (6) and cat (M. Miyaoka, J. Magnes, C. Kennedy, and L. Sokoloff, unpublished data).

The λ, φ, and the enzyme kinetic constants are grouped together to constitute a single, lumped constant (Fig. 3B). It can be shown mathematically that this lumped constant is equal to the asymptotic value of the product of the ratio of the cerebral extraction ratios of $[^{14}C]DG$ and glucose and the ratio of the arterial blood to plasma specific activities when the arterial plasma $[^{14}C]DG$ concentration is maintained constant (40). The lumped constant is also determined in a separate group of animals from arterial and cerebral venous blood samples drawn during a programmed intravenous infusion which produces and maintains a constant arterial plasma $[^{14}C]DG$ concentration (40). An example of such a determination in a conscious monkey is illustrated in Figure 4. Thus far the lumped constant has been determined only in the albino rat, monkey, cat, dog, sheep, and man (TABLE 2). Under normal conditions the lumped constant appears to be characteristic of the species and does not appear to change significantly in a wide range of physiological conditions (TABLE 2) (39). It has been found to change in pathophysiological conditions, markedly in severe hypoglycemia (42), slightly in hyperglycemia (33), and whenever the rate of glucose utilization becomes limited by glucose supply.

Despite its complex appearance, the operational equation in Figure 3B is really nothing more than a general statement of the standard relationship by which rates of enzyme-catalyzed reactions are determined

TABLE 1

VALUES OF RATE CONSTANTS IN THE NORMAL CONSCIOUS ALBINO RAT

Structure	Rate Constants (min^{-1})			Distribution Volume (ml/g)	Half-Life Precursor P (min)
	K_1^*	k_2^*	k_3^*	$K_1^*/(k_2^*+k_3^*)$	$Log_e 2/(k_2^*+k_3^*)$
Gray Matter					
Visual Cortex	0.189 ± 0.048	0.279 ± 0.176	0.063 ± 0.040	0.553	2.03
Auditory Cortex	0.226 ± 0.068	0.241 ± 0.198	0.067 ± 0.057	0.734	2.25
Parietal Cortex	0.194 ± 0.051	0.257 ± 0.175	0.062 ± 0.045	0.608	2.17
Sensory-Motor Cortex	0.193 ± 0.037	0.208 ± 0.112	0.049 ± 0.035	0.751	2.70
Thalamus	0.188 ± 0.045	0.218 ± 0.144	0.053 ± 0.043	0.694	2.56
Medial Geniculate Body	0.219 ± 0.055	0.259 ± 0.164	0.055 ± 0.040	0.697	2.21
Lateral Geniculate Body	0.172 ± 0.038	0.220 ± 0.134	0.055 ± 0.040	0.625	2.52
Hypothalamus	0.158 ± 0.032	0.226 ± 0.119	0.043 ± 0.032	0.587	2.58
Hippocampus	0.169 ± 0.043	0.260 ± 0.166	0.056 ± 0.040	0.535	2.19
Amygdala	0.149 ± 0.028	0.235 ± 0.109	0.032 ± 0.026	0.558	2.60
Caudate-Putamen	0.176 ± 0.041	0.200 ± 0.140	0.061 ± 0.050	0.674	2.66
Superior Colliculus	0.198 ± 0.054	0.240 ± 0.166	0.046 ± 0.042	0.692	2.42
Pontine Gray Matter	0.170 ± 0.040	0.246 ± 0.142	0.037 ± 0.033	0.601	2.45
Cerebellar Cortex	0.225 ± 0.066	0.392 ± 0.229	0.059 ± 0.031	0.499	1.54
Cerebellar Nucleus	0.207 ± 0.042	0.194 ± 0.111	0.038 ± 0.035	0.892	2.99
Mean ± S.E.M.	0.189 ± 0.012	0.245 ± 0.040	0.052 ± 0.010	0.647 ± 0.073	2.39 0
White Matter					
Corpus Callosum	0.085 ± 0.015	0.135 ± 0.075	0.019 ± 0.033	0.552	4.50
Genu of Corpus Callosum	0.076 ± 0.013	0.131 ± 0.075	0.019 ± 0.034	0.507	4.62
Internal Capsule	0.077 ± 0.015	0.134 ± 0.085	0.023 ± 0.039	0.490	4.41
Mean ± S.E.M.	0.079 ± 0.008	0.133 ± 0.046	0.020 ± 0.020	0.516 ± 0.171	4.51 0

From Sokoloff, et al. (29)

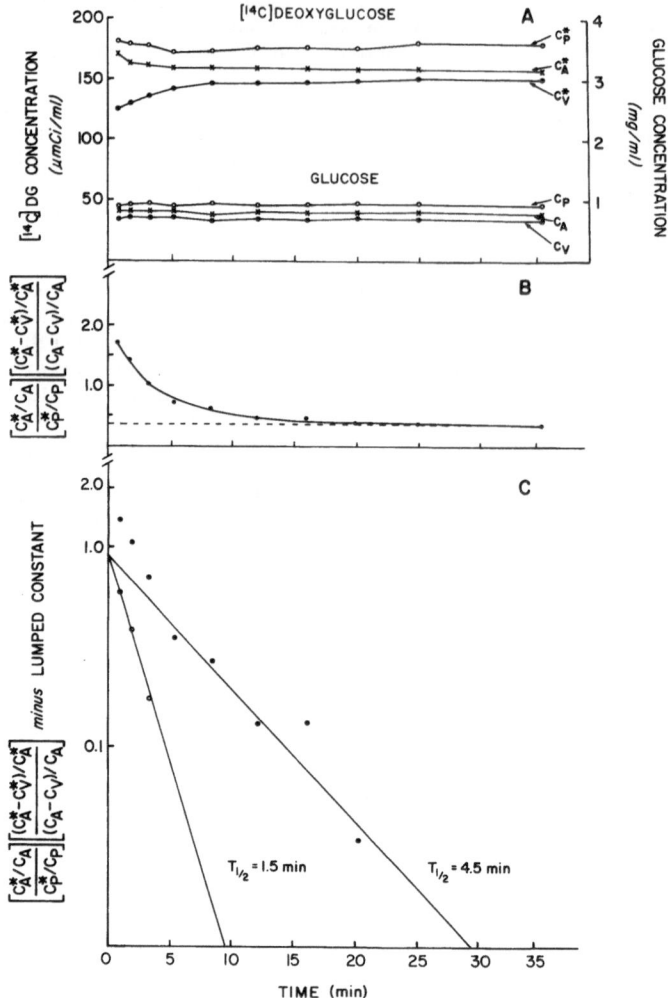

Fig. 4. Data obtained and their use in determination of the lumped
constant and the combination of rate constants, $(k_2^* + k_3^*)$, in a
representative experiment in a monkey. (A) Time courses of arterial
blood and plasma concentrations of $[^{14}C]DG$ and glucose and cerebral
venous blood concentrations of $[^{14}C]DG$ and glucose during programmed
intravenous infusion of $[^{14}C]DG$ designed to achieve and maintain a
constant arterial plasma $[^{14}C]DG$ concentration. (B) Arithemetric plot
of the function derived from the variables in (A) and combined as
indicated in the formula on the ordinate against time. This function
declines exponentially, with a rate constant equal to $(k_2^* + k_3^*)$, until
it reaches an asymptotic value equal to the lumped constant, 0.35, in
this experiment (dashed line). (C) Semilogarithmic plot of the curve in
(B) less the lumped constant, i.e., its asymptotic value. Solid circles
represent actual values. This curve is analyzed into two components by
a standard curve-peeling technique to yield the two straight lines
representing the separate comnponents. Open circles are points for the
fast component, obtained by substracting the values for the slow
component from the solid circles. The rate constants for these two

(continued)

from measurements made with radioactive tracers (Fig. 2). The numerator of the equation represents the amount of radioactive product formed in a given interval of time; it is equal to C_i^*, the combined concentrations of $[^{14}C]DG$ and $[^{14}C]DG\text{-}6\text{-}P$ and its further products in the tissue at time, T, measured by the quantitative autoradiographic technique, less a term that represents the free unmetabolized $[^{14}C]DG$ still remaining in the tissue. The denominator represents the integrated specific activity of the precursor pool times a factor, the lumped constant, which is analogous to a correction factor for an isotope effect. The term with the exponential factor in the denominator takes into account the lag in the equilibration of the tissue precursor pool with that of the plasma.

By the use of labeled 2-deoxyglucose as a probe, a single biochemical reaction, the first step in the pathway of glucose metabolism has been isolated in vivo (Fig. 5). This is the hexokinase-catalyzed phosphorylation of glucose. The total amount of radioactive product formed and the integrated specific activity of the

TABLE 2. *Values of the lumped constant in several species*

Animal	Number of animals	Mean ± SD	SEM
Albino rat			
Conscious	15	0.464 ± 0.099[a]	±0.026
Anesthetized	9	0.512 ± 0.118[a]	±0.039
Conscious (5% CO_2)	2	0.463 ± 0.122[a]	±0.086
Combined	26	0.481 ± 0.119	±0.023
Rhesus monkey			
Conscious	7	0.344 ± 0.095	±0.036
Cat			
Anesthetized	6	0.411 ± 0.013	±0.005
Dog (beagle puppy)			
Conscious	7	0.558 ± 0.082	±0.031
Sheep			
Fetus	5	0.416 ± 0.031	±0.014
Newborn	4	0.382 ± 0.024	±0.012
Mean	9	0.400 ± 0.033	±0.011
Humans			
Conscious	6	0.568 ± 0.105	±0.043

[a]No statistically significant difference between normal conscious and anesthetized rats ($0.3 < p < 0.4$) and conscious rats breathing 5% CO_2 ($p > 0.9$).

Fig. 4 (continued)

components represent the values of $(k_2^* + k_3^*)$ for two compartments; the fast and slow compartments are assumed to represent gray and white matter, respectively. In this experiment the values for $(k_2^* + k_3^*)$ were found to equal 0.462 (half-time = 1.5 min) and 0.154 (half-time = 4.5 min) in gray and white matter, respectively. From Kennedy et al. (18).

precursor at the enzyme site can be determined. From these data and the use of a correction factor, the "lumped constant," for the difference in the kinetic behavior of deoxyglucose and glucose, the net rate of glucose phosphorylation can be calculated by the operational equation. In a steady state the net rate through any step in a pathway equals the net rate through the overall pathway. The deoxyglucose method,

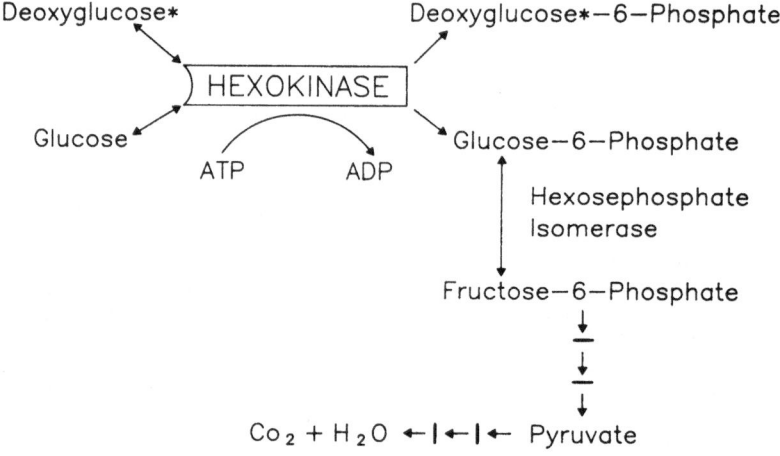

Fig. 5. Schema illustrating the fundamental principle of the radioactive deoxyglucose method for the measurement of local cerebral glucose utilization. Glucose utilization commences with the hexokinase-catalyzed phosphorylation of glucose by ATP, but the product of this reaction, glucose-6-phosphate, is not retained in the tissues. Instead, it is metabolized further to products, like CO_2 and H_2O, that leave the tissue. Deoxyglucose, an analogue and competitive substrate with glucose in the hexokinase reaction, leads to a product, deoxyglucose-6-phosphate, that does accumulate in the tissue quantitatively for a reasonable length of time. By putting a label on the deoxyglucose, it is possible to measure the rate of labeled deoxyglucose-6-phosphate formation. From a knowledge of the time course of the relative concentrations of labeled deoxyglucose and glucose in the tissue at the enzyme site and the relative Michaelis-Menten constants of hexokinase for the two substrates, it is possible to calculate how much glucose must have also been phosphorylated during the production of the measured amount of deoxyglucose-6-phosphate. The integrated relative concentrations of labeled deoxyglucose and glucose in the tissue are calculated from the measured time courses of the two compounds in the arterial plasma by subtracting from the integrated plasma specific activity a term that corrects for the lag of the tissue behind the plasma.

therefore, measures in vivo the net rate of glucose phosphorylation and in a steady state the net rate of the entire glycolytic pathway.

Procedure for Measurement of Local Cerebral Glucose Utilization

Theoretical considerations in the design of the procedure. The operational equation of the method specifies the variables to be measured in order to determine R_i, the local rate of glucose consumption in the brain. The following variables are measured in each experiment: 1) the entire history of the arterial plasma [14C]deoxyglucose concentration, C_p*, from zero time to the time of killing, T; 2) the steady state arterial plasma glucose level, C_p, over the same interval; and 3) the local concentration of 14C in the tissue at the time of killing, $C_i*(T)$. The rate constants, K_1*, k_2*, and k_3*, and the lumped constant, $\lambda V_m* k_m/\phi V_m k_m*$, are not measured in each experiment; the values for these constants that are used are those determined separately in other groups of animals as described above and presented in Tables 1 and 2.

The operational equation is generally applicable with all types of arterial plasma [14C]DG concentration curves. Its configuration, however, suggests that a declining curve approaching zero by the time of killing is the choice to minimize certain potential errors. The quantitative autoradiographic technique measures only total 14C concentration in the tissue and does not distinguish between [14C]DG-6-P and [14C]DG. It is, however, [14C]DG-6-P concentration that must be known to determine glucose consumption. [14C]DG-6-P concentration is calculated in the numerator of the operational equation, which equals the total tissue 14C content, $C_i*(T)$, minus the [14C]DG concentration present in the tissue, estimated by the term containing the exponential factor and rate constants. In the denominator of the operational equation there is also a term containing an exponential factor and rate constants. Both these exponential terms have the useful property of approaching zero with increasing time if C_p* is also allowed to approach zero. The rate constants, K_1*, k_2*, and k_3*, are not measured in the same animals in which local glucose consumption is being measured, and the standard normal rate constants in Table 1 may not be equally applicable in all physiological, pharmacological, and pathological states. One possible solution is to redetermine the rate constants for each condition to be studied. An alternative solution, and the one chosen, is to administer the [14C]DG as a single intravenous pulse at zero time and to allow sufficient time for the clearance of [14C]DG from the plasma and the terms containing the rate constants to fall to levels too low to influence the final result. To wait until these terms reach zero is impractical because of the long time required and the risk of effects of the small but finite rate of loss of [14C]DG-6-P from the tissues. A reasonable time interval is 45 minutes; by this time the plasma level has fallen to very low levels, and, on the basis of the values of $(k_2* + k_3*)$ in Table 1, the exponential factors have declined through at least 6-10 half-lives, at least under physiological conditions (Fig. 6).

Experimental Protocol. The animals are prepared for the experiment by the insertion of polyethylene catheters in an artery and vein. Any convenient artery or vein can be used. In the rat the femoral or the tail arteries and veins have been found satisfactory. In the monkey and the cat the femoral vessels are probably most convenient. The catheters are inserted under anesthesia, and anesthetic agents without long-lasting after-effects should be used. Light halothane anesthesia

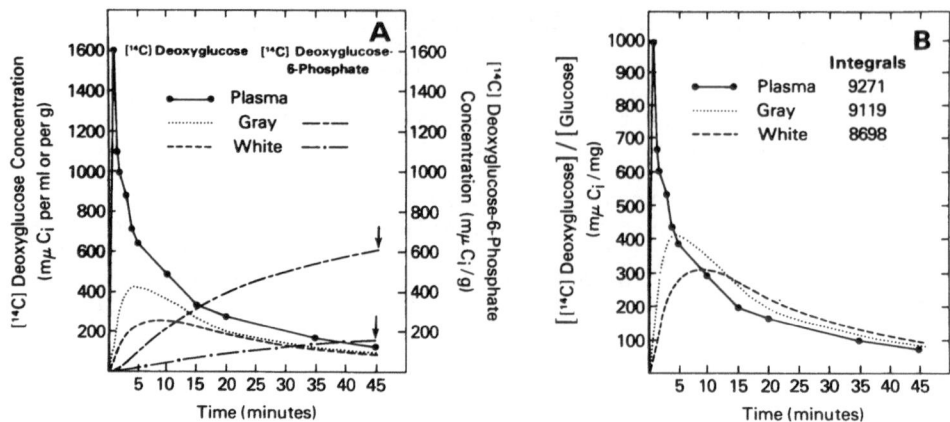

Fig. 6. Graphical representation of the significant variables in the
operational equation used to calculate local cerebral glucose
utilization. (A), Time courses of [14C]deoxyglucose concentrations in
arterial plasma and in average gray and white matter and
[14C]deoxyglucose-6-phosphate concentrations in average gray and white
matter following an intravenous pulse of 50 μCi of [14C]deoxyglucose.
The plasma curve is derived from measurements of plasma
[14C]deoxyglucose concentrations. The tissue [14C]deoxyglucose
concentrations were calculated from the plasma curve and the mean values
of K_1^*, k_2^*, and k_3^* for gray and white matter in Table 1 according to
the second term in the numerator of the operational equation. The
[14C]deoxyglucose-6-phosphate concentrations in the tissues were
calculated from the mean values of k_3^* and the integral of the
[14C]deoxyglucose concentrations in the tissues. The arrows point to
the concentrations of [14C]deoxyglucose and
[14C]deoxyglucose-6-phosphate in the tissues at the time of killing; the
autoradiographic technique measures the total 14C content (i.e. the sum
of these concentrations) at that time, which is equal to $C_i^*(T)$, the
first term in the numerator of the operational equation. Note that at
the time of killing, the total 14C content represents mainly
[14C]deoxyglucose-6-phosphate concentration, especially in gray matter.
(B), Time courses of ratios of [14C]deoxyglucose to glucose
concentrations (i.e. specific activities) in plasma and average gray and
white matter. The curve for plasma was determined by division of the
plasma curve in (A) by the plasma glucose concentrations. The curves
for the tissues were calculated by the derivative of the function in the
right set of brackets in the denominator of the operational equation.
The integrals in (B) are the integrals of the specific activities with
respect to time and represent the areas under the curves. The integrals
under the tissue curves are equivalent to all of the denominator of the
operational equation, except for the lumped constant. Note that by the
time of killing, the integrals of the tissue curves approach equality
with each other and with that of the plasma curve. From Sokoloff et al.
(40).

with or without supplementation with nitrous oxide has been found to be quite satisfactory. At least two hours are allowed for recovery from the surgery and anesthesia before initiation of the experiment.

The design of the experimental procedure for the measurement of local cerebral glucose utilization is based on the theoretical considerations discussed above. At zero time a pulse of 125 μCi (no more than 2.5 μmoles to adhere to tracer conditions) of [14C]deoxyglucose per kg of body weight is administered to the animal via the venous catheter. Arterial sampling is initiated with the onset of the pulse, and timed 50-100 μl samples of arterial blood are collected consecutively as rapidly as possible during the early period so as not to miss the peak of the arterial curve. Arterial sampling is continued at less frequent intervals later in the experimental period but at sufficient frequency to define fully the arterial curve. The arterial blood samples are immediately centrifuged to separate the plasma, which is stored on ice until assayed for [14C]DG concentrations by liquid scintillation counting and glucose concentrations by standard enzymatic methods. At approximately 45 minutes the animal is decapitated, and the brain is removed and frozen in Freon XII or isopentane maintained between -50° and -75°C with liquid nitrogen. When fully frozen, the brain is stored at -70°C until sectioned and autoradiographed. The experimental period may be limited to 30 minutes. This is theoretically permissible and may sometimes be necessary for reasons of experimental expediency, but greater errors due to possible inaccuracies in the rate constants may result.

Autoradiographic Measurement of Tissue 14C Concentration. The 14C concentrations in localized regions of the brain are measured by a modification of the quantitative autoradiographic technique previously described (29). The frozen brain is coated with chilled embedding medium (Lipshaw Manufacturing Co., Detroit MI) and fixed to object-holders appropriate to the microtome to be used. Brain sections, precisely 20 μm in thickness, are cut in a cryostat maintained at -20°C to -22°C. The brain sections are picked up on glass cover slips, dried on a hot plate at 60°C for at least 5 minutes, and placed sequentially in an X-ray cassette. A set of [14C]methylmethacrylate standards (Amersham Corp., Arlington Heights, IL), which include a blank and a series of progressively increasing 14C concentrations, are also placed in the cassette. These standards must previously have been calibrated for their autoradiographic equivalence to the 14C concentrations in brain sections, 20 μm in thickness, prepared as described above. The method of calibration has been previously described (29).

Autoradiographs are prepared from these sections directly in the X-ray cassette with Kodak single-coated, blue-sensitive Medical X-ray Film, Type SB-5 (Eastman Kodak Co., Rochester, NY). The exposure time is generally 5-6 days with the doses used as described above, and the exposed films are developed according to the instructions supplied with the film. The SB-5 X-ray film is rapid but coarse grained. For finer grained autoradiographs and, therefore, better defined images with higher resolution, it is possible to use mammographic films, such as DuPont LoDose or Kodak MR-1 films, or fine grain panchromatic film, such as Kodak Plus-X, but the exposure times are 2-3 times longer. The autoradiographs provide a pictorial representation of the relative 14C concentrations in the various cerebral structures and the plastic standards (Fig. 7). A calibration curve of the relationship between optical density and tissue 14C concentration for each film is obtained by densitometric measurements of the portions of the film representing the various standards. The local tissue concentrations are then determined from the calibration curve and the optical densities of the film in the regions representing the cerebral structures of interest. Local cerebral glucose utilization is calculated from the local tissue concentrations of ^{14}C and the plasma [^{14}C]DG and glucose concentrations according to the operational equation (Fig. 3B).

Theoretical and Practical Considerations

The design of the deoxyglucose method is based on an operational equation, derived by the mathematical analysis of a model of the biochemical behavior of [^{14}C]deoxyglucose and glucose in brain (Fig. 3). Although the model and its mathematical analysis are as rigorous and comprehensive as reasonably possible, it must be recognized that models almost always represent idealized situations and cannot possibly take into account all the known, let alone unknown, properties of a complex biological system. Several years have now passed since the introduction of the deoxyglucose method, and numerous applications of it have been made. The results of this experience generally establish the validity and worth of the method, but there are some potential problems in special situations which require further theoretical and practical considerations.

The main potential sources of error are the rate constants and the lumped constant. The problem with them is that they are not determined in the same animals and at the same time when local cerebral glucose

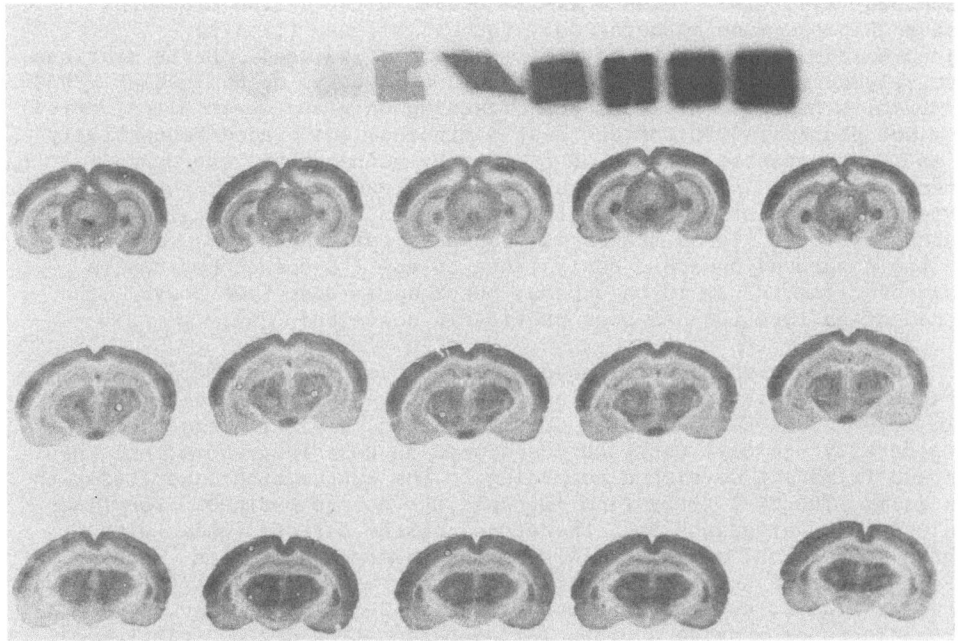

Fig. 7. Autoradiographs of sections of conscious rat brain and of calibrated [^{14}C]methylmethacrylate standards used to quantify ^{14}C concentration in tissues. From Sokoloff et al. (40).

utilization is being measured. They are measured in separate groups of comparable animals and then used subsequently in other animals in which glucose utilization is being measured. The part played by these constants in the method is defined by their role in the operational equation of the method (Fig. 3).

Rate Constants. The rate constants, K_1^*, k_2^*, and k_3^*, for deoxyglucose have been fully determined for various cerebral tissues in the normal conscious albino rat (40) (TABLE 1), but they appear to be similar in other species. All the rate constants vary considerably from tissue to tissue, but the variation among gray structures and among white structures is considerably less than the differences between the two types of tissues (Table 1). The rate constants, k_2^* and k_3^*, appear in the equation only as their sum, and $(k_2^* + k_3^*)$ is equal to the rate constant for the turnover of the free $[^{14}C]$deoxyglucose pool in the tissue. The half-life of the free $[^{14}C]$deoxyglucose pool can then be calculated by dividing $(k_2^* + k_3^*)$ into the natural logarithm of 2 and has been found to average 2.4 minutes in gray matter and 4.5 minutes in white matter in the normal conscious rat (TABLE 1).

The rate constants vary not only from structure to structure but can be expected to vary with the condition. For example, K_1^* and k_2^* are influenced by both blood flow and transport of $[^{14}C]$deoxyglucose across the blood-brain barrier, and because of the competition for the transport carrier, the glucose concentrations in the plasma and tissue affect the transport of $[^{14}C]$deoxyglucose and, therefore, also K_1^* and k_2^*. The constant, k_3^*, is related to phosphorylation of $[^{14}C]$deoxyglucose and will certainly change when glucose utilization is altered. To minimize potential errors due to inaccuracies in the values of the rate constants used, it was decided to sacrifice time resolution for accuracy. If the $[^{14}]$deoxyglucose is given as an intravenous pulse and sufficient time is allowed for the plasma to be cleared of the tracer, then the influence of the rate constants, and the functions that they represent, on the final result diminishes with increasing time until ultimately it becomes zero. This relationship is implicit in the structure of the operational equation (Fig. 3); as C_p^* approaches zero, the terms containing the rate constants also approach zero with increasing time. The significance of this relationship is graphically illustrated in Fig. 8. From typical arterial plasma $[^{14}C]$deoxyglucose and glucose concentration curves obtained in a normal conscious rat, the portion of the denominator of the operational equation underlined by the heavy bar was computed with a wide range of values for $(k_2^* + k_3^*)$ as a function of time. The values for $(k_2^* + k_3^*)$ are presented as their equivalent half-lives calculated as described above. The values of $(k_2^* + k_3^*)$ vary from infinite (i.e., T 1/2 = 0 min) to 0.14 per min (i.e., T 1/2 = 5 min) and more than cover the range of values to be expected under physiological conditions. The portion of the equation underlined and computed represents the integral of the precursor pool specific activity in the tissue. The curves represent the time course of this function, one each for every value of $(k_2^* + k_3^*)$ examined. It can be seen that these curves are widely different at early times but converge with increasing time until at 45 minutes the differences over the entire range of $(k_2^* + k_3^*)$ equal only a small fraction of the value of the integral. These curves demonstrate that at short times enormous errors can occur if the values of the rate constants are not precisely known, but only negligible errors occur at 45 minues, even over a wide range of rate constants of several fold. In fact, it was precisely for this reason that $[^{14}C]$deoxyglucose rather than $[^{14}C]$glucose was selected as the tracer for glucose metabolism. The relationships are similar for glucose. Because the products of [14C]glucose metabolism are so rapidly lost from the tissues, it is necessary with [14C]glucose to limit the experimental period to short times when enormous errors can occur if the rate constants are not precisely known for each individual structure. There is a common misconception that [14C]deoxyglucose can be used only with long experimental periods. Quite the contrary, [14C]deoxyglucose can be used with experimental periods as short as those with [14C]glucose, but with the same degree of imprecision. The advantage of [14C]deoxyglucose is that, because of its stable products, it permits the prolongation of the experimental period to times when inaccuracies in rate constants have little effect on the final result.

It should be noted, however, that in pathological conditions, such as severe ischemia or hyperglycemia, the rate constants may fall far

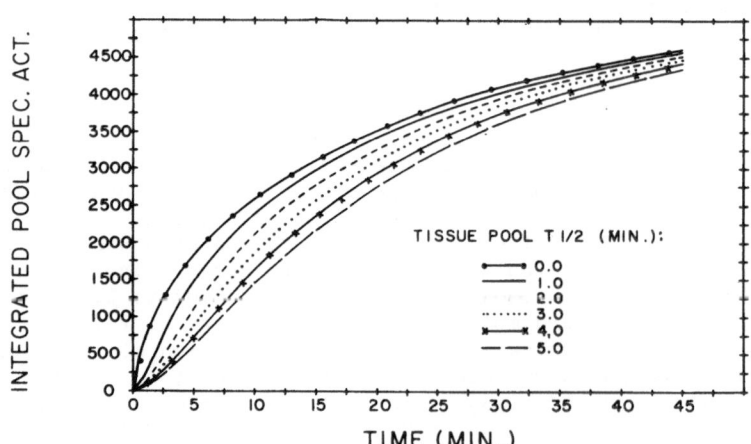

OPERATIONAL EQUATION:

$$R_i = \frac{C_i^*(T) - k_1^* e^{-(k_2^* + k_3^*)T} \int_0^T C_p^* e^{(k_2^* + k_3^*)t} \, dt}{\left(\dfrac{\lambda V_m^* K_m}{\phi V_m K_m^*}\right)\left[\int_0^T (C_p^*/C_p)\,dt - e^{-(k_2^* + k_3^*)T} \int_0^T (C_p^*/C_p) e^{(k_2^* + k_3^*)t} \, dt\right]}$$

TISSUE POOL T 1/2 (MIN.):
- 0.0
- 1.0
- 2.0
- 3.0
- 4.0
- 5.0

INTEGRATED POOL SPEC. ACT.

TIME (MIN.)

Fig. 8. Influence of time and rate constants, $(k_2^* + k_3^*)$, on integrated precursor pool specific activity in a normal conscious rat given an intravenous pulse of 50 μCi of [14C]deoxyglucose at zero time. The time courses of the arterial plasma [14C]deoxyglucose and glucose concentrations were measured following the pulse. The portion of the equation underlined, corresponding to integrated precursor pool specific activity, was computed as a function of time with different values of $(k_2^* + k_3^*)$, as indicated by their equivalent half-lives, calculated according to T 1/2 = $0.693/(k_2^* + k_3^*)$. From Sokoloff (37,39).

below the range examined in Figure 8. There is evidence, for example, that this occurs with hyperglycemia and ischemia (11). Also in conditions of markedly depressed cerebral glucose utilization and/or slow clearance of the $[^{14}C]DG$ from the plasma, the second term in the numerator which contains the rate constants may not be much smaller than $C_i^*(T)$, the first term in the numerator. Inaccuracies in the rate constants may then produce considerable errors in the numerator. In such abnormal conditions it is necessary and feasible to redetermine the rate constants for the particular condition under study. Alternatively the experimental period can be prolonged to allow more time for the exponential terms to diminish further; it may then be necessary to correct for the loss of $[^{14}C]DG$-6-P that might then occur (39).

Lumped Constant. The lumped constant is composed of six separate constants. One of these, ϕ, is a measure of the steady-state hydrolysis of glucose-6-phosphate to free glucose and phosphate. Because in normal brain tissue there is little such phosphohydrolase activity (12), ϕ is normally approximately equal to unity. The other components are arranged in three ratios: λ, which is the ratio of distribution spaces in the tissue for deoxyglucose and glucose; V_m^*/V_m; and K_m/K_m^*. Although each individual constant may vary from structure to structure and condition to condition, it is likely that the ratios tend to remain the same under normal physiological conditions. For reasons described in detail previously (40) and verified experimentally (5), it is reasonable to believe that the lumped constant is the same throughout the brain and more or less characteristic of the species of animal under normal physiological conditions. Empirical experience thus far indicates that it is generally so, except in special pathophysiological states. The greatest experience has been accumulated in the albino rat. In this species the lumped constant for the brain as a whole has been determined under a variety of conditions (40). In the normal conscious rat local cerebral glucose utilization, determined by the $[^{14}C]$deoxyglucose method with the single value of the lumped constant for the brain as a whole, correlates almost perfectly ($r = 0.96$) with local cerebral blood flow, measured by the $[^{14}C]$iodoantipyrine method, an entirely independent method (36). It is generally recognized that local blood flow is adjusted to local metabolic rate, but if the single value of the lumped constant did not apply to the individual structures studied, then errors in local glucose utilization would occur that might be expected to obscure the correlation. Also, the lumped constant has been directly determined in the albino rat in the normal conscious state, under barbiturate anesthesia, and during the inhalation of 5% CO_2; no significant differences were observed (TABLE 2). The lumped constant does vary with the species of animal. It has now also been determined in the rhesus monkey (18), cat (M. Miyaoka, J. Magnes, C. Kennedy, M. Shinohara, and L. Sokoloff, unpublished data), Beagle puppy (6), sheep (1), and man (28) and each species has a different value (Table 2). The values for local rates of glucose utilization determined with these lumped constants in these species are very close to what might be expected from measurement of energy metabolism in the brain as a whole by other methods (TABLE 3).

Although there is yet no experimental evidence of more than negligible changes in the lumped constant under physiological conditions, it certainly does change in pathophysiological states. For example, in severe hypoglycemia there is a progressive and appreciable increase in the lumped constant with falling plasma glucose concentration (42), and in severe hyperglycemia there is a small decrease (33). In severe seizure activity, when the rate of glucose consumption is so increased that it exceeds the rate of resupply, the

TABLE 3

REPRESENTATIVE VALUES FOR LOCAL CEREBRAL GLUCOSE UTILIZATION
IN THE NORMAL CONSCIOUS ALBINO RAT AND MONKEY (μmoles/100g/min)

Structure	Albino Rat (10)	Monkey (7)
Gray Matter		
Visual Cortex	107 ± 6	59 ± 2
Auditory Cortex	162 ± 5	79 ± 4
Parietal Cortex	112 ± 5	47 ± 4
Sensory-Motor Cortex	120 ± 5	44 ± 3
Thalamus: Lateral Nucleus	116 ± 5	54 ± 2
Thalamus: Ventral Nucleus	109 ± 5	43 ± 2
Medial Geniculate Body	131 ± 5	65 ± 3
Lateral Geniculate Body	96 ± 5	39 ± 1
Hypothalamus	54 ± 2	25 ± 1
Mamillary Body	121 ± 5	57 ± 3
Hippocampus	79 ± 3	39 ± 2
Amygdala	52 ± 2	25 ± 2
Caudate-Putamen	110 ± 4	52 ± 3
Nucleus Accumbens	82 ± 3	36 ± 2
Globus-Pallidus	58 ± 2	26 ± 2
Substantia Nigra	58 ± 3	29 ± 2
Vestibular Nucleus	128 ± 5	66 ± 3
Cochlear Nucleus	113 ± 7	51 ± 3
Superior Olivary Nucleus	133 ± 7	63 ± 4
Inferior Colliculus	197 ±10	103 ± 6
Superior Colliculus	95 ± 5	55 ± 4
Pontine Gray Matter	62 ± 3	28 ± 1
Cerebellar Cortex	57 ± 2	31 ± 2
Cerebellar Nuclei	100 ± 4	45 ± 2
White Matter		
Corpus Callosum	40 ± 2	11 ± 1
Internal Capsule	33 ± 2	13 ± 1
Cerebellar White Matter	37 ± 2	12 ± 1
Weighted Average for Whole Brain		
	68 ± 3	36 ± 1

Note: The values are the means ± standard errors from measurements made in the number of animals indicated in parentheses.

lumped constant rises (5). Indeed, theoretically the lumped constant could be expected to change whenever there is an alteration in the balance between glucose supply to the tissue and the tissue's rate of glucose utilization. It is only when the rate of glucose utilization becomes limited by the supply, however, that a major change in the lumped constant occurs. Also, tissue damage may disrupt the normal cellular compartmentation, and there is no assurance that λ, the ratio of the distribution spaces for [^{14}C]deoxyglucose and glucose, is the same in damaged tissue as in normal tissue. In pathological states there may be release of lysosomal acid hydrolases that may hydrolyze glucose-6-phosphate and thus alter the value of ϕ. It is necessary, therefore, to determine the lumped constant in each pathological or pathophysiological state.

Glucose-6-Phosphatase. Glucose-6-Pase activity is known to be very low in brain (12), and almost all textbooks of biochemistry attest to this fact. Although low, some activity is present, but it appears to have no influence on the deoxyglucose method if the experimental period is limited to the prescribed duration, 45 minutes. Beyond this time its effects begin to appear, increasing in magnitude with increasing time. Significant glucose-6-Pase would affect the results by hydrolyzing the $[^{14}C]DG$-6-P and causing loss of product, and its effect would be accumulative with increasing time. If its activity were significant, it would cause the calculated rates of glucose utilization to be too low and to become progressively lower with increasing time. None of this happens in the first 45 minutes. In groups of rats studied over 20, 30, and 45 minute periods, the calculated rates of glucose utilization remain constant (39,40). Also, by determining the average glucose utilization of all the structures in the brain weighted for their relative sizes, it is possible to obtain the weighted average glucose utilization of the brain as a whole and to compare this with the values obtained by the Kety-Schmidt method, the recognized standard method for measuring blood flow and energy metabolism of the brain as a whole by the Fick Principle (20). The deoxyglucose method provides values that are almost exactly those obtained with the Kety-Schmidt method (TABLE 3). There is, therefore, no detectable influence of glucose-6-Pase during the first 45 minutes after the pulse of deoxyglucose because the values for glucose utilization that are obtained are not too low nor do they decrease with time over that interval. After 45 minutes, however, effects of glucose-6-Pase begin to appear and become progressively greater with increasing time (39). The time course of the effects is compatible with the intracellular distributions of the $[^{14}C]DG$-6-P and the phosphatase. The $[^{14}C]DG$-6-P is formed in the cytosol, but the phosphatase is on the inner surface of the cisterns of the endoplasmic reticulum. The $[^{14}C]DG$-6-P must first be formed in the cytosol and then transported across the endoplasmic reticular membrane by a specific carrier before the phosphatase can act to hydrolyze it. The kinetics of this process, namely, a lag with zero phosphatase activity followed by progressively increasing activity, are exactly those to be expected from the separate compartmentalization of substrate and enzyme and a rate limiting transport of the substrate across the membrane to the enzyme. This compartmentalization, therefore, allows a period of grace before the phosphatase can act on the DG-6-P, and it is a prolonged period of grace because the carrier that transports glucose-6-phosphate and DG-6-P across the endoplasmic reticular membrane in other tissues is absent in brain (7).

If one extends the experimental period beyond 45 minutes, then correction must be made for the effects of phosphatase. It is a common misconception that phosphatase activity would invalidate the deoxyglucose method. It would not. In the original version, the model and the operational equation did not take phosphatase activity into account because the evidence described above indicated that there was no reason to do so, as long as the experimental period was limited to 45 minutes, which was the maximum period contemplated. It is a simple matter, however, to modify slightly the model and the derived operational equation to include a k_4^*, a rate constant for the phosphatase activity (13,39) which can be used in studies with prolonged experimental periods beyond 45 minutes.

The evidence above, and other evidence as well, clearly indicate that phosphatase activity is of no significance to the deoxyglucose method if it is carried out within the prescribed 45 minutes. The issue of phosphatase has been kept alive, however, by the efforts of some, who are promoting their own alternative methods, to find something that might discredit the deoxyglucose method. They have published allegations that phosphatase activity is a major source of error in the deoxyglucose method. None of these reports addressed the fact that the

[^{14}C]deoxyglucose technique provides correct, not low, values for
cerebral glucose utilization which are unchanged from 20 to 45 minutes,
all of which would be impossible if there were significant effects of
glucose-6-Pase activity during that period.

In the first of these reports Hawkins and Miller (10) applied the
[^{14}C]deoxyglucose technique to a series of rats but killed them at
various times up to 45 minutes after the intravenous pulse of [^{14}C]DG
and measured the [^{14}C]DG-6-P concentrations in the brains as a function
of time by direct chemical analyses. They also measured the time
courses of the arterial plasma [^{14}C]DG and glucose concentrations up to
the time of killing, and with these data they used a transposed version
of the operational equation (Fig. 3B) of the method to calculate the
theoretical time course of the brain [^{14}C]DG-6-P concentration (i.e.,
the numerator of the operational equation) predicted by the equation,
which assumes no loss of [^{14}C]DG-6-P due to phosphatase activity or any
other process. The computed theoretical curve that they published was
2-3 times higher than their measured concentrations, and they
interpreted their results as proof that there was enormous loss of
[^{14}C]DG-6-P during the experimental period. Their computed theoretical
curve was erroneous, however, because of a succession of errors,
compounded one upon another, that led to their misinterpretation. To
compute the [^{14}C]DG-6-P concentrations with the transposed equation it
was necessary for them to assume values for the rate of glucose
utilization and the lumped constant. They assumed a reasonable value,
0.7-0.8 μmoles/g/min for the glucose utilization. For the lumped
constant, however, they used a value of 1.1-1.25, which was 2-3 times
greater than 0.48, the value experimentally measured in rats and
reported (40). The use of the erroneous lumped constant was the direct
result of an earlier error. They calculated the lumped constant at
different times by dividing into the measured [^{14}C]DG-6-P concentration
the product of the rate of glucose utilization and the integrated plasma
specific activity corrected for the lag of the tissue behind the plasma.
This calculation is based on another transposition of the operational
equation of the method and is in principle valid. Again they assumed
the reasonable value of 0.7-0.8 μmoles/g/min for glucose utilization.
The calculated plasma specific activity was wrong, however, because they
did not measure the [^{14}C]deoxyglucose concentration in the plasma during
the first minute after the intravenous pulse of [^{14}C]deoxyglucose.
Instead, they extrapolated the arterial curve to zero time
semi-logarithmically from the measured values at 1 minute and 2 minutes
after the pulse. By doing so, they cut off the peak under the arterial
curve that occurs within the first minute after the pulse and thereby
lost a significant portion of the arterial specific activity curve
during the first 5 minutes. In calculating the lumped constant they,
therefore, divided by too small a value for the integrated plasma
specific activity during the first 5 minutes, but with increasing time
the lost portion under the plasma specific activity curve became less
significant and had less effect. This led to their erroneous conclusion
that the lumped constant decreases with time; the decrease is due to
diminishing influence of the error introduced by their failure to
include the early peak in the arterial curve. The lumped constant that
they used in their calculation of the theoretical tissue [^{14}C]DG-6-P
curve was the most erroneous one of all, the one obtained at 5 minutes
and equal to 1.1-1.25. When the theoretical curve for brain [^{14}C]DG-6-P
concentrations are calculated from their data with the correct lumped
constant, 0.48, the one that has been directly measured and equal to the
one they also found at 45 minutes, then the curve fits their measured
values remarkably well, proving, contrary to their published conclusion,
that there is no detectable loss of [^{14}C]DG-6-P from brain due to

phosphatase activity or any other cause during the first 45 minutes after the pulse of $[^{14}C]DG$.

In a second report Sacks et al. (32) infused $[^{14}C]$glucose-6-P intravenously and measured cerebral arteriovenous differences of $[^{14}C]$glucose-6-P and $[^{14}C]$glucose during the infusion. They reported positive arteriovenous differences for the $[^{14}C]$glucose-6-P and negative arteriovenous differences for the $[^{14}C]$glucose, which they interpreted as evidence that the brain was taking up the $[^{14}C]$glucose-6-P, hydrolyzing it, and releasing $[^{14}C]$glucose as a result of phosphatase activity. It is generally recognized that hexose phosphates cannot cross the blood-brain barrier, and this has been confirmed by recent experiments (24). There must have been some artifact in the experiments of Sacks et al. (32) because it would be an incredible feat for phosphatase in brain to hydrolyze a substance circulating in the blood that cannot cross the blood-brain barrier and have access to the enzyme.

A third report is by Huang and Veech (14), who injected into rats a mixture of $[2-^3H]$glucose and $[U-^{14}C]$glucose as a pulse into a carotid artery, removed the brains at various times up to 5 minutes later, and assayed the tissue glucose content for the $^3H/^{14}C$ ratio in it. The principle is as follows. Any $[2-^3H]$glucose which reaches the fructose-6-P step in glycolysis loses the ^3H label, but $[^{14}C]$fructose-6-P does not lose the ^{14}C label. By the back-reaction some fructose-6-P is converted to glucose-6-P, and if there is phosphatase activity, the glucose moiety enters the glucose pool without losing the ^{14}C label, but without the ^3H label. The $^3H/^{14}C$ ratio in the glucose pool should then fall in case of phosphatase activity, and this is what they claim to have observed. Indeed, they concluded from their results that the rate of dephosphorylation of glucose-6-P by glucose-6-Pase in brain was at least 35% of the rate of glucose phosphorylation by hexokinase. This incredibly high estimate of glucose-6-Pase activity in brain is at variance with all previous estimates and extraordinarily higher than the most liberal former claims of cerebral phosphatase activity. It is at a level one would expect from organs with known high rates of gluconeogenesis, like liver and kidney, and in conflict with results of decades of research showing very low rates of gluconeogenesis in brain.

The studies of Huang and Veech (14) have been found, however, to be grossly defective. An absolute requirement of their experiments is the separation of radioactive glucose from all the other possible labeled products of glucose metabolism in the brain tissue to insure that the ^3H and ^{14}C measurements are truly in glucose and nothing else; other labeled products, if present, could lead to the same results without the need to invoke phosphatase activity. In fact, when the experiments of Huang and Veech (14) were repeated with careful attention to the separation of the labeled glucose by Dowex-50, Dowex-1, and paper chromatography, there was no change in the $^3H/^{14}C$ ratio of the glucose in brain beyond that observed in the glucose in the plasma supplied to the brain (23,24).

In short, there is positive evidence of no influence of glucose-6-phosphatase in the deoxyglucose method if the experimental period is confined to 45 minutes. There is small but progressively increasing effect beyond that time, but it can easily be corrected for if necessary. Of the reports alleging significant phosphatase activity at all times during the deoxyglucose procedure, all are demonstrably false.

Computerized Color-Coded Image-Processing

The autoradiographs produced by the [^{14}C]DG method provide pictorial representations of only the relative concentrations of the isotope in the various tissues. Because of the use of a pulse followed by a long period before killing, the isotope is contained mainly in [^{14}C]DG-6-P which reflects the rate of glucose metabolism (Fig. 6). The autoradiographs are, therefore, pictorial representations also of the relative but not the actual rates of glucose utilization in all the structures of the nervous system. These autoradiographs often provide dramatic images of local changes in cerebral glucose utilization associated with local changes in functional activity (Fig. 9). Resolution of differences in relative rates, however, is limited by the ability of the human eye to recognize differences in shades of gray. Manual densitometric analysis permits the computation of actual rates of glucose utilization with a fair degree of resolution, but it generates enormous tables of data which fail to convey the tremendous heterogeneity of local cerebral metabolic rates, even within anatomic structures, or the full information contained within the autoradiographs. Goochee et al. (9) have developed a computerized image-processing system to analyze and transform the autoradiographs into color-coded maps of the distribution of the actual rates of glucose utilization exactly where they are located throughout the central nervous system. The autoradiographs are scanned automatically by a computer-controlled scanning microdensitometer. The optical density of each spot in the autoradiograph, from 25 to 100 μm, as selected, is stored in a computer, converted to ^{14}C concentration on the basis of the optical densities of the calibrated ^{14}C plastic standards, and then converted to local rates of glucose utilization by solution of the operational equation of the method. Colors are assigned to narrow ranges of the rates of glucose utilization, and the autoradiographs are then displayed on a color TV monitor in color along with a calibrated color scale for identifying the rate of glucose utilization in each spot of the autoradiograph from its color. These color maps add a third dimension, the rate of glucose utilization on a color scale, to the spatial dimensions already present on the autoradiographs.

Rates of Local Cerebral Glucose Utilization

Local rates of cerebral glucose utiliation determined by the 2-[^{14}C]deoxyglucose method in normal conscious adult animals are presented in TABLE 3. The rates of local cerebral glucose utilization in the normal conscious rat vary widely throughout the brain. The values in white structures tend to group together and are always considerably below those of gray structures. The average value in gray matter is approximately three times that of white matter, but the individual values vary from approximately 50 to 200 μmoles of glucose/100 g/min. The highest values are in the structures involved in auditory functions with the inferior colliculus clearly the most metabolically active structure in the brain.

The rates of local cerebral glucose utilization in the conscious monkey exhibit similar heterogeneity, but they are generally one-third to one-half the values in corresponding structures of the rat brain (TABLE 3). The differences in rates in the rat and monkey brain are consistent

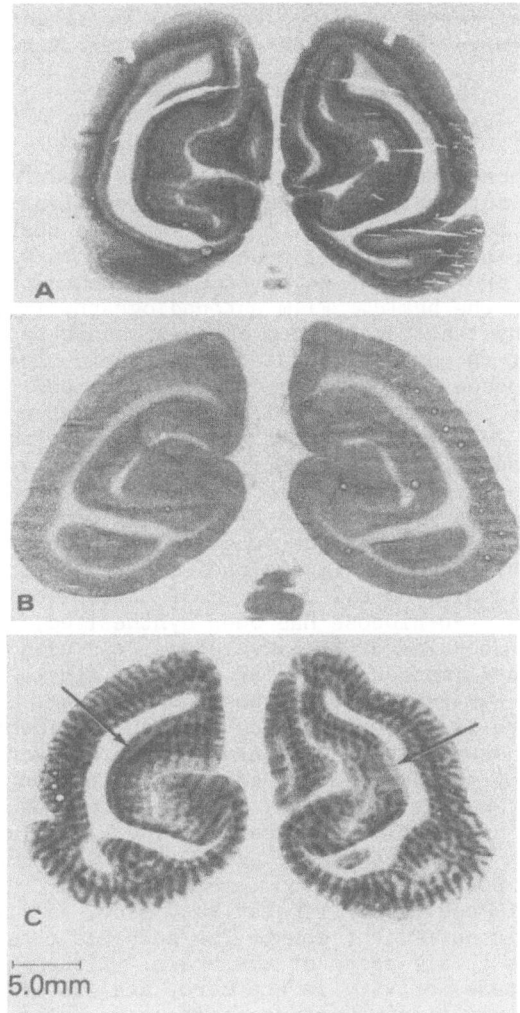

5.0mm

Fig. 9. Autoradiographs of coronal brain sections from Rhesus monkeys
at the level of the striate cortex. (A) Animal with normal binocular
vision. Note the laminar distribution of the density; the dark band
corresponds to Layer IV. (B) Animal with bilateral visual occlusion.
Note the almost uniform and reduced relative density, espectially the
virtual disappearance of the dark band corresponding to Layer IV. (C)
Animal with right eye occluded. The half-brain on the left side of the
photograph represents the left hemisphere contralateral to the occluded
eye. Note the alternate dark and light striations, each approximately
0.3-0.4 mm in width that represent the ocular dominance columns. These
columns are most apparent in the dark band corresponding to Layer IV,
but extend through the entire thickness of the cortex. The arrows point
to regions of bilateral asymmetry where the ocular dominance columns are
absent. These are presumably areas with normally only monocular input.
The one on the left, contralateral to occluded eye, has a continuous
dark lamina corresponding to Layer IV which is completely absent on the

(continued)

57

with the different cellular packing densities in the brains of these two
species. In both cases, the average glucose utilization of the brain as
whole is in good agreement with the values to be expected from
measurement of average cerebral energy metabolism with the Kety-Schmidt
method (TABLE 3).

The [^{18}F]Fluorodeoxyglucose Technique

The deoxyglucose method was originally designed for use in animals
with quantitative autoradiography and the radioactive isotopes most
suitable for film autoradiography, ^{14}C or ^{3}H. Its basic physiological
and biochemical principles apply, however, to man as well, and it is
applicable to man provided the local tissue concentrations of isotope
can be measured in the brain. Film autoradiography is a type of
emission tomography that for obvious reasons cannot be used in man, but
recent development in computerized tomography have made it possible to
determine local concentrations of γ-emitting isotopes in the cerebral
tissues. The only possible γ-emitting isotopes that could be
incorporated into 2-deoxyglucose are ^{11}C or ^{15}O, but the short
half-lives of these isotopes present problems in the synthesis of the
compounds. Alternatively, an analogue of 2-deoxyglucose with another
γ-emitting isotope but with similar biochemical properties could be used.
It is a common experience that the substitution of the very small atom,
F, in place of a hydrogen at a judicious site in the molecule often does
not alter the basic biochemical behavior of metabolic substrates.
2-[^{18}F]Fluoro-2-deoxy-D-glucose has been synthesized, found to retain
the biochemical properties of 2-deoxyglucose, and used to measure
cerebral glucose utilization in man by means of single photon emission
tomography (30). ^{18}F is actually a positron-emitter, and the absorption
of positrons in the tissues gives rise to two coincident annihilation
γ-rays of equal energy traveling at almost 180° to each other. Positron
emission tomography takes advantage of these coincident annihilation
γ-rays and, therefore, is inherently capable of better spatial
resolution than single photon tomography. The [^{18}F]fluorodeoxyglucose
method is, therefore, now generally used with positron emission
tomography (27). Positron emission tomography with
[^{18}F]fluorodeoxyglucose is still relatively slow, and it may take
2-3 hours to obtain sufficient counts for accurate measurements of local
^{18}F concentrations in all parts of the brain. Although low in brain,
glucose-6-phosphatase activity is not zero, and its effect becomes
significant after the first 45 minutes after the pulse of tracer (39).
It has, therefore, been necessary to modify the model to include a rate
constant for the hydrolysis of the phosphorylated product by
glucose-6-phosphatase and to derive a new operational equation that
takes this activity into account (13,27,39). The [^{18}F]fluorodeoxyglucose
technique for the measurement of local cerebral glucose utilization in
man is now operational and in use for studies of the human brain in
health and disease in a number of laboratories. MacGregor et al. (21)

Fig. 9 (continued)

side ipsilateral to the occluded eye. These regions are believed to be
the loci of the cortical representations of the blind spots. From
Kennedy et al. (17).

and Reivich et al. (31) have recently succeeded in synthesizing [^{11}C]deoxyglucose and applied it to the measurement of local cerebral glucose utilization in man with positron-emission tomography. Because of the short half-life of ^{11}C this development should be very useful for sequential measurements in the same subject in a short time period.

Potential Usefulness of [^{14}C]Deoxyglucose Method

The deoxyglucose method provides the means to determine quantitatively the rates of glucose utilization simultaneously in all structural and functional components of the central nervous system and to display them pictorially superimposed on the anatomical structures in which they occur. Because of the close relationship between local functional activity and energy metabolism, the method makes it possible to identify all structures with increased or decreased functional activity in various physiological, pharmacological, and pathophysiological states. The images provided by the method resemble histological sections of nervous tissue, and the method is, therefore, sometimes misconstrued to be a neuroanatomical method and contrasted with physiological methods, such as electrophysiological recording. This classification obscures the most significant and unique feature of the method. The images are not of structure but of a dynamic biochemical process, glucose utilization, which is as physiological as electrical activity. In most situations changes in functional activity result in changes in energy metabolism, and the images can be used to visualize and identify the sites of altered activity. The images are, therefore, analogous to infra-red maps; they record quantitatively the rates of a kinetic process and display them pictorially exactly where they exist. The fact that they depict the anatomical structures is fortuitous; it indicates that the rates of glucose utilization are distributed according to structure, and specific functions in the nervous system are associated with specific anatomical structures. The deoxyglucose method represents, therefore, in a real sense, a new type of encephalography, metabolic encephalography. At the very least, it should serve as a valuable supplement to more conventional types, such as electroencephalography. Because, however, it and its derivative [^{18}F]fluorodeoxyglucose and [^{11}C]deoxyglucose methods provide a new means to examine another aspect of function simultaneously in all parts of the brain, it is hoped that they will open new roads to the understanding of how the brain works in health and disease in animals and man.

DEVELOPMENT OF AN AUTORADIOGRAPHIC METHOD FOR MEASUREMENT OF LOCAL CEREBRAL PROTEIN SYNTHESIS

The basic biochemical principles for the measurement of metabolic rates in vivo, which were so effectively applied in the deoxyglucose technique, also apply to other metabolic processes. A metabolic activity of broad interest is protein synthesis. This biochemical activity is not likely to reflect directly functional activity, at least not acutely, but it may well be involved in slower, more gradual processes in the nervous system, such as growth and development plasticity, regeneration and repair, response to drugs and hormones, and possibly learning and memory. Protein synthesis may also be altered in pathological states, such as brain tumors, mental retardation due to metabolic errors, aging and senility, Alzheimer's disease, Huntington's disease, endocrine diseases, etc.

A method for the measurement of local rates of protein synthesis in the nervous system is under development (34,35). Like the deoxyglucose method, it is designed to achieve localization by quantitative autoradiography, but it is also being adapted to PET (26). Also, like the deoxyglucose method, it is based on the same biochemical principles as those described above, but their application to the measurement of local cerebral protein synthesis may be far more complex because of still undefined properties and kinetics of the equilibration of the precursor amino acid pool in the tissue with the circulating amino acid pool in the plasma.

Theory

The two essential variables that must be determined in any quantitative radioactive assay of the rate of a reaction are the amount of radioactive product formed and the integrated precursor specific activity over a measured interval of time. Because autoradiographic or emission tomographic techniques measure the concentration only of the radioisotope and not that of the radioactive product itself, it is essential that the method be so designed that the label in the tissue remains confined only to a specific product of the reaction or, at least, in well-defined chemical species that can be quantified separately. This problem is mitigated by the use of carboxyl-labeled L-leucine, an essential amino acid that is prevalent in most proteins and that, if not incorporated into protein, follows a single simple pathway of metabolic degradation. In the pathway of degradation the amino acid is first transaminated and then rapidly decarboxylated; the label is then lost as radioactive CO_2, which because of dilution by the large pool of unlabeled CO_2 constantly generated by carbohydrate metabolism, the relatively slow rate of CO_2 fixation, and the rapid removal of CO_2 from brain tissue by the blood flow, is removed from the tissues. The label is then retained only in the product of the reaction, labeled protein, and in the residual unincorporated amino acid. The concentration of free labeled amino acid remaining in the tissue can be calculated from the history of the plasma concentration and the kinetic constants for the equilibration of the tissue free amino acid pools with the plasma. The concentration of free labeled amino acid in the tissue at the time of killing can be minimized by its intravenous administration as a pulse at zero time followed by the allowance of a sufficiently long time for its clearance from the plasma and tissue. One hour after the pulse the fraction of total radioactivity in the tissue that is in the free amino acid pool is small (less than 10%) in the adult rat. In adult monkey and apparently in humans (26), however, it is large (30-50%) because of the relatively slow rates of protein synthesis and of clearance of the free amino acid pools. The error in subtracting a poorly defined free amino acid pool concentration from a total concentration that is not a great deal larger may, of course, be large. With autoradiography the error can be eliminated by histological fixation, which washes out all water-soluble radioactivity, including the free [^{14}C]leucine pool and any of its initial metabolite, α-[^{14}C]ketoisocaproate, that might be present while leaving the labeled protein product intact. This simple procedural solution of a complex theoretical problem cannot, however, be used in humans with PET.

An even more difficult problem is the determination of the integrated precursor pool specific activity in the tissue from measurements in the plasma. To accomplish this, it is necessary to know the kinetics of the equilibration of the precursor pool in the cells with that of the plasma. What makes this problem particularly perplexing is that there is evidence that amino acids in the cells are compartmentalized with only a fraction of the total intracellular amino acid content representative of the pool that serves as precursor for protein synthesis. Therefore, although it is relatively simple to measure the rate constants for the equilibration of the total amino acid pool with that of the plasma, there is little assurance that this pool reflects the kinetic behavior of the fraction of it that is the precursor for protein synthesis. A further complication is the possibility of significant dilution of the intracellular precursor amino acid pool by unlabeled amino acid derived from the slow but continuous turnover of the protein components of the cell. The magnitude of this potential dilution is even more difficult to evaluate. Nevertheless, studies are currently in progress in our laboratory that are designed to resolve these problems. If successful, they will not only determine the rate constants for the turnover of the true precursor pool but also the degree of admixture of unlabeled amino acids derived from protein breakdown with the labeled amino acids in the precursor pool.

Applications of [^{14}C]Leucine Method

Although the method is not yet fully developed, experiments with our first and most primitive model already indicate the potential usefulness of a technique that measures local cerebral protein synthesis. This simple model is essentially the same as that for the deoxyglucose method; it assumes a single tissue pool of free amino acid, all of which equilibrates uniformly with the plasma and serves as the precursor pool for protein synthesis with no dilution by unlabeled amino acids derived from protein degradation (34,35). Although the quantitative values obtained with this version of the method are probably of limited accuracy and may be underestimates of the true rates of L-leucine incorporation into protein, the results demonstrate that the rates of protein turnover do change in specific regions of the brain in response to altered function. For example, in the rat sectioning of one hypoglossal nerve is followed by increased protein synthesis in the ipsilateral hypoglossal nucleus after a lag of close to 2 to 4 days (34). The increase reaches a maximum between 20 and 30% above the contralateral hypoglossal nucleus, and protein synthesis does not return to normal until after regeneration and restoration of functional activity in the hypoglossal nerve are complete.

The method has also been used to study plasticity in the binocular visual system of the newborn rhesus monkey (19). The outputs from the retinae of the two eyes are crossed approximtely 50%, and the optic tracts terminate in the lateral geniculate nuclei in six discrete laminae: 1, 4, and 6, the sites of termination of the pathways from the contralateral retina, and 2, 3, and 5, the laminae supplied by the ipsilateral eye. The cells in these laminae project via the geniculocalcarine tracts to the ipsilateral striate cortex in such a way that the two retinal outputs for each point in the visual field converge to two adjacent cortical columns, one for each eye, for the same spot in the visual field. These are the ocular dominance columns, first described by Hubel and Wiesel (15), and demonstrated autoradiographically by the [^{14}C]DG method (Fig. 9) (17).

The visual system of the newborn monkey exhibits plasticity. Chronic occlusion of one eye in a newborn monkey leads to widening of the ocular dominance columns for the functioning eye at the expense of the columns for the deprived eye until eventually the columns disappear, and the entire striate cortex is taken over to subserve the function of the undeprived eye (4,16). Presumably the axonal terminals of the

geniculocalcarine pathway for the functioning eye grow into and take
over the synaptic connections in the adjacent columns normally reserved
for the deprived eye. If so, changes in protein synthesis required
for axonal growth and sprouting could be involved, and the protein
synthesis used for this process occurs in the cell bodies of origin of
the pathway, i.e., in the lateral geniculate nuclei. Illustrated in
Fig. 10 are the results of application of the local protein synthesis
technique to this question. The laminae in the geniculate bodies are
clearly visible and relatively uniform in the autoradiographs of the
25-day-old normal monkey. Acute monocular deprivation for 3 hr does not
alter the rates of protein synthesis in any of the laminae, including
those for the deprived eye. Chronic monocular deprivation from 2 to 25
days of age results in marked reduction in protein syntehsis in the
laminae served by the deprived eye. Apparently chronic reduction in
functional activity, in contrast to the acute state, results in a
lowering of protein synthesis in the affected pathway. These results
suggest that the loss of ocular dominance columns in the striate cortex
for the chronically deprived eye may be a consequence of depressed

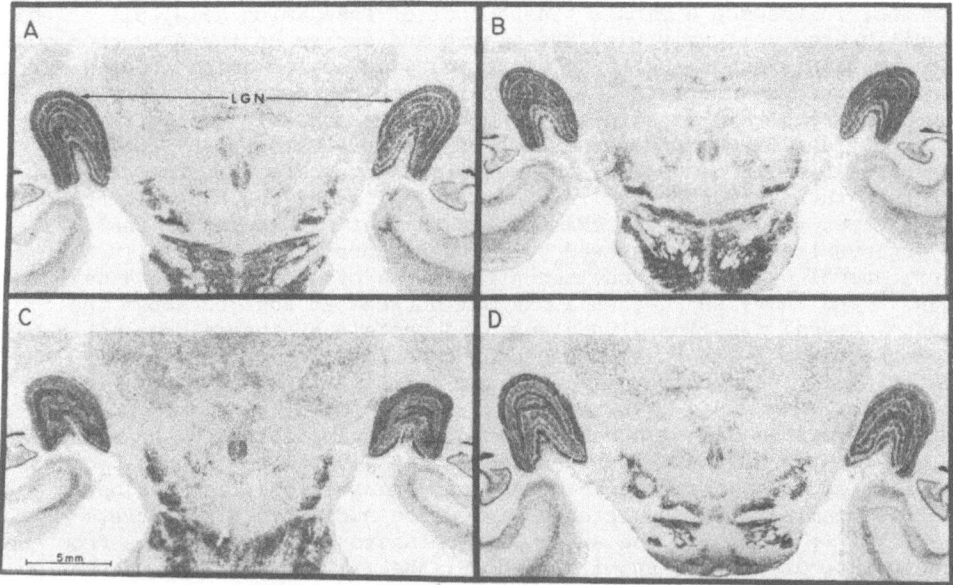

Fig. 10. Autoradiographs of coronal sections of monkey lateral
geniculate nuclei (LGN) obtained with the [^{14}C]leucine method for the
measurement of local cerebral protein synthesis. The left side of the
brain is on the left side of the autoradiographs. A:
Twenty-five-day-old monkey with intact binocular vision. B:
Twenty-five-day-old monkey with acute occlusion (3 hr) of the left eye.
C: Twenty-five-day-old monkey with chronic occlusion of right eye
initiated on second day of life. D: Twenty-five-day-old monkey with
chronic occlusion of left eye initiated on second day of life. Note the
decreased labeling and therefore decreased rate of protein synthesis in
laminae 1, 4, and 6 of the lateral geniculate nucleus contralateral to
the deprived eye and in laminae 2, 3, and 5 of the nucleus ipsilateral
to the deprived eye. From Kennedy et al. (19).

protein synthesis and deficient axonal growth in the geniculocalcarine pathways for the deprived eye.

It is hoped that measurement of local rates of protein synthesis in the nervous system may be useful to study normal and abnormal neural processes that may not be accessible to other autoradiographic and tomographic procedures--processes such as growth, maturation, plasticity, and actions of hormones. The full potential of this approach must, however, await the development of a fully accurate and reliable method.

REFERENCES

1. Abrams, R., Ito, M., Frisinger, J. E., Patlak, C. S., Pettigrew, K. D., and Kennedy, C., 1984, Local cerebral glucose utilization in fetal and neonatal sheep, Am. J. Physiol., 246:R608.
2. Bachelard, H. S., 1971, Specificity and kinetic properties of monosaccharide uptake into guinea pig cerebral cortex in vitro, J. Neurochem., 18:213.
3. Bidder, T. G., 1968, Hexose translocation across the blood-brain interface: configurational aspects, J. Neurochem., 15:867.
4. Des Rosiers, M. H., Sakurada, O., Jehle, J., Shinohara, M., Kennedy, C., and Sokoloff, L., 1978, Functional plasticity in the immature striate cortex of the monkey shown by the [^{14}C]deoxyglucose method, Science, 200:447.
5. Diemer, N. H. and Gjedde, A., 1983, Autoradiographic determination of brain glucose content and visualization of the regional lumped constant, J. Cerebral Blood Flow Metab., 3(Suppl. 1)S79.
6. Duffy, T. E., Cavazzuti, M., Cruz, N. F., and Sokoloff, L., 1982, Local cerebral glucose metabolism in newborn dogs: effects of hypoxia and halothane anesthesia, Ann. Neurol., 11:233.
7. Fishman, R. S. and Karnovsky, M. L., 1986, Apparent absence of a translocase in the cerebral glucose-6-phosphatase system, J. Neurochem., 46:371.
8. Frackowiak, R. S. J., Lenzi, G.-L., Jones, T., and Heather, J. D., 1980, Quantitative measurement of regional cerebral blood flow and oxygen metabolism in man using ^{15}O and positron emission tomography, J. Comput. Assist. Tomgr., 4:727.
9. Goochee, C., Rasband, W., and Sokoloff, L., 1980, Computerized densitometry and color coding of [^{14}C]deoxyglucose autoradiographs, Ann. Neurol., 7:359.
10. Hawkins, R. A. and Miller, D. L., 1978, Loss of radioactive 2-deoxy-D-glucose-6-phosphate from brains of implications for quantitative autoradiographic determination of regional glucose utilization, Neuroscinece, 3:251.
11. Hawkins, R., Phelps, M., Huang, S. C., and Kuhl, D., 1981, Effect of ischemia upon quantification of local cerebral metabolic rates for glucose with 2-(F-18)fluoro-deoxyglucose (FDG). J. Cereb. Blood Flow Metab., 1(Suppl 1)S9.
12. Hers, H. G., 1957, "Le Métablisme du Fructose, Bruxelles, Editions Arscia, 102.
13. Huang, S. C., Phelps, M. E., Hoffman, E. J., Sideris, K., Selin, C. J., and Kuhl, D. E., 1980, Noninvasive determination of local cerebral metabolic rate of glucose in man, Am. J. Physiol., 238:E69.
14. Huang, M.-T. and Veech, R. L., 1982, The quantitative determination of the in vivo dephosphorylation of glucose 6-phosphate in rat brain, J. Biol. chem., 257:11358.

15. Hubel, D. H. and Wiesel, T. N., 1968, Receptive fields and functional architecture of monkey striate cortex, J. Physiol., 195:215.
16. Hubel, D. H., Wiesel, T. N., and LeVay, S., 1977, Plasticity of ocular dominance columns in monkey striate cortex, Phil. Trans. R. Soc. Lond. (Biol.), 278:377.
17. Kennedy, C., Des Rosiers, M. H., Sakurada, O., Shinohara, M., Reivich, M., Jehle, J. W., and Sokoloff, L., 1976, Metabolic mapping of the primary visual system of the monkey by means of the autoradiographic [^{14}C]deoxyglucose technique, Proc. Natl. Acad. Sci. USA, 73:4230.
18. Kennedy, C., Sakurada, O., Shinohara, M., Jehle, J., and Sokoloff, L., 1978, Local cerebral glucose utilization in the normal conscious Macaque monkey, Ann. Neurol., 4:293.
19. Kennedy, C., Suda, S., Smith, C. B., Miyaoka, M., Ito, M., and Sokoloff, L., 1981, Changes in protein synthesis underlying functional plasticity in immature monkey visual system, Proc. Natl. Acad. Sci. U.S.A., 78:39.
20. Kety, S. S. and Schmidt, C. F., 1948, The nitrous oxide method for the quantitative determination of cerebral blood flow in man: theory, procedure, and normal values, J. Clin. Invest., 27:500.
21. MacGregor, R., Fowler, J. S., Wolfe, A. P., Shiue, C. Y., Lade, R. E., and Wan, C. N., 1981, A synthesis of ^{11}C-2-deoxy-D-glucose for regional studies, J. Nucl. Med., 22:800.
22. Nelson, T., Kaufman, E. E., and Sokoloff, L., 1984, 2-Deoxyglucose incorporation into rat brain glycogen during measurement of local cerebral glucose utilization by the 2-deoxyglucose method, J. Neurochem., 43:949.
23. Nelson, T., Lucignani, G., Atlas, S., Crane, A. M., Dienel, G. A., and Sokoloff, L., 1985, Reexamination of glucose-6-phosphatase activity in the brain in vivo: no evidence for a futile cycle, Science, 229:60.
24. Nelson, T., Lucignani, G., Goochee, J., Crane, A. M., and Sokoloff, L., 1986, Invalidity of criticisms of the deoxyglucose method based on allged glucose-6-phosphatase activity in brain, J. Neurochem., 46:905.
25. Oldendorf, W. H., 1971, Brain uptake of radiolabeled amino acids, amines, and hexoses after arterial injection, Amer. J. Physiol., 221:1629.
26. Phelps, M. E., Barrio, J. R., Huang, S.-C., Keen, R. E., Chugani, H., and Mazziotta, J. C., 1985, Measurement of cerebral protein synthesis in man with positron computerized tomography: Model, assumptions, and preliminary results, in: "The Metabolism of the Human Brain Studied with Positron Emission Tomography," J. Greitz, D. H. Ingvar, and L. Widén, eds., Raven Press, New York,
27. Phelps, M. E., Huang, S. C., Hoffman, E. J., Selin, C., Sokoloff, L., and Kuhl, D. E., 1979, Tomographic measurement of local cerebral glucose metabolic rate in humans with (F-18)2-fluoro-2-deoxy-d-glucose: validation of method, Ann. Neurol., 6:371.
28. Reivich, M., Alavi, A., Wolf, A., Fowler, J., Russell, J., Arnett, C., MacGregor, R. R., Shiue, C. Y., Atkins, H., Anand, A., Dann, R., and Greenberg, J. H., 1985, Glucose metabolic rate kinetic model parameter determination in humans: the lumped constants and rate constants for [^{18}F]fluorodeoxyglucose and [^{11}C]deoxyglucose, J. Cereb. Blood Flow Metab. 5:179.
29. Reivich, M., Jehle, J., Sokoloff, L., and Kety, S. S., 1969, Measurement of regional cerebral blood flow with antipyrine-^{14}C in awake cats, J. Appl. Physiol., 27:296.

30. Reivich, M., Kuhl, D., Wolf, A., Greenberg, J., Phelps, M., Ido, T., Cassella, V., Fowler, J., Hoffman, E., Alavi, A., Som, P., and Sokoloff, L., 1979, The [^{18}F]fluoro-deoxyglucose method for the measurement of local cerebral glucose utilization in man, Circ. Res., 44:127.

31. Reivich, M., Alavi, A., Wolf, A., Greenberg, J. H., Fowler, J., Christman, D., MacGregor, R., Jones, S. C., London, J., Shiue, C., and Yonekura, Y., 1982, Use of 2-deoxy-D[1-^{11}C]glucose for the determination of local cerebral glucose metabolism in humans: variation within and between subjects, J. Cerebral Blood Flow Metab., 2:307.

32. Sacks, W., Sacks, S., and Fleischer, A., 1983, A comparison of the cerebral uptake and metabolism of labeled glucose and deoxyglucose in vivo in rats, Neurochem. Res., 8:661.

33. Schuier, F., Orzi, F., Suda, S., Kennedy, C., and Sokoloff, L., 1981, The lumped constant for the [^{14}C]deoxyglucose method in hyperglycemic rats. J. Cereb. Blood Flow Metab., 1(1):S63.

34. Smith, C. B., Crane, A. M., Kadekaro, M., Agranoff, B. W., and Sokoloff, L., 1984, Stimulation of protein synthesis and glucose utilization in the hypoglossal nucleus induced by axotomy, J. Neurosci., 4:2489.

35. Smith, C. B., Davidsen, L., Deibler, G., Patlak, C., Pettigrew, K., and Sokoloff, L., 1980, A method for the determination of local rates of protein synthesis in brain, Trans. Am. Soc. Neurochem., 11:94.

36. Sokoloff, L., 1978, Local cerebral energy metabolism: its relationships to local functional activity and blood flow, in: "Cerebral Vascular Smooth Muscle and Its Control," M. J. Purves and K. Elliott, eds., Elsevier/Excerpta Medica/North-Holland, Amsterdam.

37. Sokoloff, L., 1979, The [^{14}C]doxyglucose method: four years later, Acta. Neurol. Scand. (Suppl. 70), 60:640.

38. Sokoloff, L., 1981, Localization of functional activity in the central nervous system by measurement of glucose utilization with radioactive deoxyglucose, J. Cereb. Blood Flow Metab., 1:7.

39. Sokoloff, L., 1982, The radioactive deoxyglucose method: theory, procedure, and applications for the measurement of local glucose utilization in the central nervous system, in: "Advances in Neurochemistry, Vol. 4," B. W. Agranoff and M. H. Aprison, eds., Plenum Publishing Corp, New York.

40. Sokoloff, L., Reivich, M., Kennedy, C., Des Rosiers, M. H., Patlak, C. S., Pettigrew, K. D., Sakurada, O., and Shinohara, M., 1977, The [^{14}C]deoxyglucose method for the measurement of local cerebral glucose utilization: theory, procedure, and normal values in the conscious and anesthetized albino rat, J. Neurochem., 28:897.

41. Sols, A. and Crane, R. K., 1954, Substrate specificity of brain hexokinase, J. Biol. Chem., 210:581.

42. Suda, S., Shinohara, M., Miyaoka, M., Kennedy, C., and Sokoloff, L., 1981, Local cerebral glucose utilization in hypoglycemia, J. Cerebral Blood Flow Metab., 1(1):S62.

MULTI-TRACER AUTORADIOGRAPHY FOR THE SIMULTANEOUS MEASUREMENT OF CEREBRAL BLOOD FLOW AND METABOLISM

Günter Mies

Max-Planck-Institut für neurologische Forschung
Abteilung für experimentelle Neurologie
Ostmerheimer Straße 200
D-5000 Cologne 91, FRG

INTRODUCTION

As early as 1890, Roy and Sherrington hypothesized a close relationship between functional activity, cerebral metabolism and cerebral blood flow. Convincing evidence was provided by subsequent experimental and clinical observations in which cerebral blood flow was shown to respond to functional activation (Alexander, 1912; Cobb and Talbott, 1927; Schmidt and Hendrix, 1937; Serota and Gerard, 1938; Penfield et al., 1939). With the development of the nitrous oxide technique for the quantitative determination of global cerebral blood flow and metabolism (Kety and Schmidt, 1945) it became possible to measure both parameters in animals and man and this yielded the first general correlations between the level of consciousness, blood flow and oxidative metabolism (Kety, 1949). Exploiting the principle of inert gas exchange in tissues (Kety, 1951), Landau et al. (1955) introduced an autoradiographic method for the in-vivo measurement of blood flow by which it was possible to recognize the heterogeneity in local cerebral perfusion. For example, visual stimulation was associated with an increase in local blood flow which was restricted to regions of the visual pathway (Freygang and Sokoloff, 1958) thus indicating an adjustment of local circulation to functional activity in the brain. The autoradiographic blood flow technique has since been improved further (Reivich et al., 1969; Sakurada et al., 1978) and nowadays, labeled iodoantipyrine is used as the standard blood flow tracer. The most remarkable development was the establishment of an autoradiographic technique using labeled deoxyglucose which allows local cerebral glucose utilization to be measured in-vivo (Sokoloff et al., 1977). In the past, the single tracer autoradiographic techniques have been applied to investigate the relationship between functional activity and local cerebral blood flow (see Reivich, 1972) on the one hand, and local cerebral glucose consumption (see Sokoloff, 1981), on the other. The interdependency between tissue perfusion and glucose metabolism, however, was estimated indirectly by linear correlation analysis of averaged regional flow and glucose utilization data derived from separate animal groups. In order to examine coupling between neuronal activity, local blood flow and local glucose metabolism in more detail all three parameters should be assessed in the same subject. This has become possible with the introduction of double tracer autoradiographic techniques for the simultaneous measurement of local cerebral blood flow and glucose consumption employing either ^{131}I-iodoantipyrine (Mies et al., 1981A; 1981B) or ^{123}I-iodoantipyrine (Lear et al., 1981) and

67

^{14}C-deoxyglucose, respectively. More recently, a triple tracer autoradiographic technique has been developed which in addition to cerebral blood flow and glucose utilization allows local cerebral protein synthesis to be measured using ^{3}H-labeled amino acids (Mies et al., 1983; 1986).

In the following, theoretical considerations and practical procedures involved in quantitative autoradiography and in the autoradiographic quantitation of three isotopes in brain tissue will be presented as well as the application of double and of triple tracer autoradiography to the investigation of local coupling between cerebral blood flow and metabolic rates.

PRINCIPLE AND PROCEDURES OF QUANTITATIVE AUTORADIOGRAPHY

Autoradiographic Quantitation of Single Isotope Tissue Radioactivity

Photographic film usually consists of plastic carrier material on top of which is a thin homogeneous silver salt emulsion. In order to avoid mechanical damage to the film emulsion the latter is covered by a 5 - 7 um thick protective layer of gelatine. When the photographic film is exposed to light, energy-rich photons interact with the silver halogenide molecules and form grains of silver metal during the chemical developing procedure. In general, any photons or corpuscular particles within or outside the range of visible wavelengths, stemming for example from radioactive disintegration of atoms, can blacken the film as long as the particles emitted are of sufficient kinetic energy to reach the film emulsion.

Let us assume that a radioactively labeled compound is allowed to circulate throughout the body of an animal. An organ, for example the forebrain, is removed and cut into a 20 um thick tissue sections (see Figure 1). When these radioactive brain sections are exposed to a conventional X-ray film (composed as described above) for a certain period of time increased blackening of the film will result only from those sites of the tissue section at which the isotope is present. No blackening of the film will occur were the labeled compound is not present. These 'autoradiograms' of brain sections thus reflect the qualitative local distribution of the radioactive compound. Regional blackening of the film can be assessed by measuring the percentage transmission of light intensity I distally from the autoradiogram which was transluminated with a constant source of light with the intensity I_o, i.e. transmission is 100 % ($I = I_o$) in the absence of silver grain deposits (exceeding those of film background) or 0 % ($I \ll I_o$) when there is a relative maximum in silver grain deposits within the observation field. Another way of expressing film blackening quantitatively is by taking the logarithm of the I/I_o ratio which then is called optical density (OD). In order to relate a local OD reading in an autoradiogram to isotope radioactivity the corresponding total isotope disintegrations per weight of tissue must be determined. The OD/radioactivity relationship is usually established in the following way. When a freely diffusible radioactive compound is allowed to equilibrate within tissue i.e. the radioactivity is homogeneously distributed, the organ of interest is removed and divided into two portions. One part of the organ is cut into 20 um sections (reference thickness of tissue sections) and exposed to photographic film for a cerain period of time. A weighed fraction of the other part of the organ is homogenized, digested with hyamine hydroxide and the number of disintegrations per minute (dpm) are determined in a liquid scintillation counter or a gamma counter depending on the isotope employed. From the exact weight of the analysed tissue sample, the radioactivity can be expressed in units of dpm/gram tissue wet weight (w.w.) or nCi/g tissue w.w.. The optical density of the autoradiogram corresponds to the quantitated tissue isotope radioactivity. When the same procedure is repeated for various amounts of radioactivity applied systemically to each animal the optical density induced by

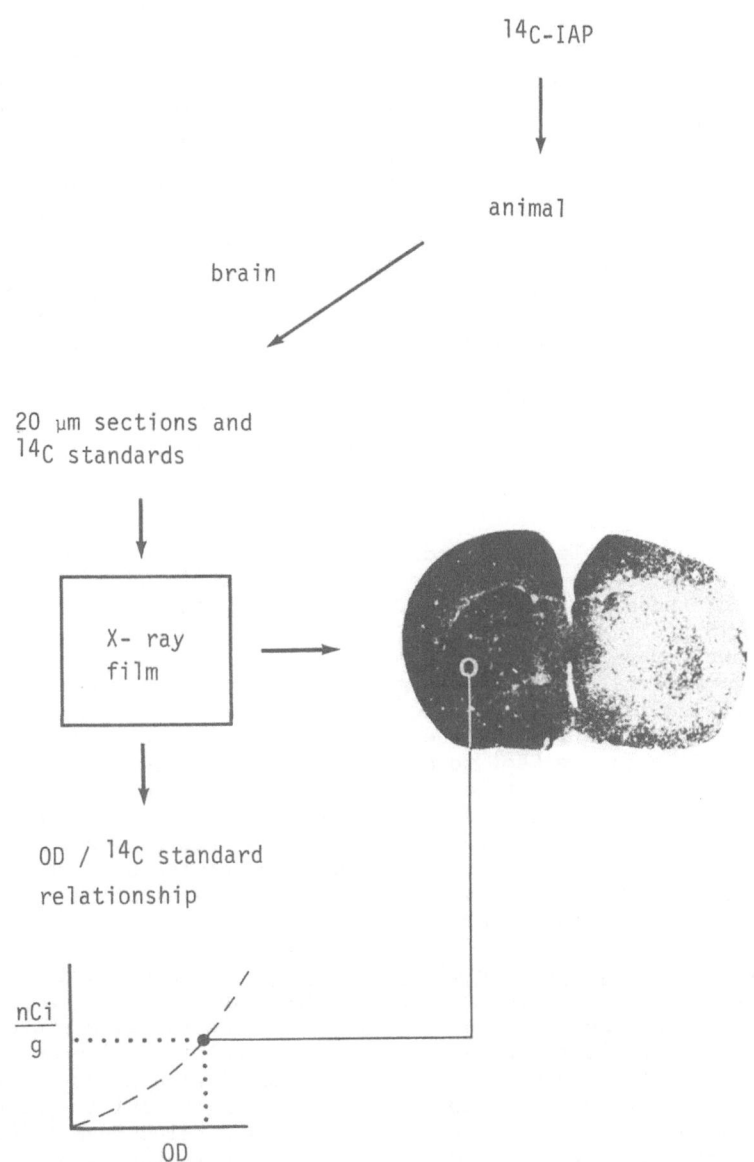

Fig. 1. Schematic illustration of quantitative autoradiographic determina-
nation of isotope tissue radioactivity. For explanation, see text.

isotope brain standards exposed simultaneously to photographic film can be related to the corresponding tissue tracer radioactivity thus yielding a calibration curve for the OD/tissue radioactivity relationship (see Figure 1). It has to be emphasized that this calibration curve is specific for the isotope and the tissue reference thickness which were used for the preparation of isotope standard brain sections. As shown in Figure 1, the OD value measured within a discrete area of the autoradiogram can now be read off from the calibration curve allowing the equivalent radioactive tissue tracer concentration to be calculated. The permanent use of isotope tissue standards however, is inconvenient in that the organic material is subject to autolytic changes and such brain standards would have to be renewed within short time intervals. For this reason plastic standards of a stable isotope incorporated into a polymer structure are now most commonly used; these, however, have initially to be calibrated in terms of radioactivity per gram tissue wet weight. A set of plastic standards containing various amounts of radioactive label is exposed to the film together with brain tissue isotope standards. Quantitation of plastic standards is achieved by inserting the corresponding optical density measured in a plastic standard into the brain standard calibration curve which then yields for each isotope the tissue equivalent radioactivity of each of the single plastic standards.

Autoradiographic Differentiation of Several Isotopes in a Tissue Section

The following possibilities of assessing local tissue radioactivity of more than one isotope in a brain section using quantitative autoradiography exist:

Half-life principle. The radiocompounds administered systemically may be such that one of the isotopes is short-lived while the others are stable, i.e. decay of those isotopes is negligible during the autoradiographic procedures. Since the intensity of film blackening can be taken as depending on the product of radioactivity (dpm or nCi) and time of apposition to photographic material (t), the energy deposits may be altered by varying either tissue radioactivity or exposure time. When it is possible to obtain high tissue radioactivities of the unstable isotope, subsequent exposure times to the film may be kept to a minimum. This reduces possible cross-contamination of stable isotopes especially when the corresponding tissue radioactivity can be adjusted so as not to be detectable autoradiographically during short-term autoradiography of the unstable isotope. After complete decay of the unstable isotope a second exposure of the same brain section to the film will allow quantitative autoradiographic evaluation of tissue radioactivity of the other stable isotopes.

Principle of photographic differentiation. As mentioned above, radionuclide-induced film blackening requires that during decay the particles emitted possess sufficient kinetic energy to reach the film emulsion which is most commonly protected by a micro-layer of gelatine. Since penetration of alpha- or beta-particles is thereby associated with a loss in kinetic energy the penetration depth depends primarily on the initial kinetic energy. Sine the maximal emission energies of beta-emission from ^{14}C-carbon and ^{3}H-tritium are about 0.2 and 0.02 MeV, respectively, the kinetic energy of ^{3}H-beta particles is reduced to zero after travelling a much shorter distance in a material such as a photographic film than ^{14}C-beta particles. This explains why the self-absorption thickness of ^{14}C-beta is about 80 - 100 um whereas that of ^{3}H-beta ranges from 4 - 5 um. Assuming that our stable isotopes remaining in the tissue section after complete decay of the short-lived isotope, for example ^{131}I-iodine, consist of ^{14}C-carbon and ^{3}H-tritium it is apparent that emission from ^{3}H-tritium will be prevented from reaching the film emulsion by the 4 - 5 um gelatine protection layer whereas those from ^{14}C-carbon do possess adequate kinetic energy. Thus, a second exposure to X-ray film will autoradiograph exclusively one of the two stable isotopes, namely ^{14}C-carbon. Tissue ^{14}C-radioactivity must necessarily be limited such that there is no cross-contamination during the first short-term exposure (^{131}I-autoradiography); consequently now the apposition time must be prolonged in order to achieve sufficient energy deposits, i.e. adequate blackening of the photographic film.

With conventional film material such as X-ray or photographic film ^{3}H-autoradiography is hardly possible and it would require at least several months in order to obtain reasonably intensive autoradiograms. In the meantime, film has been developed which lacks a gelatine protection layer. With the aid of such '^{3}H-sensitive' film it is now possible to speed up ^{3}H-autoradioradiography since the particles emitted are capable of reaching the film emulsion. But in addition this film will of course also detect disintegrations stemming from any other radionuclide present in the tissue section. As a consequence, the total blackening of the '^{3}H-sensitive' film, i.e. the optical density, has to be corrected for the energy deposits resulting from cross-contamination by subtraction of energy deposits stemming from the corresponding radionuclide.

Principle of tissue section wash-treatment. In order to improve autoradiographic differentiation between two stable isotopes with different maximal energy emission spectra, removal of that radionuclide with the higher energy, i.e. ^{14}C-carbon, is likely to facilitate the depiction of the OD-related energy deposits of ^{3}H-tritium when exposed to '^{3}H-sensitive' film. The idea behind wash-treating tissue sections is as follows: the ^{14}C-labeled compound should be a tracer molecule which is initially metabolized but is not a further substrate in the metabolic pathway. This criterion is fulfilled by ^{14}C-deoxyglucose which is used to measure cerebral glucose utilization. The accumulating label is neither built into structural elements nor fixed by reagents in the wash solution. On the other hand, ^{3}H-labeled amino acids incorporated into protein structures appear to be fixed due to the chemical binding of proteins whereas both free ^{3}H-labeled amino acids and ^{14}C-deoxyglucose, and ^{14}C-deoxyglucose-6-phosphate are essentially removed from the tissue sections.

By choosing a solvent which reacts chemically with the radioactive substrate the wash time can be significantly reduced since removal is not only a question of simple diffusion. For example, 2,2-dimethoxypropane (DMP) hydrolyses iodoantipyrine and a two minute incubation of tissue sections is sufficient to remove labeled iodoantipyrine whilst not affecting the other radioactive substrate. By employing this procedure it is possible to differentiate autoradiographically between ^{14}C-carbon and ^{3}H-tritium by apposing one untreated tissue section to a single-coated X-ray film and the section adjacent to it to '^{3}H-sensitive' film after wash treatment.

Autoradiographic Quantiation of Three Isotope in a Tissue Section

Choice of isotopes. The use of iodoantipyrine (IAP), deoxyglucose (DG) and of amino acids (AS) for the determination of cerebral blood flow (CBF), cerebral metabolic rates of glucose (CMRG) and of cerebral protein synthesis (CPS) respectively, is given by the corresponding methods which were established and validated previously. However, these method-specific compounds may be labeled in such a way that triple isotope differentiation in brain tissue sections is possible with the above mentioned approaches. For example, IAP can be labeled with ^{131}I-iodine with the following advantages: first, ^{131}I-IAP can be applied systemically at such concentration that the first short-term exposure of sections to film induces blackening which stems exclusively from ^{131}I-IAP; second, decay of ^{131}I ($T_{1/2}$ = 8.02 days) occurs within a reasonable time period which eliminates one isotope from brain tissue. When DG and AS are labeled with ^{14}C and ^{3}H, respectively, a second exposure of the same brain sections to conventional X-ray film will result in ^{14}C-autoradiograms since the gelatine layer prevents ^{3}H-disintegrations from reaching the film emulsion. When the ^{14}C-labeled DG and DG-6-phosphate (DGP) and, in addition, free ^{3}H-AS are completely removed from the tissue sections by wash-incubation, the third exposure to ^{3}H-sensitive film yields an autoradiogram depicting ^{3}H-labeled cerebral proteins.

Estimation of optimal tissue radioactivity. In order to establish the optimal range of tissue radioactivity with respect to autoradiographic sensitivity of X-ray film and isoptope cross-contamination, isotope brain standard with varying amounts of tracer radioactivity were prepared. For this purpose, ^{131}I-brain standards were obtained from four rats which received a 1 min iv. infusion of 80 - 200 uCi ^{131}I-IAP/100g body weight under halothane anesthesia. Brains were rapidly removed from the skull and divided into both hemispheres, one of which was cut into 20 um cryostat sections. The cortex of the other hemisphere was dissected out, and the tissue sample weighed, digested in hyamine hydroxide and the ^{131}I-radioactivity was measured in a gamma counter. ^{14}C-brain standards were prepared from seven animals which received 2.5 to 40 uCi ^{14}C-DG/100g body weight which was allowed to circulate for 45 min. The procedure of tissue sectioning and tissue sampling was identical to that described above. ^{14}C-radioactivity was measured by liquid scintillation counting. ^{3}H-brain standards were obtained similarly: animals received 0.2 - 2 mCi ^{3}H-DG intravenously which was allowed to circulate for 45 minutes.

The various isotope tissue standards were exposed to X-ray film (Kodak NMB) for 24 hours (^{131}I-autoradiography). After complete decay of ^{131}I (ten half-lives) all sections were re-exposed to Kodak NMB film for two weeks to obtain ^{14}C-induced blackened film (^{14}C-autoradiography) as well as to ^{3}H-sensitive film (LKB Ultrofilm) in order to estimate the interdependency between ^{14}C and ^{3}H related energy deposits on different photographic material. ^{14}C brain standards were then wash-incubated in an aqueous solution containing 10 % trichloroacetic acid, 70 % ethanol, and 1 % polyvinylpyrrolidone for 12 hours (Mies et al., 1983; 1986). Any remaining ^{14}C radioactivity, which may contaminate during ^{3}H-autoradiography, was assessed by a further two weeks' apposition of wash-treated brain sections to Kodak NMB film. Calibration curves were obtained by plotting the optical density of the cortex against the isotope radioactivity (uCi/g tissue w.w.) measured.

Estimation of isotope cross-contamination. Isotope standards were exposed to Kodak NMB film for 24 hours in order to verify contamination of ^{131}I autoradiograms by ^{14}C and ^{3}H. As illustrated in Figure 2, no blackening of the film occurred below 0.2 uCi ^{14}C/g tissue w.w.. Above 0.45 uCi ^{14}C/g tissue w.w. the optical density turned out to be significantly different from the film background. No blackening of the film was obtained from ^{3}H-standards up to 25 uCi ^{3}H/g tissue w.w. thus confirming that there was no cross-contamination by ^{3}H-tritium.

Percentage ^{14}C cross-contamination during ^{131}I-autoradiography of brain standards amounted to less than 5 % at less than 0.45 uCi ^{14}C/g tissue w.w. As shown in Figure 3, in which ^{14}C cross-contamination in ^{131}I-autoradiograms for various experimental conditions was plotted against the corresponding blood ^{131}I/^{14}C ratio, undamaged brain tissue exhibited a percentage ^{14}C cross-contamination of less than 5 % when the integral ratio of both isotopes applied systemically were met between 2 and 9. Such a range was achieved when animals received at least 100 uC/100g b.w. ^{131}I-IAP and less than 10 uCi/100g b.w. ^{14}C-DG. However, ^{14}C cross-contamination turned out to be as high as 20 % in pathological brain areas with a low-flow/high-metabolism state (brain tumor tissue) or when the integral ratio of intravenous

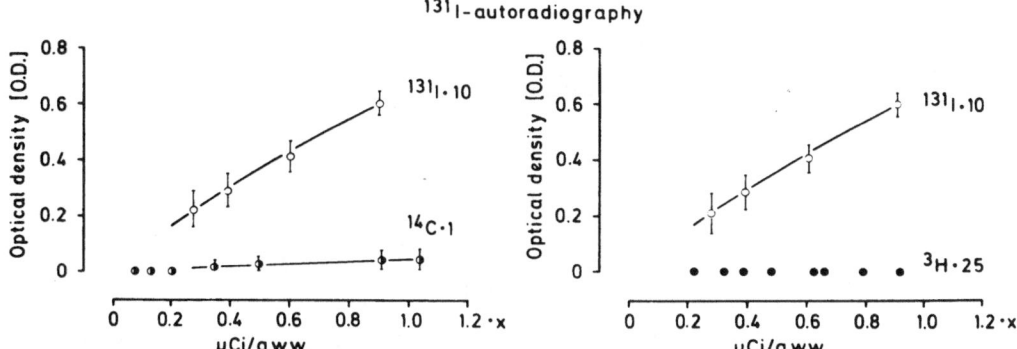

Fig. 2. Exposure of ^{131}I brain standards together with ^{14}C (left) and ^{3}H brain standards (right) for 24 hours to Kodak NMB film. Note the absence of cross-contamination by ^{3}H and negligible blackening of the film by ^{14}C-radioactivity below 0.45 uCi/g tissue w.w. and the linearity of the OD/^{131}I-radioactivity relationship ($F_{2,12}$ = 0.1342).

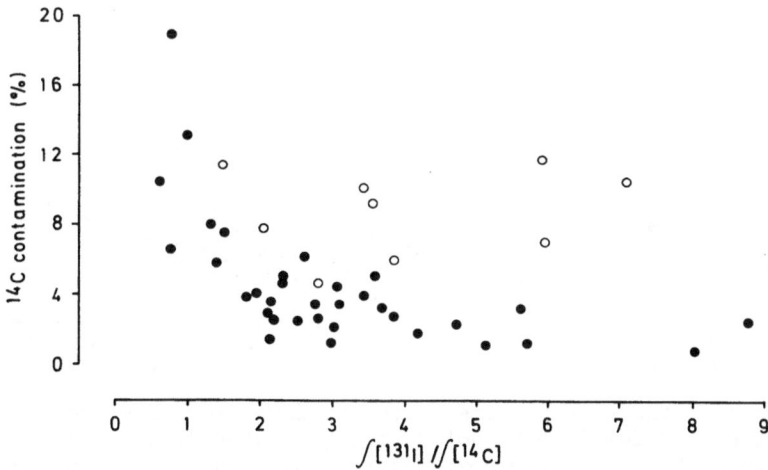

Fig. 3. Relationship between tracer input function and cross-contamination of ^{131}I-autoradiograms by ^{14}C-carbon in normal brain regions (closed circles) and in brain tumor tissue (open circles). Note that in normal brain areas cross-contamination is less than 5 % at ^{131}I/^{14}C integral ratios larger than 1.8. In tumor tissue, significant ^{14}C cross-contamination occurs independently of the integral ratios due to the high metabolic rates of glucose in tumor tissue.

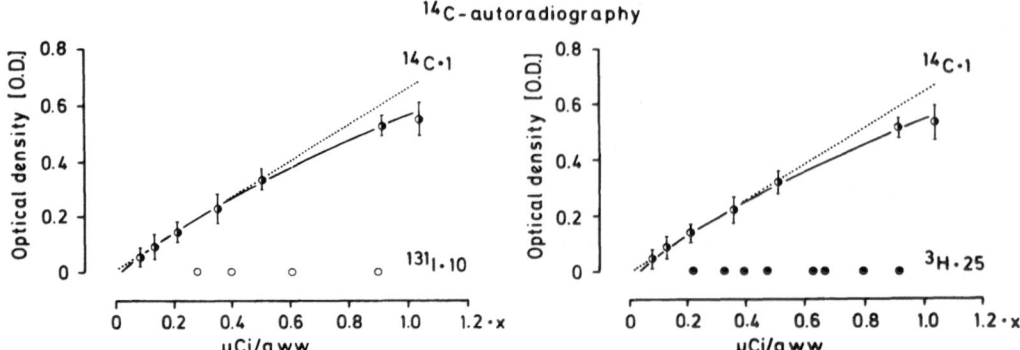

Fig. 4. Exposure of the same isotope brain standards to Kodak NMB film for two weeks after decay of ^{131}I-iodine with ^{14}C (left) and ^{14}C-carbon with ^{3}H-tritium (right). No significant blackening of the X-ray film occurred with ^{131}I and ^{3}H brain standards. Note the upper limit of the linear OD/^{14}C-radioactivity relationship up to 0.5 uCi/g tissue w.w. ($F_{3,10} = 0.1114$; dotted line).

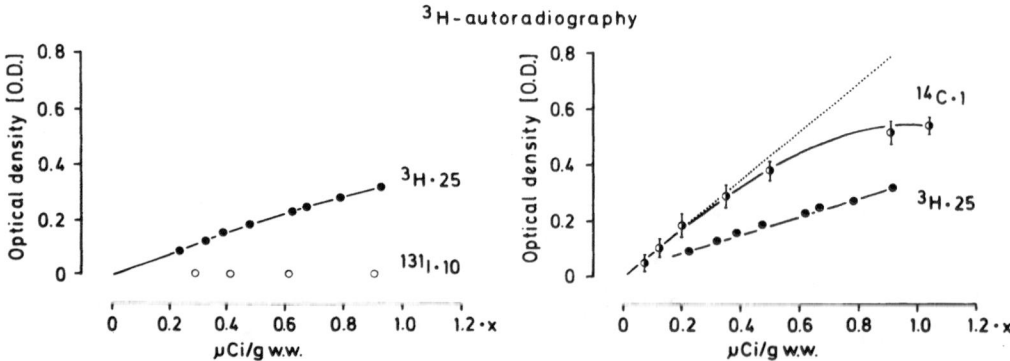

Fig. 5. Exposure of the same isotope brain standards to LKB Ultrofilm for two weeks. ^{131}I-iodine is not detectable autoradiographically but blackening by ^{14}C-carbon compared to ^{3}H-tritium turned out to be about 20 % more intensive. Selective ^{3}H-autoradiography, in consequence, either requires complete removal of ^{14}C-carbon or digital subtraction of ^{14}C cross-contamination. The extinction/^{3}H-radioactivity relationship is linear up to 23 uCi/g tissue w.w. ($F_{6,12} = 0.2847$; dotted line).

^{131}I and ^{14}C input decreased below 2 as the result of too low amounts of ^{131}I-IAP being infused systemically

Contamination of ^{14}C-autoradiograms by ^{131}I and ^{3}H did not play a role in autoradiographic differention of isotopes as illustrated in Figure 4. ^{131}I brain standards of originally 2 - 10 uCi ^{131}I/g tissue w.w. could not be detected autoradiographically after two weeks of exposure to Kodak NMB film; this shows that decay is complete and that no radionuclide impurities are present. On the other hand, the optical density of ^{3}H tissue radioactivity as high as 25 uCi/g was not found to differ significantly from the film background. As a consequence, ^{14}C-autoradiography produced images derived exclusively from ^{14}C-disintegrations.

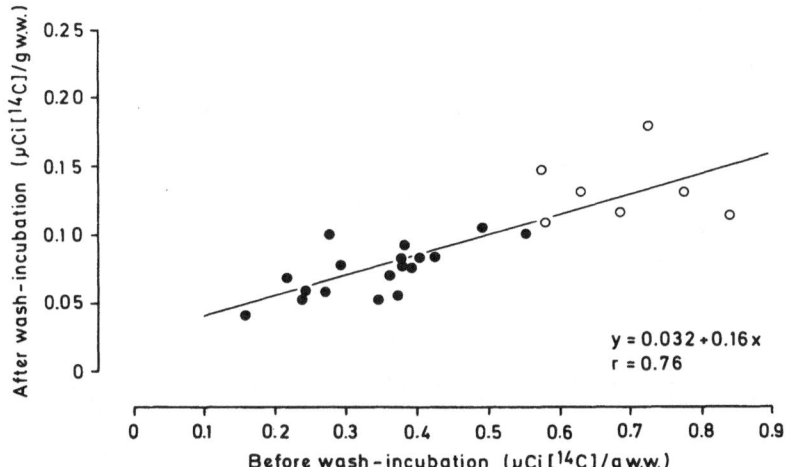

Fig. 6. Removal of ^{14}C-radioactivity from brain tissue sections by means of wash incubation. ^{14}C-tissue radioactivity was determined autoradiographically in the same brain section before and after incubation for 12 hours in normal (closed circles) and pathological (open circles) brain tissue.

However, contamination of ^{3}H-autoradiograms by ^{14}C proved to be significant whereas the exposure of ^{131}I brain standards which had decayed to ^{3}H-sensitive LKB Ultrofilm for two weeks did not produce any measurable extinction. In contrast, blackening from ^{14}C-carbon was intense even at low radioactive concentrations of ^{14}C-carbon (Figure 5). The systemic application of 10 uCi ^{14}C-DG/ 100g b.w. produced a ^{14}C radioactivity of about 0.4 uCi/g w.w. in normal brain tissue. As shown in Figure 6, a substantial amount of the ^{14}C-label is removed from the tissue section after wash-treatment and only amounts to about 0.075 uCi/g. However, this tissue radioactivity was found to be sufficient to produce significant cross-contamination, especially when the remaining ^{3}H tissue radioactivity was low after section wash-treatment. In order to improve the ^{3}H/^{14}C tissue ratio, an intensive ^{3}H-labeling of proteins is necessary; this may be achieved either by prolonging the duration of ^{3}H-labeled amino acid incorporation (Dwyer et al., 1982) or by increasing the amount of ^{3}H-labeled amino acids applied systemically.

Correction of isotope cross-contamination. Film characteristics such as grain size of silver halogenides, the thickness and density of the emulsion and/or the kinetic energy of emitted particles from an isotope determine whether the OD/radioactivity relationship may be described mathematically by linear or non-linear regression analysis. In Figure 4 and 5, the dotted lines indicate the linear range range of the ^{14}C/OD interdependency. As evident, linearity of that relationship can be assumed for ^{14}C tissue radioactivity ranging from 0 to 0.5 uCi/g w.w. using Kodak NMB film for autoradiography or from 0 to 0.4 uCi/g w.w. using ^{3}H sensitive LKB Ultrofilm after an exposure period of two weeks. The isotope/OD relationship of ^{131}I and of ^{14}C exposed to film for 24 hours and of ^{3}H (2 weeks' exposure) can be shown to be linear over the whole investigated range of isotope tissue radioactivities (Figure 2 and 5). In this case, the correction of contaminating isotope radioactivity can be performed by subtracting corresponding optical densities which, however, is not allowed with non-linear OD/radioactivity relationships.

When the contaminating ^{14}C tissue radioactivity, ^{14}C$_{cont}$, is determined after exposure of wash-treated brain sections to Kodak NMB film for two weeks (^{3}H is not detectable as shown above) it may be expressed in terms of contaminating optical density, OD$_{cont}$, in LKB Ultrofilm when a set of ^{14}C standards was exposed in addition

$$^{14}C_{cont} = a * OD_{LKB} \qquad (1).$$

Autoradiography of these wash-treated brain sections with ^{3}H sensitive LKB Ultrofilm will provide the OD/radioactivity relationship of both ^{3}H and ^{14}C and may be written

$$^{14}C_{cont} + {^3H} = b * OD_{tot} \qquad (2)$$

where total blackening of the LKB Ultrofilm OD$_{tot}$ is derived from both isotopes. Inserting equation 1 into 2 will yield

$$^3H = b * OD_{tot} - a * OD_{LKB} \qquad (3)$$

and allow the calculation of ^{3}H tissue radioactivity from the OD/radioactivity relationship of the ^{3}H standard set.

The calculation of contamination-corrected autoradiograms is greatly facilitated by the use of an image processing system. Reconstruction of single isotope 'autoradiograms' can be performed from the digitalized autoradiograms representing local blackening from both isotopes and from the contaminating radionuclide. Subtraction of local extinction values on a pixel-by-pixel basis of exactly aligned autoradiograms will yield a reconstructed image in which the optical density can be converted to tissue isotope radioactivity (for details, see Mies et al., 1986).

APPLICATION OF MULTIPLE TRACER AUTORADIOGRAPHY

As mentioned already in the introduction, multiple tracer autoradiography for the simultaneous measurement of CBF, CMRG and CPS is suggested to represent a tool allowing the determination of the precise interrelationship betwen hemodynamics and metabolism in the same subject. Since the mechanisms of linking local perfusion to metabolic activity are only partly understood it was of interest to investigate these local relationships during metabolic suppression and activation of the central nervous system. In order to study the interdependency between tissue perfusion and metabolism in pathological tissues, multiple tracer autoradiography was performed in experimental rat brain tumors.

Experimental Groups

Awake animal group. Five rats of the BD IX strain were partially immobilized with a loose fitting pelvet plaster cast in addition to femoral artery and vein cannulation. Rats were allowed to recover from anesthesia for two hours before CMRG and CBF were measured.

Anesthesia/analgesia groups. Rats of the BD IX strain were kept immobilized and were mechanically ventilated under pentobarbital (50 mg/kg ip.; n=5) or halothane anesthesia (0.8 %; n=5) or with nitrous oxide (70 % N_2O + 30 % O_2; n=5) after termination of animal surgery. One hour later, CMRG and CBF were determined.

Spreading depression group. Cortical spreading depression (CSD; n=8) was elicited by continuous topical application of KCl on to the exposed cortex in pentobarbital-anesthetized rats (50 mg/kg ip.) whereas hippocampal spreading depression (HSD; n=5) was initiated by microapplication of KCl via a glasscapillary. The maintenance of spreading depression was ascertained by the appearance of repetitive negative DC potential shifts recorded from topically positioned micro-calomel electrodes or from a microglasspipette inserted into the hippocampus. Fifteen minutes after the onset of spreading depression, the measurement of glucose utilization was started and was followed by the CBF measurement thirty minutes later.

Tumor-bearing animal group. Intracerebral tumors were produced in BD IX rats (n=5) by stereotactical implantation of RG12.2 glioma cells into the right caudatus-putamen. After 3 - 4 weeks, triple tracer autoradiographic investigations were performed on these tumor-bearing rats in which autoradiographic measurements were carried out under 0.8 % halothane anesthesia.

Experimental Conditions

Regardless of the experimental groups the following surgical standard procedure was employed in all animal preparations. Rats of both sexes and of the BD IX strain were initially anesthetized with 3 % halothane (30 % O_2 and 70 % N_2); this was reduced to 1.5 % halothane after 5 min. Pentobarbital (50 mg/kg body weight) was given intraperitoneally without prior halothane anesthesia. Polyethylene tubing was inserted into both femoral arteries and veins in order to monitor arterial blood pressure, to sample arterial blood for determination of tracer radioactity, for arterial blood gas analysis and injecting drugs or tracers. Other than the awake rat group, rats were tracheotomized, immobilized with tubo-curarrine chloride (1.5 mg/kg body weight intravenously) and mechanically ventilated with a rodent respirator. Rectal body temperature was maintained at 37°C by means of a thermistor-controlled infrared heating lamp. Arterial blood pressure was monitored throughout the experiment. Prior to tracer administration, arterial acid-base parameters were determined and taken as acceptable only when p_aH was in the range 7.35 to 7.45, p_aCO_2 35 to 45 mmHg, and p_aO_2 greater than 100 mmHg.

Tracer Application

Animals received an intravenous bolus of ^{14}C-DG (10 uCi/100g body weight) to measure CMRG as described by Sokoloff et al. (1977). Arterial blood samples were collected at increasing time intervals ranging from 20 seconds to 10 minutes after ^{14}C-DG bolus application. Blood samples were immediately centrifuged to separate plasma from blood cells. In triple tracer studies, a mixture of ^3H-labeled amino acids (50 uCi/100g body weight: phenylalanine, tyrosine, isoleucine, leucine and methionine) was administered intravenously for the measurement of CPS twenty-five minutes after ^{14}C-DG injection. Arterial blood samples were collected in order to determine the time course of ^3H specific activity of plasma amino acids. 45 minutes later, CBF was determined by intravenous infusion of 100 uCi ^{131}I-IAP/100g body weight while arterial blood samples were taken onto filter paper disks every 7 - 10 seconds. Arterial circulation was arrested by an intravenous bolus of a saturated KCl solution. In double tracer experiments, CBF was measured as early as 30 min after intravenous ^{14}C-DG administration.

Autoradiographic Procedures

Brains were removed rapidly from the skull immediately after circulatory arrest and frozen in methylbutane (-70°C). 20 um thick cryostat sections were cut at -20°C, dried on a hot plate (60°C) to prevent diffusion of labeled tracers and exposed to Kodak NMB film together with calibrated 20 um thick ^{131}I-brain standards for 24 hours. ^{131}I-iodine was allowed to decay for 10 - 12 weeks, the sections being stored at 0 - 1°C. Then sections and corresponding ^{14}C-standards were autoradiographed again for 14 days using Kodak NMB film. In case of triple tracer autoradiography, brain sections were wash-incubated in an aqueous solution containing 10% trichloroacetic acid, 70% ethanol, and 1% polyvinylpyrrolidone for 12 hours (Mies et al., 1983; 1986). Any remaining ^{14}C-radioactivity, which may contaminate ^3H-autoradiography, was assessed by a further two weeks' apposition of wash-treated brain sections to Kodak NMB film. ^3H tissue radioactivity of labeled cerebral proteins was determined by exposing brain sections and ^3H-standards to ^3H-sensitive film (LKB Ultrofilm).

Evaluation of optical densities in corresponding autoradiograms from the same brain sections and brain standards was performed using an image processing system which consisted of a microdensitometer, an image display system and a laboratory computer. Correction of 'contaminated' autoradiograms was achieved by digital subtraction on a pixel-by-pixel basis of respective optical densities, since isotope radioactivity proved to be related linearly to extinction at least over the range of isotope tissue concentrations attained under the present experimental conditions.

Calculations

^{131}I-radioactivity of weighed arterial blood samples was determined in a gamma counter. Local CBF was calculated according to Sakurada et al., (1978) with λ equal to 0.8. ^{14}C-radioactivity in arterial plasma was measured on 10 ul in a liquid scintillation counter and arterial plasma glucose content in an automated glucose analyser. Local CMRG was quantified according to Sokoloff et al., (1977) using the rate constants and the lumped constant determined in the awake rat. Quantitation of local CPS employing the suggested ^3H-labeled amino acid mixture is in progress.

Results on Local Cerebral Coupling of CBF and CMRG in Awake, Anesthetized and Analgetized Rats

Systemic application of anesthetics is known to suppress cerebral metabolism but exerts a variable influence on cerebral blood flow (Smith and Wollman, 1972). Double tracer autoradiography was employed to investigate the local CBF/CMRG couple in pentobarbital- and halothane-anesthetized rats and, in addition, under the influence of nitrous oxide. The relationship between local CBF and CMRG assessed under these experimental conditions was compared to that in awake rats. In Table 1, CBF/CMRG ratios measured in 22 individual brain structures of animals which were awake, pentobarbital- or halothane-anesthetized or under nitrous oxide analgesia are summarized.

In awake rats local CBF/CMRG ratios ranged from 1.77 ± 0.22 to 2.44 ± 0.1 (ml/umol; means $\pm$ SEM) suggesting that the local coupling between CBF and CMRG is not as close as has been previously proposed (Des Rosiers et al., 1976) and that the adjustment of perfusion to meet the metabolic demand is also probably regulated additionally by local factors other than metabolic mediators.

In pentobarbital-anesthetized rats, local CBF/CMRG ratios covered a range of 1.41 ± 0.07 to 2.58 ± 0.13 (ml/umol; means $\pm$ SEM). In spite of the general metabolic depressant effect of barbiturates, local perfusion matched the metabolic rates of glucose less than consistently. Statistical comparisons between local cortical or limbic CBF/CMRG ratios and those from awake animals were not found to differ significantly. However, in the thalamus and geniculate bodies, local CBF/CMRG ratios were significantly elevated indicating that these brain regions are oversupplied hemodynamically with respect to the local CMRG.

In halothane-anesthetized rats, on the other hand, statistical evaluation revealed that in the cortex the CBF/CMRG ratios were significantly elevated in comparison to those measured in awake controls. Although not reaching a significance level of 5 %, subcortical CBF/CMRG ratios appeared to increase distinctly probably indicating a direct vasodilatory effect of halothane on cerebral vasculature (Wollman et al., 1964; Smith, 1973). In the hippocampal stratum moleculare, however, a significantly decrease in the CBF/CMRG ratio, from $1.98 \pm .08$ to 1.52 ± 0.12 (ml/umol), was observed accounting for the absence of vasodilation mediated by halothane itself or by metabolism-related regulators.

In nitrous oxide treated animals, local CBF/CMRG ratios ranged from 1.70 ± 0.14 to 2.15 ± 0.16 (ml/umol; means $\pm$ SEM) and did not differ significantly from those determined in awake controls. Thus, analgesia by nitrous oxide did not alter local coupling between CBF and CMRG.

In order to examine whether evaluation of CBF/CMRG coupling from averaged CBF and CMRG data is sensitive enough to detect the observed differences in local CBF/CMRG ratios, correlation analysis on corresponding data was performed. The correlation coefficients were as follows: - awake: $r=0.958$, $p<0.001$; pentobarbital: $r=0.869$, $p<0.001$; halothane: $r=0.849$, $p<0.001$; nitrous oxide: $r=0.935$, $p<0.001$ - indicating a highly significant stochastic relationship between CBF and CMRG. Statistical comparison of the correlation coefficient of the awake control and of barbiturate or halothane group in which local CBF/CMRG uncoupling was observed, did not reach a significance level of 5 %.

This suggests that double tracer autoradiography for the simultaneous measurement of CBF and CMRG is suitable for the assessment of the local CBF/CMRG ratio in order to detect any local differences in coupling between CBF and CMRG under various experimental conditions. Anesthetics were shown

Table 1

Structure	Awake (n=5)	Barbiturate (n=5)	Halothane (n=5)	Nitrous oxide (n=5)
Frontal cortex	1.98 ± .1	1.98 ± .1	2.07 ± .2	1.95 ± .2
Sensory-motor cortex	2.08 ± .1	2.01 ± .1	2.74 ± .2 ***	1.94 ± .1
Parietal cortex	2.00 ± .1	2.05 ± .1	2.58 ± .2 *	2.28 ± .2
Auditory cortex	2.35 ± .2	2.10 ± .2	2.43 ± .2	2.55 ± .2
Visual cortex	1.95 ± .1	1.81 ± .2	2.53 ± .1 *	1.96 ± .2
Entorhinal cortex	1.94 ± .1	1.88 ± .1	2.18 ± .2	2.22 ± .2
Medial geniculate body	2.06 ± .1	2.55 ± .1 *	2.34 ± .2	1.95 ± .1
Lateral geniculate body	1.94 ± .1	2.53 ± .1 **	2.37 ± .2	1.89 ± .2
Superior colliculus	2.31 ± .1	2.39 ± .2	2.11 ± .2	2.21 ± .2
Thalamus: lat. nucl.	2.11 ± .1	2.58 ± .1 **	2.33 ± .1	1.99 ± .1
Thalamus: vent. nucl.	2.10 ± .1	2.57 ± .2 ***	2.20 ± .1	1.98 ± .1
Caudatus-putamen	2.11 ± .1	1.97 ± .1	2.23 ± .2	1.74 ± .1
Globus pallidus	2.08 ± .2	1.84 ± .1	2.48 ± .3	2.10 ± .1
Substantia nigra	2.19 ± .1	2.25 ± .2	2.23 ± .2	2.32 ± .2
Hippocampus:				
- stratum moleculare	1.98 ± .1	1.97 ± .1	1.52 ± .1 **	1.70 ± .1
- dentate gyrus	2.44 ± .1	2.27 ± .2	2.31 ± .1	2.34 ± .2
Septal nucleus	2.32 ± .1	1.82 ± .1	2.41 ± .2	1.88 ± .2
Habenula	2.08 ± .2	1.98 ± .2	1.74 ± .1	1.98 ± .1
Hypothalamus	2.18 ± .2	2.27 ± .2	2.42 ± .2	1.80 ± .2
Cerebellar cortex	2.24 ± .1	1.86 ± .1	2.26 ± .2	2.19 ± .1
Internal capsule	1.77 ± .2	1.41 ± .1	1.64 ± .1	1.70 ± .1
White matter	1.82 ± .1	1.62 ± .2	1.53 ± .1	1.88 ± .1

n is the number of animals in each group. All values are expressed as mean ± SEM, alpha-level of significant differences from awake animal group (Bonferroni method): * $p < 0.05$; ** $p < 0.01$; *** $p < 0.001$

to induce characteristic alterations in local CBF/CMRG ratios: under pento-barbital anesthesia elevated CBF/CMRG ratios are observed in subcortical brain structures whereas under halothane anesthesia it was the cortical CBF/CMRG ratios which exceeded the CBF/CMRG couple measured in the awake control. Under metabolic suppression regional 'uncoupling' of CBF and CMRG resulted from a relative hemodynamic oversupply in affected brain areas indicating that factors other than metabolism-related mediators may play a role in the adjustment of tissue perfusion as well as regulating global cerebral blood flow.

Results on Local Cerebral Coupling of CBF and CMRG in Spreading Depression

It was further of interest to investigate local coupling between CBF and CMRG under experimental conditions in which local glucose consumption is drastically elevated instead of being suppressed. For this purpose, double tracer autoradiographic measurements of CBF and CMRG were performed under cortical and hippocampal spreading depression. Spreading depression is cha-racterized by the following electrophysiological properties: - reversible suppression of spontaneous neuronal activity which propagates at a speed of about 3 - 4 mm/min; - reversible depolarisation of neurons and glial cells as evident from transient negative DC-shifts in cortical or hippocampal steady potential; - a refractory period after occurrence of spreading de-ression during which spreading depression cannot be evoked (see Marshall, 1959; see Ochs, 1962). Ion-sensitive micro-electrode measurements revealed that the negative shift in DC-potential is accompanied by a transient re-lease of intracellular K^+ and intracellular uptake of extracellular Ca^{++} ions (Vyskocil et al., 1972). These changes in cellular electrolyte homeos-tasis are paralleled by a transient decrease in tissue ATP (Mies and Paschen, 1984) suggesting that the affected brain regions are metabolically activated. Measurement of glucose utilization revealed a drastic increase in cortical CMRG (Shinohara et al., 1979) - the consequence of elevated energy-consuming ion pump activity to restore of physiological electrolyte homeos-tasis.

Cortical spreading depression. Typical findings of local CBF and CMRG during cortical spreading depression (CSD) are illustrated in Figure 7. Topical application of KCl on to the exposed parietal cortex induced a drastic increase in local CMRG from 39.2 ± 3.1 to 95.5 ± 5.0 umol/100g/min (p<0.001). As indicated by a more or less normal CMRG value in the cingu-late gyrus (ipsilateral: 41.2 ± 3.4; contralateral: 37.5 ± 3.2 umol-/100g/min), CSD did not invade this area corroborating electrophysiological findings on the SD resistance of the cingulate cortex (Fifkova, 1964). In the entorhinal cortex the distinct increase in CMRG from 32.8 ± 4.0 to 45.7 ± 3.4 umol/100g/min suggests some interference by CSD. Local cortical blood flow increased parallel with the rise in local CMRG of the parietal cortex; the increase corresponded to 67.1 ± 6.1 to 120.3 ± 20.5 ml/100g/min (p < 0.05). As shown in Figure 7, CBF in the cingulate cortex rose from 64.3 ± 5.8 to 83.5 ± 11.2 ml/100g/min although local glucose utilization remained unchanged. In contrast, CBF in the entorhinal cortex was found to decrease distinctly from 60.4 ± 4.9 to 55.0 ± 8.4 ml/100g/min. Quantitative evalua-tion of local coupling between CBF and CMRG revealed that in the parietal cortex (ipsilateral: 0.95 ± 0.31; contralateral: 1.73 ± 0.34 ml/umol; p < 0.05) and in the entorhinal cortex (ipsilateral: 1.36 ± 0.19; contralateral: 1.82 ± 0.35 ml/umol; p<0.05) local CBF/CMRG ratios in the territory of the spreading depression decreased significantly compared to those in the con-tralateral cortex. In the ipsilateral cingulate cortex, the CBF/CMRG ratio increased distinctly but was found not to differ significantly from that of the homotopic cortex (ipsilateral: 2.26 ± 0.5; contralateral: 1.79 ± 0.33 ml/umol).

Restoration of repetitive intra/extracellular electrolyte derangement as observed during cortical spreading depression is associated with very high metabolic requirements. Local perfusion, however, was not adjusted to local energy demand during continuous CSD as would be expected from the control CBF/CMRG relationship. As apparent, uncoupling of CBF and CMRG resulted from a less marked rise in tissue perfusion indicating that the capacity of metabolism-related mechanisms to regulate hemodynamic adjustment to the metabolic demand appears to be limited during continuous CSD.

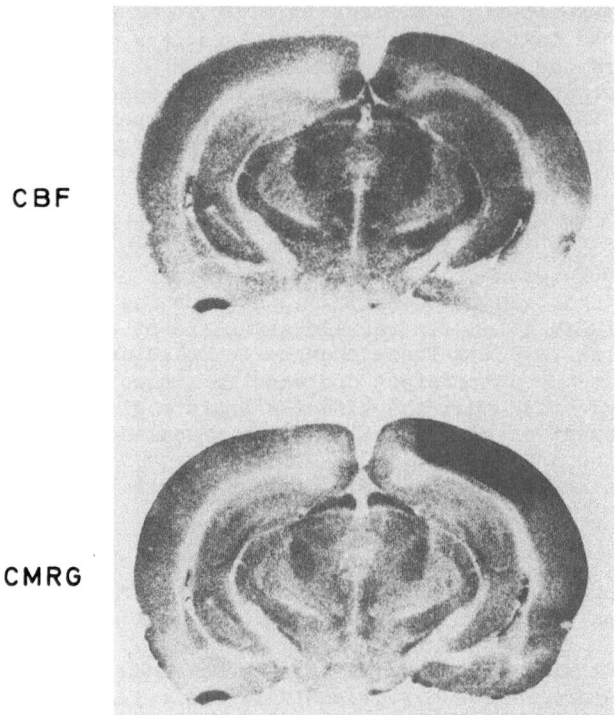

CBF

CMRG

Fig. 7. Representative autoradiograms of cerebral blood flow (CBF) and glucose utilization (CMRG) measured simultaneously by means of double tracer autoradiography in an animal submitted to continuous cortical spreading depression. Note the parallel increase in local CBF and CMRG in ipsilateral parietal cortex but mismatch of CBF and CMRG in cingulate gyrus.

Hippocampal spreading depression. The hemodynamic and metabolic situation after 30 min of continuous HSD is illustrated in Figure 8. In comparison to the unaffected hippocampus, the ipsilateral hippocampal CMRG was found to have increased drastically in the CA1-subfield (ipsilateral: 61.6 ± 3.3; contralateral: 36.0 ± 4.7 umol/100g/min; $p < 0.001$) and in the dentate gyrus (ipsilateral: 61.7 ± 5.3; contralateral: 31.9 ± 3.8 umol/100g/min; $p < 0.001$). Local CBF in the hippocampal CA1-subfield (ipsilateral: 57.2 ± 6.6; contralateral: 91.2 ± 11.5 ml/100g/min; ns) and dentate gyrus (ipsilateral:

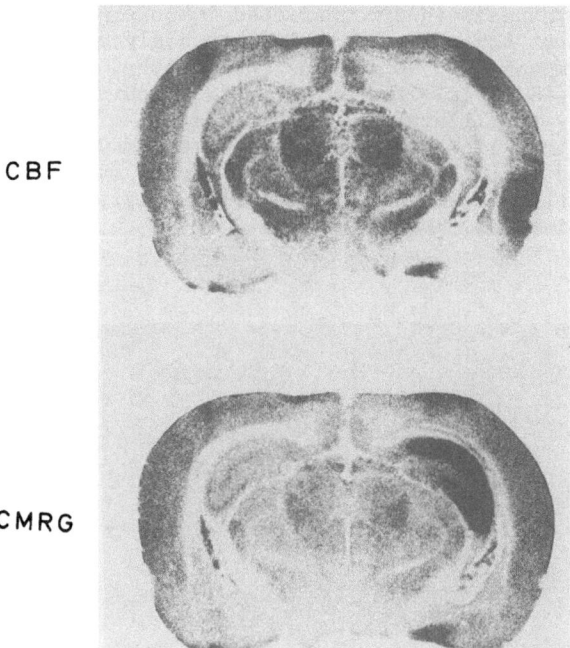

CBF

CMRG

Fig. 8. Simultaneous measurement of cerebral blood flow and glucose utilization by means of double tracer autoradiography during hippocampal spreading depression. Note the decrease in hippocampal CBF but drastically elevated CMRG in the hippocampus.

55.9 $\pm$ 2.6; contralateral: 75.0 $\pm$ 4.7 ml/100g/min; p<0.05), however, proved to decrease despite the increased metabolic demand. This mismatch in tissue perfusion and glucose metabolism led to significantly decreased CBF/ CMRG ratios in the hippocampal CA1-subfield (ipsilateral: 1.14 $\pm$ .23; contralateral: 2.56 $\pm$ 0.17 ml/umol; p<0.001) and dentate gyrus (ipsilateral: 0.97 $\pm$ 0.12; contralateral: 2.47 $\pm$ 0.27 ml/umol; p<0.001).

During hippocampal spreading depression the metabolism-related triggers which adjust tissue perfusion to the metabolic demand obviously fail to increase local hippocampal CBF which suggests an acute impairment of the flow-regulating mechanisms in this brain region.

Application of Triple Tracer Autoradiography to Experimental Brain Tumors

Typical findings are illustrated in Figure 9 where from the same brain section of a tumor-bearing rat the left autoradiograms are before and the right reconstructed images after digital subtraction of contaminating isotope radioactivty. As is apparent distinctly lower blood flows are observed in the solid parts of the tumor after appropriate correction of the CBF autoradiogram; this arises as a result of the high glycolytic activity of tumor tissue which will lead to ^{14}C tissue radioactivities above the threshold of isotope cross-contamination. In uncorrected ^{3}H-autoradiograms, on the other hand, incorporation of ^{3}H-AS appeared to be increased in the tumor mass as a whole but digital subtraction of ^{14}C cross-contamination revealed increased protein synthesis in circumscribed tumor areas in which CBF and CMRG were preserved at control values. Increased glycolytic activity within the tumor mass accompanied by low perfusion, however, resulted in a reduced incorporation of ^{3}H-AS. Triple tracer autoradiographic determination of CBF, CMRG and CPS greatly facilitates the interpretation of hemodynamic and metabolic observations: in tumor tissue increased CMRG but decreased CBF and reduced CPS suggests that the Pasteur effect may be symptomatic of impeding cell injury whereas the simultaneous metabolic activitation of CMRG and CPS suggests proliferative cellular activity.

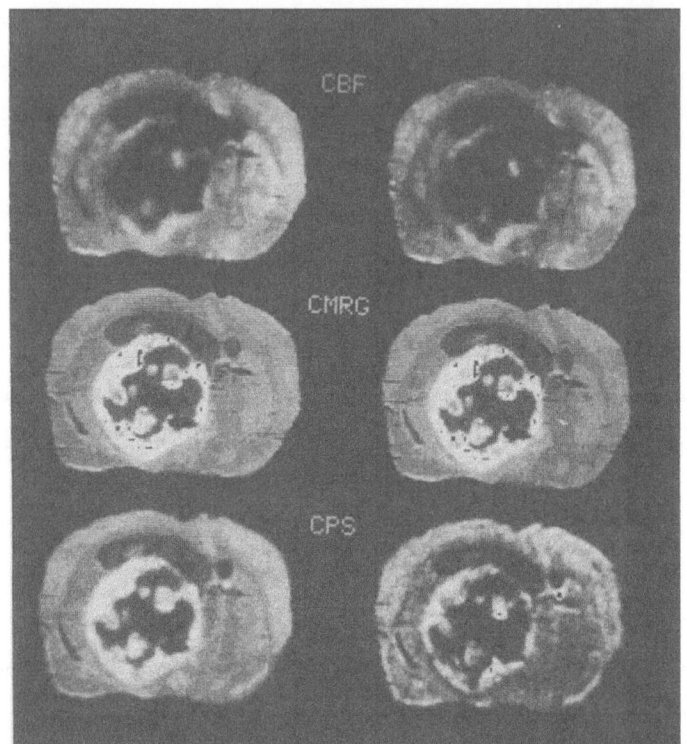

Fig. 9. Triple tracer autoradiograms of cerebral blood flow, glucose consumption and protein synthesis from the same brain section of a tumor-bearing rat. Left column: before, right column: after digital subtraction of contaminating isotope radioactivity. Note distinct cross-contamination by ^{14}C in both ^{131}I and ^{3}H autoradiograms, particularly in marginal regions of the tumor mass. After digital subtraction, a differentiation of tumor regions with low flow and high glucose utilization is possible. Circumscribed tumor regions with increased amino acid incorporation can be identified in the periphery of tumor in which glycolysis is drastically stimulated.

Discussion

Metabolic control of local cerebral tissue perfusion as hypothesized by Roy and Sherrington (1890) implies that vasoactive factors are released proportionally to local metabolic activity; this induces a decrease in cerebrovascular resistance concomitant with an increase in circulation in order to satisfy an elevated substrate demand. H^+ and adenosine have been suggested to be the major metabolism-related factors - the pH-hypothesis assumes that increased glucose consumption will provide more H^+-ions from carbon dioxide released during glucose breakdown, from the formation of lactic acid and from hydrolysis of ATP (Skinhoy, 1966; Lassen, 1968); - the adenosine-hypothesis suggests that the elevated breakdown of ATP during metabolic activation leads to a rise in the adenosine concentration which affects the vascular tonus (Berne, 1981). As shown by micro-electrode measurements of extracellular pH under experimental conditions of increased neuronal activity (Kuschinsky and Wahl, 1979) or by extravascular micro-application of adenosine (see Winn et al., 1981), both metabolic factors effect a decrease in vascular resistance with increasing substrate concentrations.

In the meantime, functional factors have been recognized as triggering vasodilatation of cerebral vessels in addition to controlling metabolically the local tissue perfusion. During increased neuronal activity intracellular K^+-ions are immediately released into the extracellular space and these have been shown to mediate a vasodilatory action at extracellular K^+ concentration less than 10 mMol. During neuronal cell depolarisation, there is a flux of Ca^{++}-ions into cells (Heinemann et al., 1977; Nicholson, 1980) and since the decrease in perivascular Ca^{++} can cause dilatation of pial arteries (Betz and Csornai, 1978), Ca^{++}-ions possibily play a role in the regulation of cerebrovascular resistance.

So far, we have addressed ourselves to the metabolic and functional factors which cause a vasodilatory effect at increasing concentrations of the metabolic and functional regulators. The question arises whether the increase in subsequent local tissue perfusion counteracts inasmuch as the mediators may be removed from the tissue by clearance. Metabolic inactivation of hemodynamically active regulators and inhibition of the metabolic compartment generating those vasoactive mediators may also occur. Although the central nervous system possesses a blood-brain-barrier at the endothelial level which prevents free exchange of most of the substrates by carrier meachanisms, a freely diffusible product of glucose breakdown such as carbon dioxide is likely to be washed out from brain tissue by increased tissue perfusion. On the other hand, increased H^+-activity inhibits glycolysis at the phosphofructokinase step (see Siesjö, 1978) thus shutting down the generation of metabolic factors responsible for the initial vasoactive interaction. Moreover, an elevated release of adenosine also leads to inhibition and in turn reduces neuronal excitability (Stone, 1981). Since the release of intracellular K^+ and the uptake of Ca^{++} supports excitability of neurons and decreases cerebrovascular resistance, the system is likely to oscillate in such a way that the balance between all factors involved is reflected in the newly re-established local perfusion/metabolism relationship.

The autoradiographic double tracer measurement of CBF and CMRG in the same animal fulfils several prerequisites which are necessary in order to evaluate the regional CBF/CMRG relationship. First of all, a CBF and CMRG autoradiogram is obtained from the same brain section in which the corresponding local parameters can be assessed in identical brain regions. Local evaluation is supported by the use of an image-processing system which ensures alignment of autoradiograms and the exact localization of the area of interest in both autoradiograms. Since the simultaneous measurements of

CBF and CMRG are performed in the same animal, stress arising from preparation of the animal, the anesthetic level and systemic parameters are identical; this is certainly difficult to achieve in separate animal groups despite standardization of the experimental set-up. Even assuming that experimental conditions are comparable, the question arises from the mathematical point of view as to what procedure is best suited for the evaluation of the interdependency between local CBF and CMRG. Conventionally, correlation analysis was applied to averaged CBF and CMRG data derived from separate animal groups and the significant linear correlation coefficient was taken as measure of the existence of a stochastic relationship between local CBF and CMRG. This statistical procedure was reviewed critically reviewed by McCulloch et al. (1982) and found not to be suitable for the the evaluation of cerebral coupling of CBF and CMRG. Regression and correlation analysis, on the other hand, do not take into account the individual change in the CBF/CMRG relationship because some local CBF/CMRG ratios may differ significantly from the control state but not alter the regression coefficient or the significance level of the correlation coefficient. The local CBF/CMRG value determined by means of quantitative double tracer autoradiography therefore, provides a measure by which a significant deviation of local coupling from the control state can be detected. This is most important for further investigations on the mechanism involved in the local control of tissue perfusion. It has to be emphasized, however, that a significant CBF/CMRG ratio itself does not permit an explicit conclusion to be drawn regarding the factor mediating dilatation or constriction of pial and parenchymal vessels but solely facilitates the interpretation of the activity level of factors suggested to regulate cerebrovascular resistance since in addition to the CBF/CMRG ratio the corresponding CMRG value is known and this allows a further estimation of the influence of metabolic factors.

As is evident from the variability of CBF/CMRG ratios obtained in awake controls and in experimental animal groups, local coupling between CBF and CMRG appears to be modulated by regional properties of hemodynamic control mechanisms over and above those generally accepted. This is most strikingly documented by the effect of pentobarbital and halothane which produced a different pattern in regional CBF/CMRG coupling: pentobarbital caused significantly increased CBF/CMRG ratios in subcortical brain regions whereas halothane induced a rise in cortical CBF/CMRG ratios. In cerebral metabolic depression by anesthetics the majority of brain regions exhibited the same hemodynamic adjustment to metabolism as in the awake control state and the significant increases in the regional CBF/CMRG couple are the result of a hemodynamic oversupply at lowered metabolic rates. Before any conclusions can be drawn regarding metabolic control of hemodynamics in cerebral metabolic depression, it must be verified that there is no direct effect of anesthetics on the cerebrovasculature. Barbiturates, for example, have been shown to constrict isolated arteries in in-vitro preparations (Price and Price, 1962) and it is difficult to decide whether the reduction in regional CMRG triggers the rise in cerebrovascular resistance or whether the direct interaction of pentobarbital is responsible for the increase in the vascular tonus. Halothane, on the other hand, turned out to mediate dilatation of cerebrovasculature in man and animals (Wollman et al., 1964; see Smith and Wollman, 1972; Smith, 1973) despite its metabolic depressant properties. As it turns out in fact, it has to born in mind that anesthetics exert a direct influence on brain vasculature in addition to the intrinsic control of cerebrovascular resistance. The more or less similar hemodynamic adaption to metabolism during cerebral metabolic depression compared to that in the awake control state could then account for the lack of significance of metabolic factors in controlling cerebrovascular resistance; the observed tissue perfusion could then result from direct effect by the drugs on the vascular response. Just how potent are the mechanisms of anesthesia-induced vascular changes in maintaining the vascular tonus? In a recent study, gamma-hydroxybutyrate was used to induce a drastic decrease in glucose metabolism in awake rats (Kuschinsky et al., 1985) but metabolic depression was not accompanied by a corresponding decline in regional tissue perfusion owing to elevated arterial carbon dioxide tensions in the treated awake animal group. Since the cerebrovascular CO_2-reactivity is mediated by H^+-ions which belong to the class of metabolic factors and which control cerebrovascular resistance, and since they are normally produced in proportion to glucose breakdown, they are obviously capable of overcoming possible vasoconstrictory effects of gamma-hydroxybutyrate. In the present study, systemic parameters were carefully controlled in order to avoid a global alteration of cerebrovascular resistance and to allow the identification of brain regions exhibiting a different CBF/CMRG relationship compared to the awake control state. Since metabolic depression and the difference in metabolic factor concentration in brain structures with elevated CBF/CMRG ratios is unlikely to be responsible for the relative hemodynamic oversupply, the sensitivity of cerebrovascular reactivity to a metabolic factor may increase. This would then result in a greater vascular response at identical concentrations of the mediator. Assuming that anesthetics exert a primary action on the vasculature a difference in vascular tonus could also result at a local level from a different sensitivity to the drug. In summary, anesthetics may affect the local CBF/CMRG relationship by direct vascular action and/or more indirectly by a metabolic depressant effect which leads to diminished generation of those metabolic factors suggested to control cerebrovascular resistance. Increased regional CBF/CMRG ratios during the exposure to anesthetics may result from a difference in regional vascular sensitivity to the drugs and/or from a hypersensitivity of vascular reactivity to metabolic factors controlling cerebrovascular resistance.

During metabolic stimulation of the cortex and hippocampus two different CBF responses were observed: cortical spreading depression (CSD) was followed by a subnormal increase in CBF, whereas in hippocampal spreading depression (HSD), regional CBF had decreased to below control values in spite of metabolic activation. Pentobarbital anesthesia had been used in both studies and this may have interfered by exerting a direct vasoconstrictory action on vascular smooth muscle cells. In CSD however, the increase in CMRG, i.e. the elevated generation of metabolic factors, was capable of dilatating the cerebrovasculature although not enough to establish the same CBF/CMRG relationship as in the homotopic cortical area. This observation may be accounted for by the fact that the upper limit of metabolism-related flow regulation was reached and/or that the increased tissue perfusion and the subsequent clearance of mediator failed to peak such that the amounts of metabolic factors were present at a concentration sufficient to establish a CBF/CMRG relationship like that in the awake control state. However, in HSD, such metabolic mediators, produced as the result of elevated metabolic activity, turned out to be ineffective in triggering a proportional vascular dilatation. On the contrary, vasoconstriction was observed in hippocampus. In comparison, an obviously identical metabolic stimulus as regards time and magnitude revealed different hemodynamic responses in the cortex and hippocampus; this suggests that in addition, there is a regional domain of vascular reactivity. The interpretation of altered CBF/CMRG ratios during spreading depression has to include those changes which may be unique to this phenomenon. Since there is no spontaneous neuronal activity during continuous spreading depression, it is most likely that the physiological fluctuation of functional factors appears to be reduced. As regards to electrolyte homeostasis, extracellular K^+-concentration in spreading depression transiently exceeds 10 mMol at which level vasoconstriction is mediated rather than vasodilatation of pial arteries. Since the timing of the electrophysiological appearance of spreading depression is the same in the cortex as in the hippocampus, the changes in the extracellular K^+ should affect the vascular diameter in a similar manner. A difference, however, may result from an alteration in the sensitivity of cerebrovascular resistance caused by K^+ and Ca^{++} ions or from a change in the sensitivity of the cerebrovasculature to metabolic factors. In summary, in cerebral metabolic stimulation elicted by spreading depression, the hemodynamic alterations appear to be a regional phenomenon. While in CSD the response in tissue perfusion consisted of an disproportional elevation which was mismatched with regional CMRG, regional tissue perfusion during HSD even decreased to below control levels, thus amplifying the discrepancy between CBF and CMRG as regards substrate availability.

This regionality in vascular response to metabolic stimulation may have some implications for pathophysiological situations. Anesthetics are commonly employed in animal surgery prior to the onset of and during recovery from experimental conditions. For example, the late postischemic recovery phase is characterized by metabolic depression and the abolishment of cerebrovascular CO_2 reactivity (Hossmann et al., 1973). An anesthetic effect may very well account for the lack of the vasculature to respond to H^+-ions or may be counterbalanced by vasoactive properties of the anesthetic drugs. Since local CBF turned out to decrease significantly during hippocampal spreading depression suggesting impairment of the regional CBF/CMRG couple by a failure of metabolic factors known to trigger at least a relative adaptation of perfusion as observed in cortical spreading depression, this phenomenon may be of importance in other pathophysiological state. For example, in short-lasting cerebral ischemia of 5 min, a delayed onset of tissue damage was observed in the hippocampal CA1-subfield (Kirino, 1982), a brain region in which a severe postischemic hypoperfusion is also observed (Suzuki et al., 1983); this may prolong the mismatch between CBF and CMRG into the post-ischemia recovery period. It appears that the onset of delayed hypoperfusion paralleled by an increase in local CMRG may be equally as responsible for the initiation of cerebral tissue damage as the neurological damage during the ischemic insult itself.

Thus quantitative double or triple tracer autoradiography for simultaneously measuring CBF and CMRG or CBF, CMRG and CPS, respectively, appears promising in extending our present knowledge and understanding of the interrelationship between hemodynamic and metabolic events taking place in the central nervous system under physiological and pathophysiological conditions.

REFERENCES

Alexander, F.G., 1912, Untersuchungen über den Blutgasaustausch des Gehirns, Biochem Z 44:127
Berne R.M., Rubio R., and Curnish R.R., 1974, Release of adenosine from ischemic brain. Effect on cerebral vascular resistance and incorporation into cerebral adenosine nucleotides, Circ Res 35:262
Cobb S., and Talbott J.H., 1927, Quantitative study of cerebral capillaries, Trans Assoc Am Physicians 42:255
Des Rosiers M.H., Kennedy C., Shinohara M., and Sokoloff L., 1978, Effect of CO_2 on local cerebral glucose utilization in the conscious rat. Neurology 26:346
Dwyer B.E., Donatoni P., and Wasterlain C.G., 1982, A quantitative autoradiographic method for the measurement of local rates of brain protein synthesis, Neurochem Res 7:563
Fifkova E., 1964, Spreading EEG depression in the neo-, paleo-, and archicortical structures of the brain of the rat, Physiol Bohemoslov 13:1
Freygang W.H., and Sokoloff L., 1958, Quantitative measurement of regional circulation in the central nervous system by the use of radioactive inert gas, Adv Biol Med Phys 6:263
Heinemann U., Lux H.D., and Gutnick M.J., 1977, Extracellular free calcium and potassium during paroxysmal activity in the cerebral cortex of the cat, Exp Brain Res 27:237

Hossmann K-A., Lechtape-Grüter H., and Hossmann V., 1973, The role of cerebral blood flow for the recovery of the brain after prolonged ischemia, Z Neurol 204:281

Kety S.S., and Schmidt C.F., 1945, Determination of cerebral blood flow in man by use of nitrous oxide in low concentrations, Am J Physiol 143:53

Kety S.S., 1949, The physiology of the human cerebral circulation, Anesthesiology 10:610

Kety S.S., 1951, The theory and application of the exchange of inert gas at the lungs and tissues, Pharm Rev 3:1

Kirino T., 1982, Delayed neuronal death in the gerbil hippocampus following ischemia, Brain Res 239:57

Kuschinsky W., and Wahl M., 1979, Perivascular pH and pial arterial diameter during bicuculline induced seizures in cats, Pflügers Arch 382:81

Kuschinsky W., Suda S., and Sokoloff L., 1985, Influence of gamma-hydroxybutyrate on the relationship between local glucose utilization and local cerebral blood flow in the rat brain. J Cereb Blood Flow Metab 5:58

Landau W.M., Freygang W.H.Jr., Roland L.P., Sokoloff L., and Kety S.S., 1955 The local circulation of the living brain: values in the unanesthetized and anesthetized cat, Trans Amer Neurol Assoc 80:125

Lassen N.A., 1968, Brain extracellular pH: the main factor controlling cerebral blood flow, Scand J Clin Lab Invest 22:247

Lear J.L., Jones S.C., Greenberg J.H., Fedora T.J., and Reivich M., 1981, Use of ^{123}I and ^{14}C in a double radionuclide autoradiographic technique for the simultaneous measurement of LCBF and LCMRglu. Theory and method, Stroke 12:589

Marshall W.M., 1959, Spreading cortical depression of Leao, Physiol Rev 39:239

McCulloch J., Kelly P.A.T., and Ford I., 1982, The effect of apomorphine on the relationship between local cerebral glucose utilization and local cerebral blood flow with an appendix on its statistical analysis, J Cereb Blood Flow Metab 2:487

Mies G., Niebuhr I., and Hossmann K-A., 1981A, Simultaneous measurement of blood flow and glucose metabolism by autoradiographic techniques, Stroke 12:581

Mies G., Niebuhr I., and Hossmann K-A., 1981B, A double tracer autoradiographic technique for the simultaneous measurement of cerebral blood flow and cerebral metabolism in rats, Eur Neurol 20:188

Mies G., Bodsch W., Paschen W., and Hossmann K-A., 1983, Experimental application of triple-label quantitative autoradiography for the measurement of cerebral blood flow, glucose metabolism, and protein biosynthesis, in:"Positron Emission Tomography of the Brain", W-D. Heiss and M.E. Phelps, eds., Springer-Verlag, Berlin

Mies G., and Paschen W., 1984, Regional changes of blood flow, glucose, and ATP content determined on brain sections during a single passage of spreading depression in rat brain cortex, Exp Neurol 84:249

Mies G., Bodsch W., Paschen W., and Hossmann K-A., 1986, Triple-tracer autoradiography of cerebral blood flow, glucose utilization and protein synthesis in rat brain, J Cereb Blood Flow Metab 6:59

Nicholson C., 1980, Modulation of extracellular calcium and its functional implications, Fed Proc 39:1519

Ochs S., 1962, The nature of spreading depression in neuronal networks, Int Rev Neurobiol 4:1

Penfield W., von Santha K., and Cipriani W., 1939, Cerebral blood flow during induced epileptiform seizures in animals and man, J Neurophysiol 2:257

Price M.L., and Price H.L., 1962, Effects of general anaesthetics on contractile responses of rabbit aortic strips, Anesthesiology 23:16

Reivich M., Jehle J., Sokoloff L., and Kety S.S., 1969, Measurement of regional cerebral blood flow with antipyrine-^{14}C in awake cats, J Appl Physiol 27:296

Reivich M., 1972, Regional cerebral blood flow in physiologic and pathophysiologic states, Prog Brain Res 35:191

Sakurada O., Kennedy C., Jehle J., Brown J.D., Carbin G.L., and Sokoloff L., 1978, Measurement of local cerebral blood flow with iodo(^{14}C)antipyrine, Am J Physiol 234:H59

Schmidt C.F., and Hendrix J.P., 1937, The action of chemical substances on cerebral blood vessels, Res Publ Assoc Res Nerv Ment Dis 18:229

Serota H.M., and Gerard R.W., 1938, Localized thermal changes in the cat's brain, J Neurophysiol 1:115

Shinohara M., Dollinger B., Brown G., Rapoport S., and Sokoloff L., 1979, Cerebral glucose utilization: local changes during and after recovery from spreading depression, Science 302:188

Siesjö B.K., 1978, "Brain Energy Metabolism", John Wiley and Sons, Chichester

Skinhoy E., 1966, Regulation of cerebral blood flow as a single function of the intestinal pH in the brain. A hypothesis, Acta Neurol Scand 42:604

Smith A.L., and Wollman H., 1972, Cerebral blood flow and metabolism: Effect of anesthetic drugs and techniques, Anesthesiology 36:378

Smith A.L., 1973, The mechanism of cerebral vasodilation by halothane, Anesthesiology 39:581

Stone T.W., 1981, Physiological roles for adenosine and adenosine-5'-triphosphate in the nervous system, Neuroscience 6:523

Sokoloff L., Reivich M., Kennedy C., Des Rosier M.H., Patlak C.S., Pettigrew K.D., Sakurada O., and Shinohara M., 1977, The (^{14}C)deoxyglucose method for the measurement of local cerebral glucose utilization: theory, procedure and normal values in the conscious and anesthetized albino rat, J Neurochem 28:897

Sokoloff L., 1981, Localization of functional activity in the central nervous system by measurement of glucose utilization with radioactive deoxyglucose, J Cereb Blood Flow Metab 1:7

Suzuki R., Yamaguchi T., Kirino T., Orzi M., and Klatzo I., 1983, The effect of 5-min ischemia in mongolian gerbils: I. Blood-brain barrier, cerebral blood flow, and cerebral glucose utilization changes, Acta Neuropathol (Berl.) 60:207

Vyskocil F., Kriz N., and Bures J., Potassium selective micro-electrodes used for measuring the extracellular brain potassium during spreading depression and anoxic depolarization in rats, Brain Res 39:255

Winn H.R., Rubio R., and Berne R.M., 1981, The role of adenosine in the regulation of cerebral blood flow, J Cereb Blood Flow Metab 1:239

Wollman H., Alexander C.S., Cohen P.J., Chase P.E., Melman E., and Behar M.G., 1964, Cerebral circulation of man during halothane. Effects of hypocarbia and of d-tubocurarine, Anesthesiology 25:180

SINGLE PHOTON EMISSION TOMOGRAPHY

INTRODUCTION TO METHODS USED FOR MEASURING REGIONAL CEREBRAL BLOOD FLOW WITH SINGLE PHOTON EMISSION TOMOGRAPHY

Giovanni Lucignani and Maria Carla Gilardi

Consiglio Nazionale Delle Ricerche
Centro Studi Fisiologia Del Lavoro Muscolare
Via Olgettina, 60
20132 Milano, Italia

INTRODUCTION

The measurement of regional cerebral blood flow (rCBF) has been pursued by methods generally based on kinetic models that require the systemic administration of tracer substances and the measurement of their concentration in arterial blood and brain tissue. Most of the kinetic models used for this purpose have been based on the Fick principle. This principle has been expanded by Kety to describe the exchange of inert gas, or other non-metabolizable, freely diffusible substances, between capillary blood and tissue [1]. The nitrous-oxide method for measurement of CBF [2] was one of the first applications of the principles of inert gas exchange. This method in its original form did not utilize a radioactive tracer, but has been modified for use with radioactive gases [3-12].

The use of radioactive substances has allowed the _in vivo_ assay of the local concentration of tracer in tissue. Several radioactive tracers have been proposed for this purpose and their concentrations are monitored either by external detectors for use in human beings or by quantitative autoradiography for studies in animals [13,14]. More recently, the measurement of local tissue concentrations of gamma-emitting radioactive tracers by single photon emission tomography (SPECT)[15,16], a technique which allows the reconstruction of three dimensional images of radioactivity distribution in human brains, has been investigated as a possible tool to determine rCBF following the administration of radioisotopes [17-19]. However, the measurement of radioactivity distribution does not _per se_ provide any information on the physiology or biochemistry of the organ or system under study unless it is combined with kinetic models of tracer distribution in arterial blood and brain tissue. For this reason the selection of a proper tracer, of the timing of the procedure, and the system used for the recording of radioactivity are of critical importance to obtain meaningful information on brain physiology.

The purpose of the present paper is to introduce the general concepts, the basic kinetic models, and the experimental strategies for the measurement of rCBF by SPECT with currently available tracers.

THEORY

The amount of blood supplied to the brain per unit time per unit mass of tissue (CBF) is a physiological parameter which can be measured following the administration of tracer substances which can cross the blood-brain barrier. According to the Fick principle the rate of accumulation of an inert diffusible tracer is equal to the difference in rates at which the tracer is brought to the tissue in the arterial blood and removed from it in the venous blood. This statement can be expressed mathematically as follows:

$$dQ_i/dt = dQ_a/dt - dQ_v/dt \qquad (1)$$

where dQ_i/dt, dQ_a/dt and dQ_v/dt represent the rate of change of the quantity of the substance in the brain, the rate of delivery in the arterial blood, and the rate of removal from brain with mixed cerebral venous blood, respectively. However, the rate at which the tracer is carried to the tissue , dQ_a/dt is equal to the arterial inflow (F_a) times the arterial concentration (C_a), and the rate at which the substance leaves, dQ_v/dt, is equal to the venous outflow (F_v) times the mixed venous concentration (C_v). Thus, if the arterial blood inflow equals the venous outflow:

$$dQ_i/dt = F(C_a - C_v) \qquad (2)$$

Dividing both sides of the equation by tissue mass W:

$$dC_i/dt = (F/W)(C_a - C_v) \qquad (3)$$

where C_i is the concentration of the tracer in the brain tissue per unit mass. The knowledge of the concentration of the tracer in the brain and of its arteriovenous difference allows, according to Equation 3, the calculation of the rate of flow, F/W, once an appropriate experimental procedure is defined.

The Fick principle expressed in the above differential form is the basis of the nitrous-oxide method for measurement of global cerebral blood flow in human subjects.

The Nitrous-Oxide Method

The nitrous-oxide technique [2] is based on the administration by inhalation of N_2O at low concentration. For a nonradioactive gas like N_2O whose concentration cannot be measured in tissue by external probes the amount of tracer in the tissue can be calculated by use of an appropriate experimental procedure. Inhalation of nitrous-oxide at a constant concentration, 15% N_2O in air, allows the eventual achievement and maintenance of a virtual equilibrium of the gas between the brain tissue and the mixed cerebral venous blood within approximately 10 minutes from the onset of the gas administration. Under these conditions, i.e. identical tension of N_2O in blood and tissue, the ratio of the concentrations of the tracer in tissue, C_i, and venous blood, C_v, is by definition the tissue:blood partition coefficient, λ, for N_2O. Thus C_i can at equilibrium also be defined as λC_v. Equation 3 can be, therefore, rearranged and integrated between time T_o when C_a = 0 and time T to obtain the operational equation of the method as follows:

$$F/W = \lambda C_v(T) \bigg/ \int_{T_0}^{T} (C_a - C_v)\, dt \qquad\qquad (4)$$

Equation 4 defines the variables which must be determined during the experimental procedure for measurement of the rate of flow. On the basis of this equation, therefore, F/W can be determined in an entire organ or region where arterial and representative venous blood are accessible to examination.

It is possible to use these same principles which govern the exchange of inert substances between blood and tissue for measurement of flow in small areas whose arterial input and venous drainage are not accessible to measurement [20]. This approach, however, is based on the assumption that the concentration of the tracer is uniform throughout the arterial system so that C_a can be measured in any convenient peripheral artery. Moreover, the method requires the direct measurement of the tissue concentration of the tracer in individual structures, C_i, made possible by the use of radioactive substances which can be detected by autoradiography or external probes.

The Tissue Equilibration Principle

For tissues that are homogeneous with respect to the rate of perfusion and solubility of tracer, the following relationship between the arterial-venous and the arterial-tissue concentration differences for freely diffusible tracer substance can be defined as follows [20]:

$$(C_a - C_v) = (C_a - C_i/\lambda) \qquad\qquad (5)$$

where λ is the tissue:blood partition coefficient for the tracer. The assumption of complete equilibrium between tissue and venous drainage is also made. Equation 5 can be used, therefore, only in the absence of arteriovenous shunts and diffusion limitations and for a homogeneous compartment. However, for those cases in which there is some diffusion limitation of the tracer, Equation 5 has been modified as follows [20]:

$$(C_a - C_v) = m(C_a - C_i/\lambda). \qquad\qquad (6)$$

In this equation m is a constant between 0 and 1 which represents the fraction of complete equilibrium achieved between blood and tissue during each passage through the tissue from the arterial to the venous end of the capillary.

A definition of m was derived in terms of the capillary permeability coefficient, P, the capillary surface area, S, and the rate of blood flow, F, as follows [1]:

$$m = 1 - e^{-(PS/F)}. \qquad\qquad (7)$$

Combining Equations 3 and 6:

$$dC_i/dt = m\, FC_a/W - m\, FC_i/\lambda W. \qquad\qquad (8)$$

Let $\qquad m\, F/W = K_1 \qquad\qquad (9)$

and $\qquad m\, F/\lambda W = k_2. \qquad\qquad (10)$

Thus $\qquad \lambda = K_1/k_2. \qquad\qquad (11)$

Substituting K_1 and k_2 in Equation 8:

$$dC_i/dt = K_1 C_a - k_2 C_i \qquad\qquad (12)$$

The solution of Equation 12 at any time T is given by:

$$C_i(T) = C_i(T_0)e^{-k_2 T} + K_1 e^{-k_2 T} \int_{T_0}^{T} C_a(t)e^{k_2 t}dt \tag{13}$$

Equation 13 can also be written as follows:

$$C_i(T) = C_i(T_0)e^{-k_2 T} + \lambda k_2 e^{-k_2 T} \int_{T_0}^{T} C_a(t)e^{k_2 t}dt \tag{14}$$

Equation 14 is the general equation for measurement of rCBF with tracers which cross the BBB. Equation 14 can be solved for k_2 from knowledge of the tissue concentration of the tracer at time, T; the partition coefficient, λ; and the time course of the arterial tracer concentration from zero time, T_0, to time, T. Then from knowledge of the diffusibility coefficient, m, and λ the value of flow, F/W, can be determined by Equation 10.

It is important to note, however, as can be readily seen from Equation 7, that only for PS/F values greater than 5 does the value of m approach 1 within 1%, i.e. almost complete diffusibility. Under this condition the quantity of tracer diffusing across the BBB is directly proportional to blood flow. On the other hand, if PS/F is not high enough to yield complete diffusibility of the tracer over the range of flows of interest, increasing values of F will result in decreases in relative diffusibility. The use of a tracer substance which is not completely diffusible leads to an underestimation of the rates of flow. Because the magnitude of the error due to diffusion limitation is flow dependent, the range of blood flow values is compressed and the evaluation of any condition that changes the rate of blood flow is distorted [21]. However, it has also been suggested that the permeability surface product, PS, may not be a fixed value, but may instead increase with blood flow because of vasodilation and/or capillary recruitment [22].

Equation 14 can be used to measure rCBF in studies carried out with experimental procedures which require the administration of the tracer by different routes and the monitoring of the tracer either during tissue saturation or during tissue desaturation.

In studies carried out during desaturation two possibilities can occur: the first is a sudden drop to zero of the arterial concentration of the tracer at the time of onset of desaturation, T_0. Under this condition, i.e. $C_i(T_0)$ different from 0 and $C_a(t) = 0$ for times greater than T_0, Equation 14 reduces to:

$$C_i(T) = C_i(T_0)e^{-k_2 T} \tag{15}$$

The second possibility occurs when the tracer recirculates after the beginning of the desaturation process. This case requires the use of Equation 14 whose second term on the right side corrects for recirculation of the tracer. This is the case of the ^{133}Xe clearance studies of rCBF in humans [8].

When the method is used during saturation, $C_i(T_0) = 0$ and Equation 14 then reduces to:

$$C_i(T) = \lambda k_2 e^{-k_2 T} \int_{T_0}^{T} C_a(t)e^{k_2 t}dt \tag{16}$$

Equation 16 is used with the autoradiographic technique in animals with iodo[14C]antipyrine [14] and had previously been used with [14C]antipyrine [13] and [131I]trifluoroiodomethane [23].

In all these cases the experimental strategy must provide the possibility of measuring the clearance rate constant of the tracer, k_2. This can be done: a) by monitoring the time course of the concentration of the tracer in tissue from a sequence of instantaneous measurements during desaturation, and/or b) by an instantaneous measurement of the tissue concentration of the tracer at a single time during the saturation process.

In certain cases the measurement of the tissue concentration of the tracer cannot be made instantaneously, and significant variations may occur during the time of measurement because of continuous uptake of recirculating tracer and/or redistribution of the tracer within the tissue. Under these conditions Equation 14 must be integrated over the time interval of the scanning procedure to obtain an integrated tissue concentration measurement as follows:

$$\int_{T_0}^{\tau} C_i(T)dT = \int_{T_0}^{\tau} C_i(T_0)e^{-k_2 T}dT + \lambda k_2 \int_{T_0}^{\tau} e^{-k_2 T} \int_{T_0}^{T} C_a(t)e^{k_2 t}dt \ dT \qquad (17)$$

Integrating by parts Equation 17 can also be written as follows:

$$\int_{T_0}^{\tau} C_i(T)dT = \int_{T_0}^{\tau} C_i(T_0)e^{-k_2 T}dT + \lambda \int_{T_0}^{\tau} C_a(t)dT - \lambda e^{-k_2 \tau} \int_{T_0}^{\tau} C_a(t)e^{k_2 T}dT \qquad (18)$$

The principle of the measurement of the wash-out rate constant, k_2, does not apply to tracers which once in the brain remain trapped irreversibly. A different model and equation are then necessary.

The Indicator Distribution Principle

With tracers which cross the BBB from blood to brain but are completely extracted and trapped in tissue during the time of the experimental procedure, i.e. $C_v = 0$, according to the Fick principle:

$$dC_i/dt = (F/W)C_a \qquad (19)$$

and

$$C_i(T) = (F/W) \int_{T_0}^{T} C_a(t) \ dt. \qquad (20)$$

With tracers which are incompletely extracted, but are still irreversibly trapped, Equation 20 can be modified as follows:

$$C_i(T) = E(F/W) \int_{T_0}^{T} C_a(t) \ dt \qquad (21)$$

where E is a value between 0 and 1 which defines the unidirectional capillary single-pass extraction fraction of the indicator[24]. It is important to notice that in this case, since the model assumes no backflux, the unidirectional extraction fraction is equivalent to the net extraction fraction.

Although E has the same mathematical definition as m, i.e. $E = 1 - e^{-(PS/F)}$, the two parameters m and E describe different processes: m is the bidirectional diffusion limitation coefficient and E the unidirectional fractional transport coefficient. The same consideration expressed for the parameter m, i.e. dependence on the rate of blood flow, applies to the parameter E as well; the relationship between the unidirectional rate of influx of tracer, EF, and the rate of flow, F, is not linear [25-27].

A widely used application of the indicator distribution principle is based on the use of radioactive microspheres. These tracers, which remain physically trapped in the capillary bed because of their size, cannot be administered intravenously and are generally used for experiments in animals by direct intracardiac injection. The measurement of tissue concentration of these tracers can be performed over an interval of time since its level remains essentially constant. The intravenous route, on the other hand, is only feasible with the so-called "chemical microspheres". These are tracers which can diffuse freely across the capillary bed and supposedly remain trapped in tissue on the basis of some chemical mechanism. These tracers are likely to recirculate for a certain time after administration unless they are sequestered in some target organ and/or metabolized into products which do not enter the brain. For these tracers the tissue concentration increases throughout the experimental time and its measurement has to be performed over a short interval. When short time measurements of tissue concentration are not feasible, Equation 21 must be integrated over the time interval of scanning to obtain an integrated tissue concentration as follows:

$$\int_{T_0}^{\tau} C_i(t) dT = \int_{T_0}^{\tau} EF/W \int_{T_0}^{T} C_a(t) \ dT \tag{22}$$

In summary, the operational equations derived for each method (Equations 4, 14 through 16, 21) define the variables which must be measured and the parameters which must be known for the measurement of rCBF. The application of these operational equations necessitates the development of adequate experimental strategies and methods to measure rCBF under different conditions and boundaries such as type of tracer used, time frame of the study, and finally techniques and instruments for monitoring the tissue concentration of the tracer in brain.

METHODS FOR SPECT STUDIES OF rCBF

The Xenon Clearance Technique

The ^{133}Xe technique has been the most common method used for the assessment of rCBF for many years. The theoretical basis of this method rests on the principles of the inert gas exchange developed by Kety. The first application of these principles to the measurement of local tissue perfusion rates was the [24]Na clearance method of Kety developed for use in muscle [28]. The ^{133}Xe was a modification of the [24]Na method for use in brain where a tracer that diffuses through the BBB is required.

Conn applied the clearance technique for measuring CBF by substituting ^{133}Xe for [85]Kr and monitoring the brain uptake and washout of tracer with an external probe [29]. The measurement of rCBF by ^{133}Xe is based on the monitoring of the clearance of the gas from the tissue; sample observations are made at intervals of one or, at most, a few seconds apart. The resulting exponential time activity curve is then used to calculate blood flow from Equation 14.

The assessment of rCBF obtained by two dimensional planar images, following the administration of the tracer by inhalation has some limitations caused by: a) representation of the three dimensional distribution of the tracer on a two dimensional plane, b) presence of tracer in the upper airways and extracerebral tissues, c) isotope recirculation. The administration of the ^{133}Xe by intracarotid injection produces images with better resolution which are less affected by the above mentioned sources of error [8-10]. Both dedicated instruments and procedures have been validated for the measurement of rCBF following both inhalation and intracarotid routes of administration [29-32]. The introduction of emission tomography has recently made possible three-dimensional cross sectional images of brain perfusion in human beings [17-19]. In particular, the use of SPECT has reduced the effects of activity superimposition. Since the washout of ^{133}Xe is an event which has a half life of less than one minute, a conventional rotating gamma camera is unsuitable for studies of rCBF because it is both too slow and has too little sensitivity. Tomographic images, however, can be obtained only by computerized reconstruction of data acquired over a certain time frame. In order to reconstruct one slice of the brain with an acceptable spatial resolution with SPECT, it is necessary to accumulate at least 300,000 counts [33]. Due to these requirements and due to the limited amount of tracer which can be administered in a routine study, it is necessary to perform the measurement of the activity over a relatively long period of time. Several scanning systems have been developed for this purpose, but two are most widely used. The first one consists of a system of 64 sodium iodide crystals in four square banks placed as close as possible to the subject's head (Tomomatic 64) [34]. The system rotates around the head at the speed of 180 degrees/5 sec. The second system (Headtome) [35] consists of a stationary ring of scintillation crystals. In this system a tungsten collimator moves so that each crystal can see different lines of brain at different times. Both systems allow recording of a series of 1-minute images of the isotope distribution in each slice of brain tissue with an approximate resolution of 17 mm [36]. Each image is obtained by accumulating counts for several rotations of the detector over a time of 60 seconds. A complete study with this technique lasts 4.5 minutes: during the initial 90 seconds ^{133}Xe is inhaled. A series of four consecutive tomographic images, one during the inhalation and three 1-minute images thereafter, are collected. The air curve of ^{133}Xe monitored over the chest is taken to represent the arterial curve [36,37].

As proposed by Kanno and Lassen [33], the calculation of the rate of blood flow can be based on either the measurement of the activity in tissue during the initial distribution or during the clearance of the tracer. The assessment of the initial distribution is the basis of the "early picture method" which requires the analysis of the first 90-second image recorded while ^{133}Xe enters the brain. The "sequence of pictures method" consists of the analysis of the shape of the time activity curve obtained from the sequence of the last three one-minute images. Both methods have some limitations [36]. The poor discrimination among different regions with high flow rates due to the nonlinearity of the distribution of the tracer with respect to the rate of flow is a limitation of the early picture method, whereas a poor sensitivity to areas of low blood flow is a limiting aspect of the sequence of pictures method. A procedure which combines both calculation methods has been proposed by Celsis et al. [36] to improve estimation of blood flow in areas of reduced perfusion.

The "Chemical Microsphere" Technique

New radiopharmaceuticals along with kinetic models for measurement of rCBF with currently available rotating gamma camera systems have been sought. These have required a different approach and the development of alternative techniques. The general purpose rotating gamma cameras have a major advantage, due to the use of large planar crystals, i.e. the possibility to obtain contiguous sections at any level of the brain, whereas the dedicated multicrystal tomographs can provide only a limited number of discrete sections with unseen interposed slices. A major disadvantage of general purpose instruments, however, is the length of time necessary for a complete rotation. The tomographic images obtained by these instruments are reconstructed on the basis of multiple views obtained by a single rotation of the gamma camera per study. It is self evident that with such an instrument only studies under steady state conditions are possible. Considerable efforts have been devoted, therefore, to synthesize radioactive tracers which remain trapped in the brain tissue after systemic administration by intravenous route. The goal has been the development of tracers suitable for currently available SPECT tomographs.

Since 1980 Winchell et al. [38] and Tramposch et al. [39] have investigated many iodophenylalkyl amines labeled with radioactive iodine in search of a flow tracer that could cross the intact BBB and remain trapped inside the brain tissue, i.e. an ideal "chemical microsphere". The two "chemical microspheres" which have been most widely tested and used for rCBF assessment are isopropyl-iodoamphetamine (IMP)and N,N',N'-trimethyl-N-(2-hydroxy-3-methyl-iodobenzyl)-1,3-propane- diamine (HIPDM). The biodistribution and the kinetics of these tracers have been studied in animals and in man in order to derive quantitative methods for rCBF measurement [19,40-43].

As with other compounds which do not have active transport systems, the penetration of the BBB by these tracer is attributable to free diffusion of the un-ionized lipophilic form of the compound. In particular, the mechanism of uptake and retention of IMP has been attributed to its lipophilicity and affinity for high-capacity, relatively nonspecific binding sites in brain and capillary endothelium [39]. For HIPDM the mechanism of trapping has been attributed instead to a change from un-ionized to ionized form in the passage across the BBB due to the difference of pH between blood and brain tissue [44].

Both tracers have been reported to be distributed in brain in proportion to blood flow. Although this represents a necessary condition for measurement of rCBF with any tracer, it is not sufficient. An additional series of conditions must be satisfied: a) the tracer must be delivered continuously to the brain in proportion to blood flow, b) it must cross freely the BBB and be completely extraced from blood into brain tissue during a single passage through the brain capillaries, c) the tracer must be distinguished in arterial blood from its polar metabolites that do not enter the brain tissue [19]. Under these conditions rates of rCBF can be calculated, assuming that there is not significant efflux of tracer from brain during the time of measurement of tissue concentration of the radioisotope, using the equation of the indicator distribution principle (Eq. 20 in theory section).

In the model proposed for rCBF measurement with IMP [19] the loss of tracer from brain is considered to be negligible and the possible error is minimized by making the measurement of the tissue concentration of tracer sufficiently early so that the amount of tracer diffusing back from brain is small. It has also been observed that IMP is not completely extracted and the kinetic model takes into account also the partial extraction [19]. Depending on the systemic metabolism of IMP, not all the circulating radioactivity is available for exchange with the brain and a separation procedure between IMP and its polar metabolites which do not cross the BBB has been proposed [19].

Also in the process of developing a kinetic model for rCBF measurement with HIPDM it was observed that the tracer is not completely extracted, it is systemically metabolized, and it is lost from brain tissue after initial delivery. An analysis of the effects of these factors on the quantitative method for measurement of rCBF has been carried out [43].

The Analysis of HIPDM Kinetics

The capillary first pass cerebral extraction, the rate of HIPDM metabolism in vivo and the extent of the backflux of [125I]HIPDM from brain tissue to blood were studied in male adult Sprague-Dawley rats. Initially the regional [125I]HIPDM distribution was compared with the rCBF measured with iodo[14C]antipyrine ([14C]IAP)[14] in studies performed over different experimental times up to 60 min after i.v. injection. The results confirmed the correlation between rCBF, as measured with [14C]IAP and [125I]HIPDM distribution in all the regions of the brain examined.

The first pass extraction fraction of [125I]HIPDM was determined experimentally by the indicator diffusion method [24] using [57Co]DTPA as an intravascular reference tracer. The [125I]HIPDM was found to be incompletely extracted by the brain tissue in one pass; the extraction fraction ranged between 75% and 85%. The results were the same whether the tracer was dissolved in saline, plasma or blood.

Therefore the microsphere model was applied, assuming complete trapping of the tracer after 80% extraction and negligible systemic metabolism, for quantitative assessment of rCBF with [125I]HIPDM. The values of rCBF estimated from the [125I]HIPDM data and the equation of the microsphere method (Eq. 21 in theory section) were lower than those measured with [14C]IAP, and the underestimation increased with longer experimental periods. One possible explanation for the time dependent underestimation of rCBF calculated with this procedure from the [125I]HIPDM uptake data was the distribution in blood of 125I labeled metabolites which do not cross the BBB. Tissue and arterial blood samples were, therefore, extracted with ethylacetate and analyzed by high performance liquid chromatography (HPLC) for unaltered [125I]HIPDM and 125I labeled breakdown products.

The [125I]HIPDM concentration was found to drop rapidly in the arterial blood after an i.v. pulse and represented only 30% of the total blood radioactivity 60 min after injection. Radioactive HIPDM metabolites in the brain tissue were minimal; more than 92% of the radioactivity in brain tissue 60 min after i.v. administration was unaltered [125I]HIPDM. Corrections for partial first pass extraction and for the presence of metabolites in the arterial blood were applied to the experimental data. Despite these corrections, the values of flow calculated by the microsphere equation from [125I]HIPDM uptake in brain and [125I]HIPDM time course in the arterial blood continued to result in a time-dependent underestimation of rCBF. A possible explanation for this time-dependent error could be the presence of substantial backflux of tracer from the brain tissue to blood.

Previous observations [40,42] showed a constant level of radioactivity in brain tissue for a long period following i.v. administration of a pulse of iodine-labeled HIPDM and a low but constant level of radioactivity in the arterial blood during the same period. In our experiments the tracer's concentration in the tissue reached a plateau within a few minutes after the i.v. pulse and remained fairly constant for up to 60 minutes. The concentration of [125I]HIPDM in arterial blood dropped to very low but still measurable levels. Inasmuch as the integral of concentration of the tracer in the arterial blood is slowly but continuously increasing with increasing experimental time, an increase in brain activity over the 10-60 min time period was not observed in the experiments. These observation suggested that a steady-state bidirectional flux of [125I]HIPDM between blood and brain tissue was occurring rather than a complete irreversible trapping of this compound in the brain.

The experimental data obtained with $[^{125}I]$HIPDM were then analyzed on the basis of a distribution model different from the indicator distribution model. Instead of the unidirectional flux the revised model, i.e. the tissue equilibration model, assumed bidirectional exchange between the tissue and the blood compartments as described by the following equation:

$$C_i(T) = K_1 e^{-k_2 T} \int_{T_0}^{T} C_a(t) e^{k_2 t} dt \qquad (23)$$

The regional rate constants for $[^{125}I]$HIPDM transport across the BBB were estimated by a nonlinear least-squares best fit of Equation 23 to the measured tissue concentrations of the tracer in discrete brain regions and the history of the arterial blood concentration in animals killed at different times up to one hour [45,46].

The overall average values for the whole brain of K_1 and k_2, are 0.915 ml/g/min and 0.021 min^{-1}, respectively. K_1 is several times larger than k_2, but k_2 is clearly not negligible. Values for rCBF were recalculated from the same data according to the tissue equilibration method (Eq. 16 in theory section), introducing the λ and m constants. The value of λ specific for each structure was calculated from the ratio K_1/k_2, and m was obtained by averaging the m values for different structures. The values of m for each individual structure were obtained using data from 1 minute experiments in which $[^{14}C]$IAP and $[^{125}I]$HIPDM were administered concurrently. Equation 9 was then used to calculate m from the value of F measured with $[^{14}C]$IAP and the constants k_2 and λ of $[^{125}I]$HIPDM.

There was a high correlation between the rates of rCBF determined with $[^{125}I]$HIPDM by the tissue equilibration method and the values obtained with $[^{14}C]$IAP at all the experimental times up to one hour after the administration of $[^{125}I]$HIPDM.

As described by the operational equation of the tissue equilibration method, the tissue concentration of $[^{125}I]$HIPDM at any time after its administration is a function of its clearance constant, k_2, which reflects both blood flow and tissue-blood partition coefficient. The concentration of $[^{125}I]$HIPDM in the tissue immediately after the administration as an i.v. pulse is mainly a reflection of the blood flow. As the distribution of the tracer between blood and tissue tends to approach equilibrium, however, the brain concentration becomes more a function of the partition coefficient. It is therefore reasonable to expect changes with time in the pattern of distribution of $[^{125}I]$HIPDM within the brain tissue, as the λ for HIPDM is not uniform throughout the brain.

These studies in animal provide a model for the analysis of the kinetic behavior of HIPDM in humans. They also indicate that the use of radioactive-labeled HIPDM for quantitative measurement of rCBF in human subjects with SPECT requires the determination of several constants. In particular, the partition coefficient and the diffusion factor must be determined before any attempt can be made to correlate measurement of tissue radioactivity with rates of regional blood flow. Knowledge of the metabolism of $[^{125}I]$HIPDM is also necessary if quantitative measurements are pursued. Furthermore, a time frame for measurements of the radioactive iodine-labeled HIPDM concentration in the brain tissue needs to be chosen according to the kinetic properties of the tracer in humans and according to the limitation imposed by the duration of the scanning procedure with a conventional rotating gamma camera. In fact, as stated before, the tissue equilibration method requires the instantaneous measurement of the tissue concentration of the tracer at a selected time, either by autoradiography or by SPECT.

The significant efflux of tracer might represent a limitation in the use of HIPDM for rCBF measurement if SPECT is carried out with a slow-rotating gamma camera. This problem could be, however, partially overcome by use of a fast-scanning device and the same approach proposed for ^{133}Xe. As with any other tracer no inferences can be made about the kinetic behavior of this compound in pathological brain tissue. The values of the constants m and λ used in the operational equation for measurement of rCBF with [^{125}I]HIPDM in physiological states are, in fact, likely to change in pathological states. Even the qualitative interpretation of the brain images obtained with SPECT requires at least knowledge of the regional partition coefficient of the tracer.

The results of these studies demonstrate that HIPDM is not suitable for quantitative measurement of rCBF with the indicator distribution method unless the scanning procedure is performed over a short time. The tissue equilibration method, although more suitable, still represents a relatively unsatisfactory alternative because of the need to know the kinetic constants and metabolic degradation of HIPDM in each study in humans. Both the kinetic constants and the rate of HIPDM metabolic degradation could vary among individuals and in pathological conditions.

CONCLUSIONS

Kinetic models and validated experimental methods are necessary for the quantitative measurement of physiological and biochemical processes with emission tomography. The ^{133}Xe SPECT technique appears to satisfy the requirements of a quantitative blood flow method because a) the choice of the tracer which is an inert diffusible gas, b) the simplicity of the kinetics, and c) adequacy of the instrumentation used. None of the presently available tracers for SPECT with conventional rotating gamma cameras seems to satisfy completely the chemical and kinetic requirements of an ideal "chemical microsphere" for measurement of rCBF, and all can be considered at most indicators of perfusion until their kinetics and biodistribution is completely understood. Some doubts arise, therefore, about the actual relationship between images of radioactivity distribution in brain following the administration of "perfusion tracers" and rates of rCBF. On the other hand, the experience of clinical routine application using semiquantitative analysis of radioactivity distribution indicates that perfusion studies with these tracers provide functional information that correlates with the clinical presentation in neurological patients beyond the anatomical damage demonstrated by CAT scan. Further refinement of SPECT techniques may be justified by its potential clinical usefulness to define the neurological status and the clinical outcome of patients with brain pathology. Although semiquantitative assessment of perfusion might be useful for clinicians, quantitative measurement should be the goal since this represents the only unequivocal parameter to compare results from different studies.

ACKNOWLEDGEMENT

We are grateful to K. Schmidt and L. Sokoloff for revising the manuscript.

REFERENCES

1. Kety, S.S., The theory and applications of the exchange of inert gas at the lungs and tissues. Pharmacol. Rev., 3:1 (1951).
2. Kety, S.S., and Schmidt, C.F., The nitrous oxide method for the quantitative determination of cerebral blood flow in man: Theory, procedure, and normal values. J. Clin. Invest., 27:476 (1948).
3. Betz, E., Cerebral blood flow: its measurement and regulation. Physiol. Rev., 52:595 (1972).
4. Freygang, W.H., Jr. and Sokoloff, L., Quantitative measurement of regional circulation in the central nervous system by the use of radioactive inert gas. Adv. Biol. Med. Physics, 6:263 (1958).
5. Gjedde, A., Caronna, J.J., Hindfelt, B., and Plum, F., Whole-brain blood flow and oxygen metabolism in the rat during nitrous oxide anesthesia. Am. J. Physiol., 229:113 (1975).
6. Gotoh, F., Meyer, J.S., and Tomita, M., Hydrogen method for determining cerebral blood flow in man. Arch. Neurol., 15:549 (1966).
7. Kety, S.S., The cerebral circulation. In: Handbook of Physiology: Neurophysiology, Vol. III, edited by J. Field, H.W. Magoun, and V.E. Hall, pp1751-1760. American Physiological Society, Washington D.C. (1960).
8. Lassen, N. A. and Ingvar, D. H., The blood flow of the cerebral blood flow of the cerebral cortex determined by radioactive krypton-85, Experientia, 17:42 (1961).
9. Lassen N.A., and Munck, O., The cerebral blood flow in man determined by the use of radioactive krypton. Acta Physiol. Scand., 33:30 (1955).
10. Ingvar, D. H. and Lassen, N. A., Quantitative determination of regional cerebral blood flow in man, Lancet, 2:806 (1961).
11. Nilsson, B., and Sjesjo B. K., A method for determining blood flow and oxygen consumption in the rat brain. Acta Physiol. Scand.,96:72 (1976)
12. Sokoloff, L., Quantitative measurement of cerebral blood flow in man. In: Methods in Medical Research, Vol VIII, edited by H.D. Bruner, pp253-261. Year Book Publishers, Chicago (1960).
13. Reivich, M., Jehle J., Sokoloff L., and Kety S.S., Measurement of regional cerebral blood flow with antipyrine-^{14}C in awake cats. J. Appl. Physiol., 27:296 (1969).
14. Sakurada, O., Kennedy, C., Jehle, J., Brown, J.D., Carbin, G.L., and Sokoloff, L., Measurement of local cerebral blood flow with [14C]iodoantipyrine. Am. J. Physiol., 239:H59 (1978).
15. Kuhl, D. E. and Edwards, R. Q., Image separation radioisotope scanning, Radiology, 80:653 (1963).
16. Kuhl, D. E., Edwards, R. Q., Ricci, A. R. and Reivich, M., Quantitative section scanning using orthogonal tangent correction, J. Nucl. Med., 14:196 (1973).
17. Lassen, N. A., Sveinsdottir, E., Kanno, E., Stokely, M., and Rommer, P., A fast moving single photon emission tomograph for regional cerebral blood flow studies in man, J. Comput. Assist. Tomogr., 2:661 (1968).

18. Fazio, F., Fieschi, C., Collice, M., Nardini, M., Banfi, F., Possa, M., and Spinelli, F., Tomographic assessment of cerebral perfusion using a single-photon emitter (Krypton-81m) and a rotating gamma camera, J. Nucl. Med., 21:1139 (1980).

19. Kuhl, D. E., Barrio, J. R., Huang, S-C., Selin, C., Ackermann, R. F., Lear, J. L., Wu, J. L., Lin, T. H., and Phelps, M. E., Quantifying local cerebral blood flow by N-isopropyl-p[123I]iodoam- phetamine (IMP) tomography, J. Nucl., 23:196 (1982).

20. Kety, S. S., Measurement of local blood flow by the exchange of an inert, diffusible substance, Methods Med. Res., 8:228 (1960).

21. Eckman, W. W., Phair, R. D., Fenstermacher, J. D., Patlak, C. S., Kennedy, C., and Sokoloff, L., Permeability limitation in estimation of local brain blood flow with [14C]antipyrine, Am. J. Physiol., 229:215 (1975).

22. Phelps, M. E., Huang, S. C., Hoffman, E. J., Selin, C., and Kuhl, D. E., Cerebral extraction of N-13 ammonia: Its dependence on cerebral blood flow and capillary permeability-surface area product, Stroke, 12:607 (1981).

23. Landau, W. M., Freygang, W. H., Jr., Rowland, L. P., Sokoloff, L., and Kety, S. S., The local circulation of the living brain: values in the unanesthetized and anesthetized cat, Trans. Am. Neurol. Assoc., 80:125 (1955).

24. Crone, C., The permeability of capillaries in various organs as determined by use of the 'indicator diffusion' method, Acta. Physiol. Scand., 58:292 (1963).

25. Renkin, E. M., Transport of potassium-42 from blood to tissue in isolated mammalian skeletal muscles, Am. J. Physiol, 197:1205 (1969).

26. Crone, C., Permeability of capillaries in various organs as determined by use of the indicator diffusion method, Acta. Physiol. Scand., 58:292 (1964).

27. Huang, S-C. and Phelps, M. E., Principles of tracer kinetic modeling in positron emission tomography and autoradiography, in: "Positron Emission Tomography and Autoradiography: Principles and Applications for the Brain and Heart," M. Phelps, J. Mazziotta, and H. Schelbert, eds., Raven Press, New York (1986).

28. Kety, S.S., Measurement of regional circulation by the local clearance of radioactive sodium, American Heart Journal, 38:321 (1949).

29. Conn, H. L., Measurement of organ blood flow without blood sampling, J. Clin. Invest., 34:916 (1955).

30. Hoedt-Rasmussen, K., Sveinsdottir, E., and Lassen, N. A., Regional cerebral blood flow in man determined by intra-arterial injection of radioactive inert gas, Circ. Res., 18:137 (1966).

31. Obrist, W.D., Thompson, H.K., Wang, H.S., Wilkinson W.E., Regional cerebral blood flow estimated by ^{133}Xe inhalation, Stroke, 6:245 (1975).

32. Sveinsdottir, E. Larsen, B., Rommer, P., and Lassen, N.A., A multidetector scintillation camera with 254 channels, J. Nucl. Med., 18:168 (1977).

33. Kanno, I. and Lassen, N. A., Two methods for calculation of regional cerebral blood flow from emission computed tomography of inert gas concentration, J. Comput. Assist. Tomogr., 3:71 (1979).

34. Stokely, E. M., Sveindottir, E., Lassen, N. A., and Rommer, P., A single photon dynamic computer-assisted tomograph (DCAT) for imaging brain function in multiple cross-sections, J. Comput. Assist. Tomogr., 4:230 (1980).

35. Kanno, I., Uemura, K., Miura, S., Miura, Y., Headtome: A hybrid emission tomograph for single photon and positron emission imaging of the brain, J. Comput. Assist. Tomogr 5:216 (1981).

36. Celsis, P., Goldman, T., Henriksen, L., and Lassen, N. A., A method for calculating regional cerebral blood flow from emission computed tomography of inert gas concentrations, J. Comput. Assist. Tomogr., 5:641 (1981).

37. Lassen, N.A., Cerebral blood flow tomography with Xenon-133, Seminars in Nuclear Medicine 14:347 (1985).

38. Winchell, H. S., Baldwin, R. M., and Lin, T. H., Development of I-123-labeled amines for brain studies: Localization of I-123 iodophenylalkyl amines in rat brain, J. Nucl. Med., 21:940 (1980).

39. Tramposch, K. M., Kung, H. F., and Blau, M., Radioiodine labeled N,N-dimethyl-N-(2-hydroxyl-3-alkyl-5- iodobenzyl)-1,3-propanediamines for brain perfusion imaging, J. Med. Chem., 28:121 (1983).

40. Holman, B., Lee, R., Hill, T., Lovett, R., and Lister- James, J., A comparison of two cerebral perfusion tracers, N-isopropyl I-123 p-iodoamphetamine and I-123 HIPDM, in the human, J. Nucl. Med. 25:25 (1984).

41. Lear, J. L., Ackermann, R. F., Kameyama, M., and Kuhl, D. E., Evaluation of 123 I-isopropyliodoamphetamine as a tracer for local cerebral blood flow using direct autoradiographic comparison, J. Cereb. Blood Flow Metabol., 2:179 (1982).

42. Fazio, F., Lenzi, G. L., Gerundini, P., Collice, M., Gilardi, M. C., Colombo, R., Taddei, G., Del Maschio, A., Piacentini, M., Kung, H.F., and Blau, M., Tomographic assessment of regional cerebral perfusion using intravenous [123HIPDM and a rotating gamma camera, J. Comput. Assist. Tomogr., 8:911 (1984).

43. Lucignani, G., Nehlig, A., Blasberg, R., Patlak, C.S., Anderson, L., Fieschi, C., Fazio, F., and Sokoloff, L., Metabolic and kinetic considerations in the use of [^{125}I]HIPDM for quantitative measurement of regional cerebral blood flow, J. Cereb. Blood Flow Metabol. 5:86 (1985).

44. Kung, H. F. and Blau, M., Regional intracellular pH shift: a proposed new mechanism for radiopharmaceutical uptake in brain and other tissues, J. Nucl. Med., 21, 147 (1980).

45. Knott, G. D. and Reece, D. K., MLAB: A civilized curve-fitting system, Proceedings of the ONLINE '72 International Conference, Vol. 1, Brunel University, England, 497 (1972).

46. Knott, G. D. and Shrager, R. I., Computer Graphics: Proceedings of SIGGRAPH Computers in Medicine Symposium, Vol. 6, No. 4, ACM, SIGGRAPH Notices, 138 (1972).

ESTIMATING BLOOD FLOW BY DECONVOLUTION OF THE INJECTION OF

RADIOISOTOPE TRACERS

Andrew Todd-Pokropek

Dept of Medical Physics
University College London U.K.

INTRODUCTION

Blood flow in vivo can be estimated by a variety of methods. This paper is concerned primarily with techniques associated with the injection of a bolus of a radioisotope tracer, making measurements using an external detector of a bolus being delivered to some target organ, and the amount of the tracer within the target organ. In practice such in vivo methods suffer from many problems, not least, in the choice of an appropriate tracer. The total amount of tracer is difficult to estimate as a result of attenuation and scatter. It is not usually possible to estimate arrival of activity directly to the target organ. The data observed are noisy. This paper is primarily concerned with techniques that may be used to handle such data, with the aim of rendering the use of such methods 'stable' if not completely precise. A brief discussion of some other single photon methods of estimating cerebral blood flow is included.

MATHEMATICAL INTRODUCTION.

Following the literature[1,2], let us assume that as estimate of the amount of radioactivity within a blood vessel can be observed as a function of time and assume also that the total amount of activity within an organ $Q(t)$ can also be estimated as a function of time. The Fick principle states that the rate of accumulation of a tracer dQ/dt is the difference between the rate of arrival and departure of the tracer. If F is the flow rate of the 'carrier' and P_A is the concentration on arrival and P_V is the concentration on departure (the subscripts being adopted according to the accepted convention of referring to arterial and venous concentrations) then

$$dQ/dt = F \, P_A(t) - F \, P_V(t) \qquad [1.]$$

such that

$$F = Q(t) \, / \int_0^t (\, P_A(t') - P_V(t') \,) \, dt' \qquad [2.]$$

from which flow F can be estimated given that the other variables can be evaluated, with the many assumptions that have been stressed elsewhere. These are, in particular, that P_A and P_V are both measured for the complete flow to the 'compartment' in which Q is being measured, and that the tracer is 'freely diffusible'.

Now consider the definitions of the transfer function of such a 'compartment', that is, a measurable sub-space, with well defined input and output, obeying the constraints of being linear and stationary. If it is assumed that the transfer function corresponding to this compartment can be represented by some function h(t) such that the convolution equation is true, then

$$P_V(t) = \int_0^t h(t-t') \; P_A(t') \; dt' \qquad\qquad [3.]$$

and also

$$Q(t) = F \int_0^t H(t-t') \; P_A(t') \; dt' \qquad\qquad [4.]$$

where

$$h(t) = -dH(t)/dt, \quad t \neq 0 \qquad\qquad [5.]$$

and

$$H(t) = 1 - \int_0^t h(t) \; dt \qquad\qquad [6.]$$

with F being as before, the flow. H(t) is often called the retention function, while transfer function h(t) is sometimes also called the spectrum of transit times. Substitution of eqns 3 and 6 in eqn 4, leads back directly to the Fick principle.

If a bolus injection of tracer is given directly on input to the compartment or, it would be better to say, organ being considered, then $P_A(t) = 0$ for $t \neq 0$. Therefore from eqn 3 therefore $P_V(t) = h(t)$, so that the output of the system to a delta function is the transfer function h. Likewise $Q(t) = F \, H(t)$, and H indicates the content of tracer within the organ.

Note also that the mean transit time t_m can be defined as

$$t_m = \int_0^\infty t \; P_V(t) \; dt \; / \; \int_0^\infty P_V(t) \; dt. \qquad\qquad [7.]$$

These are the basic mathematic relationships, all of which are essentially obvious, given the implied constraints.

When making measurements in vivo, not all of these parameters are directly assessable, in particular with respect to cerebral blood flow. In this case we can make an estimate of Q(t) the concentration of activity in the brain with time, with considerable problems associated with scaling and 'uniformity'. In addition estimates of $P_A(t)$ can be obtained at points somewhat remote, for example the heart, again with problems of normalization. Estimates of $P_V(t)$ cannot be obtained except by taking blood samples, again with considerable problems of normalization. Complete sampling of the system is in no way guaranteed. Therefore within this poorly sampled system, it remains to be seen whether F, flow, can be determined. A second issue, which of a second subject of interest in this paper, is whether that determination is mathematically stable, (or how can it be stabilized).

Many additional relationships can be determined, for example the mean transit time is also given by

$$t_m = \int_0^\infty t\ h(t)\ dt\ \ /\ \int_0^\infty h(t)\ dt \qquad\qquad [8.]$$

from which it may be shown that

$$t_m = \int_0^\infty H(t)\ dt\ \ /\ H(0) \qquad\qquad [9.]$$

Note that although a bolus injection of a tracer cannot be achieved directly to an organ, if the shape of the bolus can be measured at the position, then, by deconvolution, the effect of administering a bolus injection can be computed.

Following conventional lines, if the tracer is freely diffusible, then let s be the partition coefficient for the organ defined as

$$s = Q(t)\ /\ \{\ M\ P(t)\ \} \qquad\qquad [10.]$$

where M is the mass of the organ, $Q(t)$ is the quantity if tracer within the organ, and $P(t)$ is the concentration of the tracer in the blood within the organ. Thus, as an approximation it is assumed that $P_v(t)$ is a good estimate of $P(t)$, in other words that the tracer diffuses rapidly, and therefore that

$$F = s\ M\ P_v(t)\ \ /\ \int_0^t (\ P_A(t') - P_v(t')\ )\ dt' \qquad\qquad [11.]$$

substituting for $Q(t)$ from eqn 2. For a delta function bolus injection, as before, therefore

$$Q(t) = s\ M\ P_v(t) \qquad\qquad [12.]$$

and hence, since $dQ(t)/dt = F\ P_v(t)$ under these conditions from eqn 1, then

$$F/M = s\ (\ dQ(t)/dt\ /\ Q(t)). \qquad\qquad [13.]$$

If both sides are integrated with respect to t then and noting that for a bolus injection $Q(t) = H(t)$, then

$$F/M = s\ H(0)\ \ /\ \int_0^\infty H(t)\ dt \qquad\qquad [14.]$$

which from eqn 9 implies that

$$F/M = s/ t_m \qquad [15.]$$

Note that there is a certain circularity to all these equations. Note also the basic assumptions made throughout of linearity, good sampling, rapid diffusion of a (freely diffusible) tracer etc, (but see the notes on the choice of tracer below).

This is the mathematical background which will be used in attempting to determine blood flow by deconvolution methods in vivo.

CHOICE OF TRACER IN BRAIN

There is considerable confusion about the appropriate choice of tracer in the literature. Britton[3,4] uses labelled albumin and claims that this is necessary in order that the tracer does not cross the brain blood barrier. Lever et al[5], state that what is required is a freely diffusable tracer, but one which remains in the brain long enough (has a long enough transit time) such that adequate measurements can be made over the brain. Obrist[6] and Lassen[7,8], as examples of the many workers using ^{133}Xe, would appear to recommend a freely diffusable tracer. Kuhl[9], Knapp[10], and many others justify a tracer which is completely (or nearly) removed from the brain and has a very long retention time in brain.

Thus it seems that all possible types of tracers have been proposed. This apparent contradiction can be understood when the detection systems proposed by the different authors is considered. Two groups can be distinguished, workers looking at dynamic processes, for example the passage of a bolus etc, and a second group who cannot image as rapidly and therefore can only look at 'equilibrium' images. Thus the different methods for estimating cerebral blood flow using isotopes in vivo will be divided into three classes: those looking at the passage of a bolus of tracer (which will be called the first pass method, and is the main subject of interest here), those looking at washout of for example ^{133}Xe, and those looking at equilibrium images for example using Single Photon Emission Computerized Tomography (SPECT).

FIRST PASS METHOD

The basic measurements that can be made in vivo, using either external probes or a gamma camera, when investigating blood flow in an organ (such as the brain) are to sample tracer levels in blood at some distance from the target organ, and to estimate total amounts of tracer over parts, or possibly all, of the target organ. Such estimates are of the number of detected photons corresponding to such regions. These are not easily converted into estimates of tracer concentration. It is possible to calibrate the sensitivity of the detector with respect to known samples of the tracer. However, the amount of attenuation undergone is unknown, and is in fact a function of the distribution of the tracer within the volume of interest. In addition to unattenuated photons a considerable fraction of scattered photons are also detected, the exact amounts depending on tracer distribution. Finally, there are usually sources of contamination. Estimates of tracer concentration in the target organ include photons from both arterial and venous blood adjacent to (and within) the organ. However, clinical medicine does not always require exact measurement which is probably fortunate in this case.

From eqn 15, if it is possible to determine the mean transit time t_m, then, knowing the partition coefficient s, and M the mass of the organ, the flow can be determined[11]. Estimating transit time is less sensitive to some of the previously mentioned sources of area since it is essentially a relative measurement in which some of the errors cancel. Here the basic equations used are eqn 4, enabling us to determine H from Q and P_A, and eqn 9, which then allows us to find t_m.

Although in eqn 4, the measurement of P_A is supposed to be as close to the organ as possible, since the form of P_A will change from measurements made close to the heart to those made, for example, close to the brain. However, suppose $P_H(t)$ be an estimate made of the blood concentration of tracer close to the heart then

$$P_A(t) = \int_0^t d(t-t') \ P_H(t) \ dt \qquad\qquad [16.]$$

where d(t) is a bolus distortion function which can also be estimated. Thus by deconvolution, P_A can be estimated from P_H, after which, by a second deconvolution with respect to Q[12], the mean transit time can then be estimated, and hence, blood flow.

Temporal sampling does not present any difficulties since images of the target organ can be acquired at least 10 images per second. This is quite adequate for transit time analysis. The pulsatile nature of the input function is specifically excluded, and was not included in the basic model of the flow process. Likewise, laminar rather than turbulent flow is normally assumed. Typical sampling time for such dynamic studies is of the order of 1 image per second.

Sampling of the input function may require an external probe to the main (imaging) detector placed over the brain, for example, when the input function is sampled over the heart or when the vertex view is used. The input from such a probe needs to be normalized with respect to data from the brain detector.

ATTENUATION CORRECTION

One practical problems is related to the choice of region(s) of interest. Some workers have used a vertex view of the brain in order to get good sampling and reduce attenuation, at the expense of increasing contamination. The label for the tracer in single photons studies has most often been $^{99}Tc^m$ for convenience, and for sensitivity (since relatively large doses can be used), at the cost of increasing the problems associated with attenuation.

Since the linear attenuation correction μ for $^{99}Tc^m$ is 0.15 (narrow beam) or about 0.12 (as determined using a gamma camera), then the relative difference in sensitivity between the near side and the far side of the brain, in a lateral view is about 10:1. This difference is still greater for AP and PA views. Thus, the brain is poorly sampled in the sense that the signal is collected preferentially from certain (near) regions of the brain. This poor sampling is not as bad, when using a vertex view, but, however, in this case contamination from arterial and venous tracer is greater (although reduced by attenuation).

This problem associated with attenuation is also observed for the sampling of the input bolus. Suppose that the vessel be considered is at a distance D deep underneath tissue of attenuation coefficient μ. The correction factor associated with attenuation is therefore exp(µD). Since neither D not µ are exactly known, it is not unusual for errors of 30% to result from this correction. When trying to perform an attenuation correction for a large organ such as the brain, there is no such thing as D, the thickness of tissue interposed, and therefore some guesstimate of the 'effective thickness' of tissue must be used. This does not in anyway correct for the problem of poor sampling.

Some relief results from the use of two opposed detectors. Here for a total thickness of tissue L, for activity at a depth D from one side, the observed count density is reduce by exp(-µD) for the one detector and exp(-µ(L-D)) for the other. Often a geometric mean of the two measurements is used which should give a correction factor of exp(µL/2). It has been demonstrated by various authors that the use of the geometric mean improves both the accuracy of the attenuation correction, and (to some extent) the sampling problems, at the expense of having two detectors.

In addition to the photons detected from tracer in the correct place, photons will also be detected from other sources, so called background. There are two important types of background: constant background, and correlated background. Correlated background is a signal observed which is correlated in some way to one of the desired signals. A good example is that, when estimating activity in the brain, photons are also observed from arterial and venous blood flow in the cerebral region. Background subtraction has been an important topic in such isotopic methods.

It should also be pointed that, when a deconvolution is performed of two unscaled functions, the shape of the computed function (here the retention function) remains unchanged, although its amplitude is unknown. Thus the inability to normalise or quantitate the arterial input function, or the total amount of tracer in the brain does not as such alter the values then computed for the mean transit time.

DECONVOLUTION METHODS

The method presented here basically relies on the ability to perform accurate deconvolutions of various functions. It is possible to rewrite the convolution equation in matrix notation. If Q is a vector being the observed values, for example of Q(t), and P is the vector of values of $P_A(t)$, then we can defined a matrix A such that

$$A \ P \ = \ Q \qquad\qquad\qquad [17.]$$

where A is a band matrix generated from H, the discrete transfer function (or retention function as appropriate) H(t). To be precise, if H is of length m, and P is of length n, n>m, then Q is of length n+m-1 and A is of size n+m-1 by n such that

$$A(i,j) = h(i-j+1) \text{ for } 0<i-j+1<m+1$$
$$\qquad = 0 \text{ otherwise} \qquad\qquad [18.]$$

In nuclear medicine, the most common method used for deconvolution is a stripping method[15] given by an algorithm of the form:

$$H(i) \ = \ 1/P(1) \ [\ Q(i) - \sum_{j=1}^{m} P(i-j+1) \ H(j) \ ] \qquad\qquad [19.]$$

such that

$$H(1) = Q(1)/P(1), \tag{20.}$$

and other values for i>1 being determined subsequently. Note that, as always P Q and A must be correctly normalized, in the same units of time etc. This solution method is obvious given the band matrix form for the convolution being

$$
\begin{bmatrix}
H(1) & 0 & \ldots\ldots\ldots\ldots\ldots 0 \\
H(2) & H(1) & 0\ldots\ldots\ldots\ldots 0 \\
H(3) & H(2) & H(1) & 0\ldots\ldots 0 \\
\ldots\ldots\ldots\ldots\ldots\ldots\ldots \\
H(m) & \ldots\ldots\ldots H(1) & 0\ldots 0 \\
\ldots\ldots\ldots\ldots\ldots\ldots\ldots \\
\ldots\ldots\ldots\ldots\ldots\ldots\ldots \\
\ldots\ldots\ldots 0 & H(m)\ldots & H(1) \\
\ldots\ldots\ldots\ldots\ldots\ldots\ldots \\
0 & \ldots\ldots\ldots\ldots\ldots\ldots 0 & H(m)
\end{bmatrix}
\begin{bmatrix}
P(1) \\
P(2) \\
P(3) \\
\cdot \\
P(m) \\
\cdot \\
\cdot \\
P(n)
\end{bmatrix}
=
\begin{bmatrix}
Q(1) \\
Q(2) \\
Q(3) \\
\cdot \\
Q(m) \\
\cdot \\
\cdot \\
Q(n) \\
\cdot \\
Q(n+m-1)
\end{bmatrix}
\tag{21.}
$$

Eqn. 21 is just eqn. 17 (using 18) written out in full.

By a change in time origin in the basic convolution (eqns 3 and 4), substituting t"=t-t', we can construct a band matrix from P, of size n+m-1 by m, and a vector from H such that

$$
\begin{bmatrix}
P(1) & 0 & \ldots\ldots\ldots\ldots\ldots 0 \\
P(2) & P(1) & 0\ldots\ldots\ldots\ldots 0 \\
P(3) & P(2) & P(1) & 0\ldots\ldots 0 \\
\ldots\ldots\ldots\ldots\ldots\ldots\ldots \\
P(m) & \ldots\ldots\ldots\ldots\ldots P(1) \\
P(m+1) & \ldots\ldots\ldots\ldots\ldots P(2) \\
\ldots\ldots\ldots\ldots\ldots\ldots\ldots \\
\ldots\ldots\ldots\ldots\ldots\ldots\ldots \\
P(n) & \ldots\ldots\ldots\ldots\ldots P(n-m+1) \\
\ldots\ldots\ldots\ldots\ldots\ldots\ldots \\
0 & \ldots\ldots\ldots\ldots\ldots 0 & P(n)
\end{bmatrix}
\begin{bmatrix}
H(1) \\
H(2) \\
H(3) \\
\cdot \\
H(m)
\end{bmatrix}
=
\begin{bmatrix}
Q(1) \\
Q(2) \\
Q(3) \\
\cdot \\
Q(m) \\
Q(m+1) \\
\cdot \\
\cdot \\
Q(n) \\
\cdot \\
Q(n+m-1)
\end{bmatrix}
\tag{22.}
$$

Note that eqn 22 results just from the longhand expression for the value of Q(1). The algorithm given in eqn 19, is in fact just a solution determined from the **first m rows** of eqn 21, or eqn 22, obtained by backsubstitution. It is therefore far from optimal, neither in the least squares sense, nor in the sense of using the data efficiently.

If eqn 22 is rewritten in the form

$$B H = Q \tag{23.}$$

then it is possible to find B^{-1}, or to solve eqn 26 in some optimal sense. Incidentally the size of B is typically of the order of 100x20, and so computationally quite accessible.

Deconvolution is, in general, the process of solving either the matrix equation 17 or 23. Normally, we know Q and P (and hence B) and wish to find A (or rather H). Both eqns 17 and 23 are over determined, but not badly posed (as assessed by looking at their singular values). It does not therefore have an exact solution, and a solution must be found which minimises some criterion. For example, if the residual r is defined as being

$$r = | B H - Q |_2 \qquad\qquad [24.]$$

then we can find a solution which minimizes r. Here, this is defined as the so called L2 norm, which gives just the least squares solution. In this case r is just the sum of the squares of the differences between the observed Q(t) and the computed value obtained by convolving H and P (or matrix multiplying B and H).

Since B is a band matrix, it can be simplified by the Fourier transform. Hence the well known convolution theorem can be used which states that

$$F\{ H(t) \} F\{ P(t) \} = F\{ H(t) * P(t) \} \qquad\qquad [25.]$$

where $F\{\}$ represents the Fourier transform, and * represents the convolution operation. Thus, if $F^{-1}\{\}$ represents the inverse Fourier transform, then

$$H(t) = F^{-1}\{ F\{ Q(t)\} / F\{ P(t) \} \} \qquad\qquad [26.]$$

which is the so called Fourier solution for H. This formulation also gives insight into sources of error. For example where $F\{P(t)\}$ is close to zero, then the solution is unstable. A simple solution is to use in addition a window function W(u) where u is spatial frequency such that

$$H(t) = F^{-1}\{ F\{ Q(t)\} (W(u) / F\{ P(t)) \} \} \qquad\qquad [27.]$$

such that for small values of $F\{P(t)\}$ $(W(u) / F\{P(t)\}$ is small or zero. Suitable window functions, such Hamming, Butterworth etc, may be found in the literature. Such a window function is equivalent to preprocessing P(t) before deconvolution. Other methods based on the use of the z transform and the Laplace transform have also been proposed[15].

Finally, in addition to L2 or least squares methods, L1 and L∞ methods have also been proposed[14]. The L1 norm for the residual is just the sum of the differences and can be made equivalent to finding the median of the data (as opposed to the L2 norm which finds the mean). The L∞ method uses as its estimate of the residual just the maximum difference found. The general methods for solving the L1 problem is to use a Linear Programming or Simplex algorithm.

MODEL DRIVEN METHODS

In all the above discussion, a single model was used, that the forward operation was a convolution. The major constraints were of stationarity and linearity. However, much more is known about the system than this, and it seems desirable to include such information when trying to estimate blood flow by deconvolution. For example, if the transfer function is related to information known about the passage of tracer through an organ, then it is clear that h(t) is always positive. Indeed it is possible to set up a compartmental model of the process of the flow of a tracer through an organ[15]. This could have several compartments, with different rate constants to connect them. In addition it might be necessary to include 'delays' into the model, or indeed a whole variety of non-linear constraints. The problem is then changed from being one of linear minimization, to one of non-linear constrained minimization[14].

The basic method available to analysis of such problems is still similar to that described previously, although cannot be represented by a simple matrix equation. The 'model' can be considered at a black box (function evaluator), driven by the known input function, generating an output which can be compared with the known output such that a residual r can be computed. The minimization proceeds by adjusting the parameters of the model, while respecting any constraints (inequalities) to minimize r. Computational, such techniques are far from simple, for example, when a local minimum has been established, there is no guarantee that this is a global minimum. All such methods are iterative. Of considerable importance is the rate of convergence. An important parameter in accelerating the rate of convergence is to know the gradient of the function with respect to the various parameters of the model. For such problems the gradient may not be known analytically, and quadratic convergence may be difficult to establish. However, on the good side, the dimensions of the problem (no. of parameters by the no. of data points) is reasonable. In addition, since much information exists about the likely solution, good starting points can be incorporated, thereby limiting the problem of false minima. The method of simulated annealing has also been suggested as being appropriate for such problems, and is good at rejecting false local minima.

While a full compartmental model might be difficult to achieve in the sense of it being physiologically realistic, in particular in cases of pathology, a convolution model with constraints might be an acceptable compromise. In addition, the main aim of deconvolution (in this example) is to determine the mean transit time. The exact form of the transfer function (or retention function) is of little interest. Therefore a good algorithm should concentrate on producing the most robust estimate of this parameter.

From a more practical point of view, most users of deconvolution methods in nuclear medicine insist both on smoothing the input function, and on smoothing the transfer function, often very considerably, before calculating the mean time, This process is quite unnecessary, and in fact harmful. For example, if it is believed that a gamma function is a good representation of the input function, then the model driven approach can determine the parameters of that gamma function in an optimal (for example least squares) sense rather than distorting the input function by smoothing. The same arguments apply to the calculation of the mean transit time.

However, in order to produce a more stable solution, it may be appropriate to regularize the method. So far, the minimization process has been discussed as reducing the residual r as defined in eqn 24, for an appropriate norm. However, it is often helpful to minimize a cost function r' for example

$$r' = \| B H - Q \| + \alpha \| C H \| \qquad [28.]$$

where α is a constant and C is an operator. A common regularizing methods are to chose C is give the nth derivative of Q such that

$$r' = \| B H - Q \| + \alpha \| d^n H/dx^n \| \qquad [29.]$$

where n=0 (the so called Tikhonov method) or 2 (the Philips-Twomey method)[14]. In other words, in addition to making the solution fit the data, we also require that the solution is 'smooth' in some sense. Both of these methods can be used to generate equivalent filters (of the singular values), where α is the variable 'smoothing' parameter. Obvious for α=0, then a conventional minimization is performed. More complex cost functions are easy to imagine and can readily be included into such non-linear minimization schemes.

TESTING DECONVOLUTION TECHNIQUES

It is possible to establish a test bed for any of the previously described deconvolution methods. Thus data is simulated, and the computed values compared to the values used to generate the data. From this, the performance of the algorithm used can be assessed both in terms of accuracy and precision. However, the generation of the data implies the use of a model, and therefore, to some extent a model is being tested against itself, which is clearly artificial. It is therefore important, as far as possible, to exclude such circular methods, generating data by methods other than those to be used for analyzing them. Nevertheless, such validation cannot prove that any given algorithm will work in clinical practice. However, an additional technique is of considerable importance. The algorithm being tested should be stretched by giving it essentially 'bad' data, in which the model assumptions are not respected, in order to see how stable the algorithm is with respect to the model. For example for conventional deconvolution, the data can be generated with a known variable delay, to see how the result perform with respect to this source of error.

If algorithms tested in this manner do not perform satisfactorily with respect to test data, they cannot perform well on clinical data. Unfortunately, after such simulations, they must also be tested on clinical data, preferably validated cases, in order to assess their performance. The simulation method can be very powerful in ascertaining under what condition such algorithms are likely to fail, or indeed, with respect to which parameters are they stable. An example of the latter case is that most deconvolution methods are usually very stable with respect to background subtraction, for example when a component associated with $P_A(t)$ is included in the estimate of $Q(t)$.

ALTERNATIVE TECHNIQUES- WASHOUT

There are a number of other methods using isotopes which claim to measure cerebral blood flow, and indeed, regional cerebral blood flow. The best known of these are the methods using ^{133}Xe after intra-arterial or intra-venous injection or inhalation[6,7,8]. In this case, the concentration of Xe is measured typically during two periods of time. Firstly, it is measured as a function of time in the brain as the Xe accumulates (wash-in), while the tracer is being administered continuously. Secondly, it is measured as a function of time to estimate the rate at which it disappears (washout). The hypothesis is usually made that the washout rate τ is proportional to blood flow F

$$F = s V \tau \qquad [30.]$$

where s is the partition coefficient for Xenon, and V the volume of the brain. The assumption is made that s is constant, independent of region and, in particular, of flow. The wash-in rate is also dependent to some extent on flow. Likewise it is usually assumed that the time to peak is constant for all regions. It should be noted that Xenon diffuses very readily through tissue, but is also soluble (to different extents) in different tissues. Thus eqn 30 is likely to be a poor approximation.

While [135]Xe is the most common tracer used, [127]Xe has also employed being of higher energy and therefore suffering less from attenuation effects. [81]Krypton[m] has also been employed after injection[17].

This method has been widely used, notably by Lassen[7,8], and some remarkable results have been published in terms of localized blood flow within specific regions of the brain while various types of stimulation are employed. However, physiologists have claimed that such a method does not measure blood flow but only give an index which 'changes' with blood flow.

ALTERNATIVE TECHNIQUES- EQUILIBRIUM IMAGES

An alternative technique, the injection of labelled microspheres, suffers from the same draw-back. The microspheres are of a size to be trapped by the capillaries during the passage of blood through the brain. Thus, the number of microspheres trapped should be directly proportional to the amount of blood delivered. Again there are considerable problems of attenuation and calibration.

A whole series of tracers such as Iodo-amphetamine, HIPDM (both labelled with ^{123}I)[18], and PAO in various forms (labelled with [99]Tc[m]) have been employed[19]. Uptake as measured at some suitable time after injection is said to reflect blood flow to the brain, and exciting clinical results have been claimed. However, the physiological and pharmacological mechanisms of what is happening are poorly understood, and, although the images show clear abnormalities in cases of pathology, they cannot be considered as serious methods for estimating quantitatively cerebral blood flow as such.

THE USE OF SINGLE PHOTON EMISSION COMPUTERIZED TOMOGRAPHY (SPECT)

Single photon emission computerized tomography SPECT is a powerful method for looking at organs and investigating their function, however, it is not possible to consider this technique at any depth here. Images are collected for all angles around the patient and then reconstructed, in a manner similar to that used in X-ray CT, to give a volume image of the distribution of tracer within the patient. The minimum time for collecting such data is probably of the order of 2 minutes, and therefore cannot be used for collecting information of the passage of a bolus of some tracer. They can be use for looking at equilibrium studies, for example after the continuous adminstration of some tracer in a manner similar to the methods as used in PET, or for looking at studies such as those obtained with the tracers previously mentioned such as microspheres or PAO, in this case looking at transverse slices.

This form of imaging may be much more convenient, and many workers have studied the problems of trying to quantitate either relatively (the ratio of activity in one region relative to another) or absolutely (the concentration in $Bq.cc^{-1}$)[9,10]. The problems of attenuation, scatter, the so called partial volume effect (where the sample is not contained completely in a transverse slice) and the sampling effect (where the observed concentration of activity appears to be a function of the size of the region in which it is concentrated) all combine to make quantitation of such studies difficult[20]. However, the problem of measurements being influenced from photons from surrounding tissue is largely eliminated.

DISCUSSION AND CONCLUSION

Measurements of cerebral blood flow in vivo are much more difficult than such measurements in vitro (or in animals). However, many important clinical conditions or disease states can be studied, although the accuracy of the measured parameters cannot be guaranteed. None of the three basic types of methods discussed really give anything other than an index of blood flow. Such indices, while being reasonably reliable when determined for the whole brain, become more of an 'art form' when determined for some region within the brain. Apart from problems associate with the tracers themselves (are they freely diffusable, or is there complete extraction from the blood), the uniformity of sampling the brain is very poor, and there are many problems associated with quantitation for the tomographic and non-tomographic methods.

While deconvolution can be accurately performed, and indeed improved using model driven methods, the mathematical and computational support does not limit these methods. The basic limitation is with respect to the physiology (and pharmacology) of what is happening, and the difficulty of making 'clean' measurements. It is neither possible using external probes to measure blood flow close to the point of arrival in the brain, nor is it possible to sample all the blood arriving. What can be hoped for is that the techniques such as are being used, which indeed give good clinical information, can be validated in terms of more rigorous physiological measurements, and that cerebral blood flow determination in vivo becomes less of an art, more of a science.

REFERENCES

1. P. Meier and K.L. Zierler, On the theory of the indicator dilution method for the measurement of blood flow and volume, J. Appl. Physiol. 6:731-743, (1954).
2. D. Ingram and R. Block, eds, Tracer techniques in nuclear medicine, in "Mathematic Methods in Medicine", Wiley, Chichester, 1:361-404, (1984).
3. K.E. Britton, C.C. Nimmon, T-Y Lee, P.H. Jarritt, M. Granowska, A. Greening A and J.M. McAlister, Carotid and cerebral blood flow, in "Information Processing in Medical Imaging" A.B. Brill ed, Nashville, 499-525, (1977).
4. K. Britton, C. Nimmon, M. Granowska and T. Lee, Regional cerebral flow in cerebro vascular disease: validation of a non invasive quantitative radionuclide technique, in "Information processing in Medical Imaging", R. Di Paola and E. Kahn eds, INSERM, Paris, 469-487, (1979).
5. S.Z. Lever, H.D. Burns, T.M. Kervitsky et al, Design, preparation and biodistribution of a Tc-99m triaminedithiol complex to assess regional cerebral blood flow, J. Nucl. Med. 26:1287-1294, (1985).
6. W.D. Obrist, T.K. Thompson, C.H. King and H.S. Wang HS, Determination of regional cerebral blood flow by inhalation of ^{133}Xe. Circ. Res. 20:124, (1967).
7. N.A. Lassen and D.H. Ingvar, Radioisotopic assessment of regional cerebral blood, in "Progress in Nuclear Medicine", E. Potchen and V.R. McCready eds, Karger, Basle 1:376-409, (1972).
8. N.A. Lassen, E. Sviensdottir, I. Kanno et al, A fast moving single photon emission tomograph for regional cerebral blood flow studies in man, J. Comput. Assist. Tomogr. 2:661-662, (1978).
9. D.E. Kuhl, J-L. Barrio, S.C. Huang et al, Quantifying local cerebral blood flow by N-isopropyl-p-[^{123}I]-iodoamphetamine (IMP) tomography, J. Nucl. Med. 23:196-203, (1982).

10. W.H. Knapp, R. von Kumner and W. Kubler, Imaging of cerebral blood flow to volume distribution using SPECT, J. Nucl. Med. 27:465-470, (1986).

11. C.M. Coulam, H.R. Warner, E.H. Wood and J.B. Bassingthwaighte, A transfer function analysis of coronary and renal circulation calculated from upstream and downstream indicator dilution curves, Circ. Res. 19:879-890, (1966).

12. A. Kuruc, A. Treves, A. Parker, C. Cheng and A. Sawan A, Radionuclide angiography: an improved deconvolution technique for improvement after suboptimal bolus injection. Radiology 148:233-238 (1983).

13. M.E.. Valentinuzzi and E.M. Volachec, Discrete deconvolution, Med. Biol. Eng., 13:123-125, (1975).

14. J.L. Mohamed and J.E. Walsh JE, eds, "Numerical Algorithms", Clarendon Press, Oxford, (1986).

15. K. Godfrey, "Compartmental models and their applications", Academic Press, New York, (1983).

16. H. Bererhi, Deconvolution analysis in nuclear medicine, MSc thesis, University of London (1986).

17. F. Fazio, C. Fieschi, M. Collice et al, Tomographic assessment of cerebral perfusion using a single-photon emitter (Krypton-81m) and a rotating gamma camera, J. Nucl. Med. 21:1139-1145, (1980).

18. B.L. Holman, R.G.L. Lee, T.C. Hill et al, A comparison of two cerebral perfusion tracers, N-isopropyl I-123 p-iodoamphetamine and I-123 HIPDM in the human, J. Nucl. Med. 25:25-30, (1984).

19. I. Podreka, E. Suess, G. Goldenberg et al, Initial experiences with Tc-99m hexamethylpropyleneamine oxime (Tc-99m-HMPAO) brain SPECT, J. Nucl. Med. 27:887 (1986).

20. A. Todd-Pokropek, G. Clarke and R. Marsh, Preprocessing of SPECT data as a precursor for attenuation correction, in "Information Processing of Medical Images", F. Deconnink ed, Martinus Nijhoff, Boston, 130-150, (1984).

REGIONAL CEREBRAL BLOOD FLOW MEASUREMENTS USING THE 133-XENON INHALATION METHOD

Rodriguez G., De Carli F., Novellone G., Marenco S., and Rosadini G.

Institute of Neurophysiopathology
University of Genoa
Italy

INTRODUCTION

The perfusional condition of the human brain has been studied extensively during the last three decades, since the first measurements of cerebral blood flow (CBF) by the nitrous oxide method were reported (Kety & Schmidt, 1948).

With the gradual development of the simple, non-invasive 133-Xenon inhalation method (Mallet & Veall, 1965; Veall & Mallet, 1966; Jensen et al., 1966; Obrist et al., 1967, 1975b; Potchen et al., 1969; Blauenstein et al., 1977; Reivich et al., 1975; Risberg et al., 1975a, 1975b, 1977b; Deshmukh & Meyer, 1978; Meyer et al., 1978a; Obrist & Wilkinson 1981; Risberg & Prohovnik, 1981; reviewed in Stump & Williams, 1980) routine clinical investigations of regional cerebral blood flow (rCBF) became possible. The technique has been used in the study of cerebrovascular disease (due to its capability to ascertain the patency of the vascular bed), as well as in the evaluation of brain functions, in view of the strict coupling between CBF and cerebral metabolic rate (Raichle et al., 1976).

An exhaustive survey of all the literature being now out of place, only the basic features of the method and some of its applications to human physiological and pathological conditions will be reviewed here.

TECHNICAL FEATURES

The method involves the inhalation of a mixture of the diffusible, metabolically inert 133-Xe gas (6-7 mCi/L) and normal air through a face mask connected to a shielded spirometer by a gas delivery arm. This delivery system allows for comfortable physiological respiration, controls constantly the 133-Xe concentration and expired CO_2 partial pressure, and enables to adjust oxygen concentration in the air-gas mixture.

The subject inhales the tracer for 1 minute and room air for the ensuing ten minutes of the examination. The tracer reaches the brain through the blood stream and diffuses freely in the tissue. The radioactive emissions are recorded during 11 minutes by collimated probes placed radially to the head of the patient in fixed, reproducible positions. The head probes are 32 scintillation detectors (16 over each hemisphere) consisting of a thallium activated NaI crystal and of a photomultiplier tube. The signal collected by the probes is processed by an amplifier and a pulse height analyser; a scaler then counts the electrical pulses collected in 5-seconds epochs (for more details see Deshmukh & Meyer, 1978; Stump & Williams, 1980).

Collimators measuring 1.7 cm. in diameter and 2.0 cm. in length enclose the detectors. The geometric field of view of a probe is approximately 2 cm. at a distance of 2 cm. from the collimator surface. The method gives reliable information on cortical CBF only, and not on the deeper structures of the brain. More details about the system's spatial resolution may be found in Potchen et al. (1969); Lassen & Ingvar (1972); Risberg et al. (1975b); Risberg (1980); Bolmsjo (1984). The end-tidal 133-Xe concentration in the expired air is also recorded. This estimate of the arterial isotope concentration (Mallet & Veall, 1966), is used to correct for the recirculation of the tracer to the head. In our equipment (Novo Cerebrograph 32c, Denmark), a Digital Pc 350 computer processes the data "off-line" and stores them on a floppy disk for further analysis.

CBF ANALYSIS

The first step is to obtain a curve from the counts collected by the head detectors by a least square fit method. After correction for recirculation and background activity, a biexponential analysis is performed (Obrist et al., 1975b). In this model the calculation of flow parameters starts when 133-Xe activity of the end-tidal air curve has decreased to 20% of its maximum value due to the artifactual activity (mainly 133-Xe in the air passages) which influences the first 60-90 seconds of the curve. The coefficients P1, K1, P2, K2 corresponding to the size and clearance rate of the fast compartment (gray matter) and slow compartment (white matter and extracranial tissue) are calculated by the following equation:

Head curve(t)= [P1*exp(-K1*t)+P2*exp(-K2*t)] § A(t)

where A(t) is the end-tidal air curve, § being the convolution.

The following rCBF values are obtained from the coefficients: F1 (gray matter flow), W1 (relative weight of gray matter), K2 (white matter flow), ISI (initial slope index: Risberg et al. 1975a; Risberg, 1980) and CBF 15 (mean tissue flow: Obrist & Wilkinson, 1981). ISI and CBF 15 are non-compartmental indices, used to evaluate low flow conditions when the compartments are difficult to separate.

Correction routines for the air-passage and for other artifacts were also proposed (Jablonski et al., 1979; Risberg, 1980; Hazelrig et al., 1981; Prohovnik et al., 1983a).

DATA MANAGEMENT

The system's output is a display and a numerical print-out of all parameter values for each probe.

However, besides the numerical information intrinsically related to each specific probe, the spatial relationship among the probes must be fully taken into account. To achieve this goal various topographic imaging systems have been suggested (Takano, 1979; Matsuda et al., 1984). A system to represent data with readily interpretable bidimensional maps was developed also in our laboratory (Sandini et al., in press) and applied in clinical data evaluation (Rosadini et al., 1985; Rodriguez et al., in press).

After storage of rCBF values in a suitable data-base, a computer (Digital Pc 350) processes the information and displays a map of the distribution of the rCBF index selected by the operator. The mapping procedure is briefly described as follows:
a) the skull location and relative distance of the 32 probes is shown on a 64 x 64 matrix. The matrix within each hemisphere is formed into triangular patches by drawing lines joining triplets of nearby probes;
b) the flow value for each probe is ascribed to the corresponding vertex of each triangle. The numerical values of the pixels inside the triangle are computed from the values at the vertices by linear interpolation;
c) a relationship is established between a green to red colour scale and raw numerical data;
d) simultaneous presentation of the interpolated coloured triangular patches on a monitor gives a bidimensional map of flow values over the whole brain surface.
The standard display for the clinical records of the patients (raw data of single examinations) consists of 5 small-sized maps, one for each rCBF index, and of a larger one representing the parameter selected by the operator. Also the mean values of a population can be displayed by this output (Fig. 2 and 3).

Moreover, the system allows to statistically compare two populations, or to evaluate differences between a single subject and a population or between two single examinations. The most appropriate statistical procedure is chosen by the operator among the most common parametric and non-parametric ones. The graphic output of statistical comparisons is a map representing the topographic distribution of the statistical probability of differences, displayed as a six level pseudo-colour scale (Fig. 4, 5, 6, 7).

CBF AND THE "RESTING" CONDITION

During an examination in the resting condition, the subject lies with closed eyes, without falling asleep, in a quiet and dark room, all the surrounding noise sources being kept at a minimum level. Many data have been collected using the inhalation method in this condition (Risberg et al., 1975b; Wang & Busse, 1975; Wada et al., 1975; Meyer et al., 1978a; Maximilian et al., 1978; Ingvar, 1979; Naritomi et al. 1979; Melamed et al., 1980; Prohovnik et al., 1980; Cossu et al., 1982; Davis et al., 1983; Hagstadius & Risberg, 1983; Matsuda et al., 1984; Lassen, 1985). We found FG values of 69.9+/-9.5 (mean +/-SD) in the left hemisphere and 70.1+/-9.7 in the right one in a population of 35 normal volunteers (18 males and 17 females, mean age 39.40+/-6.73), age ranging from 30 to 50 years (Fig.2,3). The corresponding ISI values were of 53.1+/-7.2 in the left and 53.4+/-7.3 in the right hemisphere without any statistically significant asymmetry between the two hemispheres . These values are close to those recently reported by Prohovnik et al. (1980) and Melamed et al. (1980).

In fig. 2 and 3 the so called "hyper-frontal" distribution (Ingvar, 1979) is represented with higher flow values in frontal areas and lower clearance rates in posterior regions. In more detail, frontal probes had values 7-8% higher than the hemispheric mean; central regions were only slightly higher than the hemispheric mean, while parietal, temporal and occipital areas all had values below the hemispheric mean (3, 4 and 6% respectively). A possible explanation for this particular perfusional

condition during rest is the higher metabolic activity of the frontal
lobes, due to their function in controlling behaviour and alertness
(Risberg et al., 1977a; Maximilian et al., 1978). An alternative
explanation could be the arousal condition due to anxiety during the
examination, but no conclusive data on this topic have been presented.

The reproducibility of the method was investigated by many authors
(Blauenstein et al., 1977; McHenry et al., 1978; Prohovnik et al., 1980;
Meric et al., 1983). In agreement with others (Prohovnik et al., 1980), we
found in 10 normal, young subjects a flow reduction in the second run of
about 5% as compared with the first examination. This decrease never
reached statistical significance. This finding confirms the system's
sensitivity and is probably the consequence of the subject's different
psychological attitude towards the examination procedure.

It is well known that age is an important factor influencing CBF. Many
studies showed a highly significant negative correlation between advancing
age and flow values (Ishihara et al., 1977; Lavy et al., 1979; Naritomi et
al., 1979; Melamed et al., 1980; Hagstadius & Risberg, 1983). We studied
107 normal subjects, age ranging from 18 to 78 years. A significant flow
regression was found with advancing age, but, interestingly enough, when
the group was divided in two subgroups, one over 45 and the other under 45
years of age, the regression analysis showed significant CBF decreases in
the younger subgroup only, while older subjects had fairly constant flow
rates (Fig. 1). These data agree with those reported by Smith (1984) and
Dastur (1985) with other techniques. No conclusive explanation of this
finding may be given yet. While the absence of significant CBF decrease
with advancing age in the older group may be explained by careful exclusion

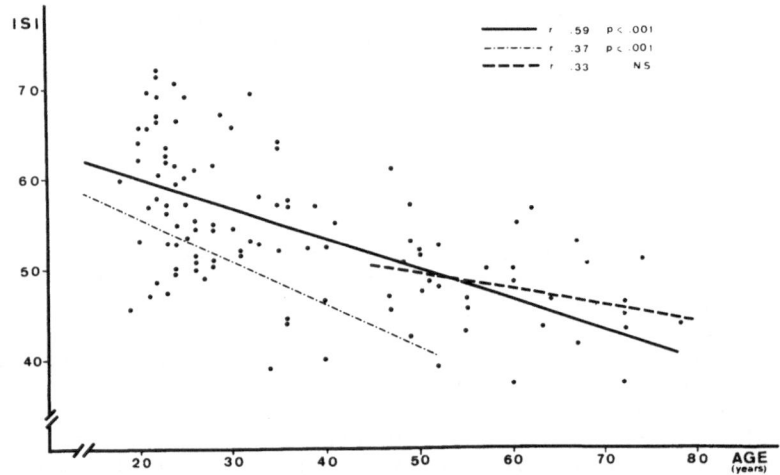

Fig. 1. Correlation of mean global ISI values with age in 107
normal right handed volunteers. The regression line for
the whole group (continuous line) and for the two
subgroups (young and old subjects respectively: dashed
lines) are reported with the corresponding statistical
significance of r values (Pearson's regression analysis).

from our sample of all patients with even mild signs of cerebrovascular
disease or related risk factors, the steep regression slope observed in
young subjects is an intriguing problem. Few and inconclusive studies
(Fang, 1976; Barron et al., 1976) suggest that anatomic changes (neuronal

124

loss) may occur during aging, which could explain CBF reduction. Also the slight non-significant CBF decrease observed in the old subjects may be due to senile changes in brain tissue (Brody, 1955; Thomlinson et al., 1968).

Another interesting point discussed by some authors is the role of the sex variable in determining rCBF. Some authors found no difference between CBF values in males and females (Melamed et al., 1980; Matsuda et al., 1984) and Lassen (1985) concluded that this variable should not influence CBF values. Other studies showed higher flow in females than in males (Gur et al., 1982; Davis et al., 1983; Shaw et al., 1984). To investigate the sex variable, a large number of females and males were examined during a cooperative study with another laboratory (Department of Psychiatry, CBF laboratory, University of Lund). Each female was paired with a male of the same age, therefore excluding the influence of aging. Fig. 4 shows data concerning 12 females and 12 males, age range from 30 to 50 years. Females had mean global ISI values 12% higher than males, and mean global F1 values 14% higher than males. Hormonal levels could play a role in determining this difference, as Shaw et al. (1984) suggested. Since the well known fluctuations of hormones during the menstrual cicle lead to changes in basal temperature, ionic and water balance and blood viscosity, they may influence also rCBF. Nevertheless, many other factors such as different habits (diet, smoking, etc.), personality differences and differing levels of anxiety and arousal during the examination should be investigated to explain this finding.

Various physiological variables like CO_2, O_2, hemoglobin, pH, blood viscosity, fibrinogen, and hematocrit influence CBF. Among them, the one most investigated is cerebrovascular reactivity to CO_2. CBF increases globally during hypercapnia (Olesen, 1974; Maximilian et al., 1980; Jarret et al., 1982; Davis et al., 1983) and hypoxia (Nordstrom & Sjesjo, 1977). During inhalation of 5% CO_2 gas-air mixtures mean flow increases of about 2-4% for each Torr rise in pCO_2 were observed (Jarret et al., 1982). Davis et al. (1983) reported that cerebrovascular response to CO_2 did not change in normal aging.

Brown et al. (1985) pointed out the importance of arterial oxygen content in determining CBF. Using the intravenous injection technique, they found a close correlation between arterial oxygen content and cerebral blood flow depending on a local autoregulative mechanism that maintained oxygen delivery to the brain constant. They also found that other factors such as whole blood viscosity were less important in determining CBF levels.

CBF IN THE WORKING BRAIN

The 133-Xe inhalation method allows "just the right degree of temporal and geographical specificity for optimal description of those brain-behavior relationships of greatest interest to neuropsychology" (Wood, 1980). Indeed, many different activation procedures have been experimented with this method, in view of its non-invasiveness and the possibility to explore all brain regions in both hemispheres at the same time. It should be noted that the activation procedure should last 5-6 minutes at least, due to the temporal resolution of the method.

Many data (some are quoted below) have been collected with the intra-arterial method, due to its higher spatial resolution. An extensive review of these results may be found in Roland (1985).

Motor tasks

Olesen (1971) first demonstrated with the intra-arterial method a

Fig. 2. "Summary image" presenting mean flow values for all rCBF parameters related to the group of normal subjects 30 to 50 years old (NOR3T5). The maximum and minimum computed value for each parameter is reported near the chromatic scale on the right side of each map. The population code, the number of studied subjects, the mean pCO2 and the mathematical model used for data analysis appear on the right hand side of the picture.

Fig. 3. The map of the F1 parameter presented in fig. 2 is magnified.

Fig. 4. Result of the statistical comparison (unpaired t-test) between the mean ISI maps of a group of males (MEN) and females (WOMEN). Mean hemispheric ISI values are written below the corresponding hemispheres. The scale in the centre of the picture is the same for both the upper maps, ranging from the minimum to the maximum computed ISI values. The lower map refers to results of the statistical comparison between the flow values of the two upper maps, and the six level colour scale beside it is graded according to the per cent probability of significance of differences.

Fig. 5. Result of the statistical comparison (z-statistics) between a control population (C$50) and a subject (1598) with a "unilateral" completed stroke on the right side. All details as in fig. 4. The age and pathology code (V66N02) of the patient are shown in the upper right corner of the display. At neurological examination a left hemiparesis was present, while CT-scan showed an ischemic lesion in the territory supplied by the middle cerebral artery. Angiography showed an occlusion of the right internal carotid artery.

Fig. 6. Result of the statistical comparison (unpaired t-test) between the mean ISI maps of a group of normal age-matched controls (CNTRL) and untreated hypertensives (ARTHT). All details as in figure 4.

Fig. 7. Result of the statistical comparison (unpaired t-test) between the mean ISI maps of a group of age-matched normal controls (CONTROLS) and cirrhotic patients with subclinic hepatic encephalopathy (SUBENC). All details as in figure 4.

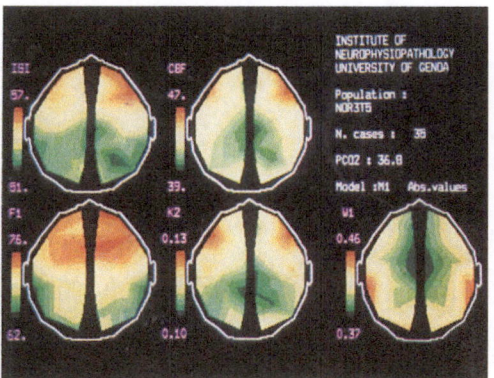

Figure 2

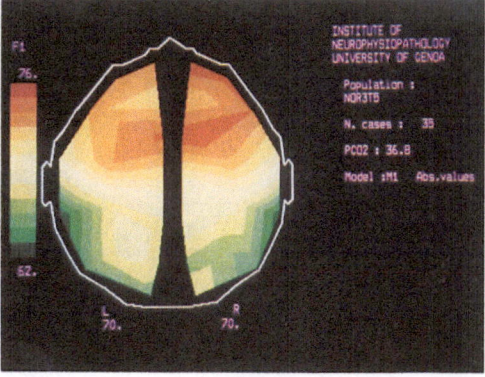

Figure 3

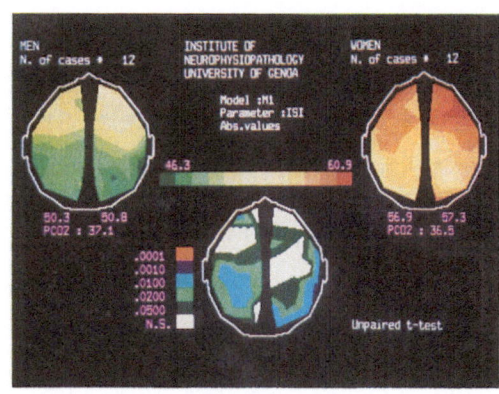

Figure 4

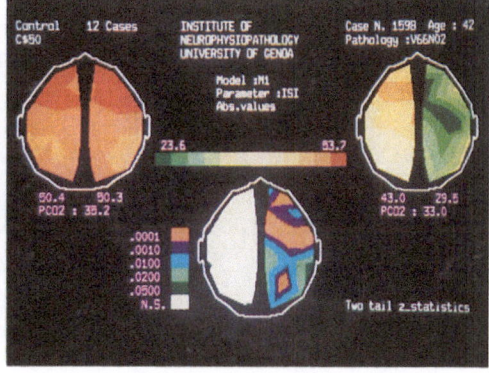

Figure 5

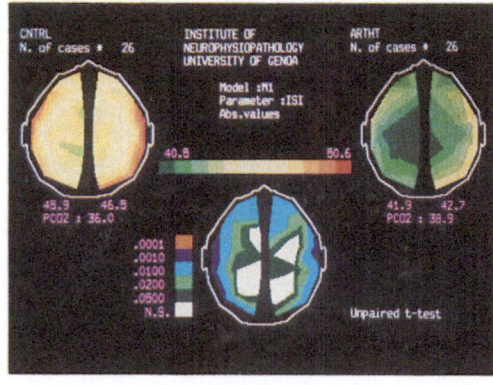

Figure 6

Figure 7

focal rCBF increase in the Rolandic hand-area during vigourous exercise of the contralateral hand. This finding was confirmed by Ingvar & Philipson (1977) and Roland et al. (1980a). Using the inhalation method, Halsey et al. (1979, 1980a) studied left and right hand movement in left and right handed normal subjects. They used a complex task requiring the rithmic opposition of the thumb to the other fingers in sequence. Flow increased in the Rolandic areas of both hemispheres, more so in those contralateral to the movement. However, the only statistically significant activation was found in the sensorimotor hand area of the non-dominant hemisphere contralateral to the side of the movement.This might be related to a higher effort required from the hemisphere which is usually not engaged with skillful finger movements or otherwise to a different organization of movement in the speech-dominant hemisphere as compared to the other one.

We studied 10 young right handed volunteers during movement of the right hand (the same activation task as Halsey et al., 1979 was adopted). A statistically significant rCBF increase was seen not only in the primary sensorimotor areas contralateral to the movement, but also in the ipsilateral ones (Rodriguez & Rosadini, 1984). rCBF activation in the ipsilateral hemisphere was also demonstrated with the intra-arterial method (Roland et al., 1980b).

During spontaneous speech, Halsey et al. (1980a) reported focal rCBF increases in the inferior frontal region (Broca's area) in both hemispheres. Perhaps no asymmetry between the response of the two hemispheres was seen because the task was too simple. Indeed, in a more complex experimental design, left hemisphere frontal and fronto-temporal rCBF increases were seen (Risberg, 1986).

Sensory tasks

Knopman et al. (1980) reported a significant increase in temporal auditory areas in the left hemisphere during a non-verbal activation study, while a verbal task produced strong increases in the left hemisphere and slighter ones in the right parietal areas. Maximilian (1982) saw no significant rCBF activation during monoaural presentation of verbal material. Nevertheless, left-right asymmetries were present: the left hemisphere had higher flow in the temporo-parietal region during left-ear as well as right-ear stimulation. During left-ear stimulation a right fronto-temporal area showed higher values than the contralateral one. Therefore, the left Wernicke area would process verbal material regardless of the side stimulated, while the right fronto-temporal region might play a role in transfer of partially processed verbal information from one hemisphere to the other.

The metabolic activation of the temporo-occipital visual cortex was first demonstrated by Melamed & Larsen (1977) with the intra-arterial technique. Inhalatory rCBF measurements showed progressively higher rCBF increases in the visual areas while looking at a still spiral, at a rotating one, or while performing a spiral-after-effect test. Activation was seen also in the frontal, temporal and parietal regions when the spiral was moving; and the spiral-after-effect test produced flow increases over the whole cortex except in the central areas (Risberg & Prohovnik, 1983). Occipital and inferior frontal activation was observed by Deutsch et al. (1986) during visual stimulation and recognition. When more complex tasks were presented visually the role of occipital regions was always confirmed (Maximilian et al., 1980; Leli et al.. 1983).

Concerning somatosensory stimuli, both intra-arterial (Roland & Larsen 1976) and inhalatory methods (Risberg & Prohovnik, 1983) confirmed contralateral activation in parietal regions corresponding to the primary sensory hand area and in some frontal regions. In the last study, when attention to the stimulation of the left hand was required, a global activation was observed, with flow increases occuring especially in the right hemisphere, interpreted as right hemisphere dominance for attention.

Complex Tasks

The most simple experimental setting of this kind was brought out by Meyer et al. (summarized in Meyer et al.,1980). The subjects had open eyes, they spoke aloud and heard music or other people speaking to them. The results were flow increases in almost all brain regions, especially in the posterior and inferior temporal regions.

There are two main groups of tasks involving abstraction: those concerning language (with verbal presentation of the stimuli and verbal reply of the subject), and those involving spatial abilities (where the stimuli are usually visually presented and motor or verbal reaction of the subject is required). A third group of experiments used tasks of both kinds to study hemispheric specialization.

In a word-pair memory test Prohovnik et al. (1983b) showed global flow increases in both hemispheres but most marked in the left temporal lobes. Risberg (1986) reported frontal and fronto-temporal rCBF increases in the left hemisphere during a word fluency and a verbal creativity test. In the latter condition the activation seemed stronger.

During habituation to a spatial reasoning test Risberg et al. (1977a) found loss of frontal activation. Maximilian et al. (1980) used the same task with subjects inhaling a 6% CO_2-air mixture. This study is quite interesting in that it demonstrates the reliability of rCBF in detecting the small regional changes occuring in the occipital regions during hypercapnia. Maximilian et al. (1978) showed less frontal activation during recall than during learning of a visually presented paired word list. Visual input was also used by Wood et al. (1980a), who found higher left hemisphere flow values during a semantic classification than during a memory task. The performance level during the recognition memory test was inversely correlated to flow in the occipital region. Bilateral flow increases in frontal, parietal and occipital regions were observed during a test of right-left discrimination by Leli et al. (1983).Deutsch et al. (1986) tested normal subjects at rest, during visual stimulation and visual memory tasks, finding that the sensory component of the visual task diminished the anterior-to-posterior resting state gradient, while the cognitive component exaggerated it.

Risberg et al. (1975b) found higher flow increases in the right hemisphere during a spatial test and stronger activations in the left one when a verbal test was performed. Gur & Reivich (1980) showed the importance of individual differences in hemispheric activation replicating Risberg's et al. (1975b) results on hemisphere specialisation. They saw that the strategy used to solve spatial tasks is not so strongly bound to right hemisphere function as verbal strategies are to the left one. Moreover, the subjects who performed better in the spatial task were those with higher flow increases in the right hemisphere. Gur et al. (1982) also investigated sex and handedness differences during cognitive activity, finding higher flow levels in females than in males. Right-handed females showed stronger CBF asimmetry (related to hemispheric specialization) than right-handed males. Moreover, females and left-handers had a greater percentage of fast perfusing tissue.

Some of these activation procedures may be useful in clinical studies.

Examples are in Halsey et al. (1980a; 1980b), Meyer et al. (1980), Wood et al.(1980b; 1980c), Yamaguchi et al. (1980).

CLINICAL APPLICATIONS

The review of all literature concerning CBF in primitive and secondary intracranial pathology being now outside our aim, only the most relevant results in major pathologies will be summarized; further details may be found in the extensive specific bibliography attached here.

Cerebrovascular Disease

Brain ischemia is one of the main fields of application for rCBF measurements. There is extensive evidence that ischemia results in rCBF reduction especially in the area of infarction (Rao et al., 1974; Donley et al., 1975; Keating et al.. 1975; Obrist et al., 1975a; Fieschi & Des Rosier, 1976; Scheinberg et al.. 1976; Tuteur et al. 1976; Halsey et al., 1977; Naritomi et al. , 1977; Norrving et al., 1982; Burke et al., 1986) but also in the whole ipsilateral hemisphere due to decreased metabolic demand (Halsey et al 1981; Thomas et al 1984). Though to a lesser extent, a flow reduction occurs also in the contralateral hemisphere (Fujishima et al., 1977; Yonekura et al., 1981; Awad et al., 1982; Norrving et al., 1982), where the areas opposite to the lesion are most affected. We found four different flow abnormality patterns in patients with a completed stroke compared statistically with an age-matched population. They were described as "focal" (presence of one significantly hypoperfused area), "multifocal" (several hypoperfused areas in one or both hemispheres), "unilateral" (significant rCBF impairment in a whole hemisphere) and "bilateral" (CBF reduction in the whole brain). An example is in fig. 5 (Rosadini et al., 1985).

Concerning the relationship between clinical symptoms and rCBF abnormalities, lower flow values were found, though not exclusively, in the patients with more severe clinical disability (Rao et al 1974 ; Norrving et al., 1982). Recently, an index for ISI asymmetries was proposed which related well to clinical symptoms and angiographic findings in ischemic patients (Mosmans et al., 1986). Burke et al. (1986) reported that in patients suffering from acute stroke initial rCBF measurements were not predictive of clinical outcome, probably due to luxury perfusion, a common although transient phenomenon in this condition, or to edema and modifications in intracranial pressure (Slater et al., 1977; Mies et al., 1983). When clinical evolution was favourable, rCBF tended to recover especially in the ischemic area (Heiss et al., 1977), but CBF values seldom returned to normal either in the zone of infarction or in the controlateral one, at least according to our experience in patients with completed stroke.

It is worth mentioning that hypoperfused areas reacted badly to activation procedures such as hyperventilation or respiration of air at higher pCO2 levels (Ingvar & Risberg, 1967; Halsey et al., 1977). These procedures were helpful in determining wether vascular reactivity had been lost, providing useful information in selecting patients for STA-MCA by-pass surgery (Norrving et al., 1982). Moreover, they increased the sensitivity of the method from 30 to 90% in patients with TIA, and from 40 to 82% in other cerebrovascular patients (Jarret et al., 1982). Several patients in the last mentioned study showed a paradoxic response to hypercarbia (flow reduction) which returned to normal after endarterectomy.

In transitory ischemic attacks (TIAs) impaired CBF areas persisted for a few days after resolution of the clinical symptoms. The time of persistence and the dimension of these areas seemed related to the clinical severity and to the level of residual risk. In general, patients suffering from TIA can have lower flow values than normals (Ackerman et al., 1979).In a follow-up study Hartmann (1985) reported CBF impairment in both hemispheres, but more marked in the one affected. In this hemisphere a significant increase during the first two days after TIA was observed.

In subarachnoid hemorrhage (SAH), an overall CBF reduction related to the severity of the symptoms was described (Heilbrun et al., 1972; Rosenstein et al. 1985).This was ascribed to the higher intracranial pressure which is a feature of this syndrome.

Regarding by-pass surgery in cerebrovascular disease, the 133-Xe method yields useful information before the operation allowing selection of those patients whose cerebrovascular reactivity is still good (Schmiedek et al., 1976; Nilson et al., 1979) and in the evaluation of its results (Heilbrun et al., 1975; Tanahashi et al., 1985).The cited paper by Tanahashi et al. is of particular interest: it reports no difference between a group of surgically treated (STA-MCA by-pass) cerebrovascular patients and a similar medically treated group after a four-year follow-up.

In patients with cerebral arteriovenous malformations (AVM), rCBF is generally increased in the zone of the malformation, where rCBF may double normal values. The increased blood flow is due to the blood by-passing the small resistance vessels (Deshmukh & Meyer, 1978).

Head Trauma

CBF reduction was demonstrated in patients with acute severe head trauma , often accompanied by hemorrhage and deep alterations of the state of consciousness (Bruce et al., 1973; Fieschi et al., 1974; Overgaard & Tweed, 1974; 1976; 1983; Obrist et al., 1979; Overgaard, 1982).

We studied 14 patients, age ranging from 14 to 18, ten with a simple cerebral contusion and four with diffuse brain swelling. The rCBF examinations were repeated 1, 4 and 24 weeks after the initial study. The group with contusion showed on first examination an impairment of flow values even in the presence of negligible neurological signs , while higher flow values were usually recorded later. In the first examination, the group with brain swelling had higher values than the other one (Arvigo et al., 1984). In another study, 17 patients with minor cerebral contusion were examined. The mean global flow on early measurement was found to be significantly lower as compared to CBF values in an age-matched group of healthy volunteers. Recovery to normal flow values began in the first week and was complete within a month's time (Arvigo et al., 1985).

We also collected other data in this field studying two groups of boxers, professionals and amateurs.All of them had normal CT-scans. A global CBF reduction was found in professional boxers as compared to amateurs and normal volunteers of the same age (Rodriguez et al.. 1983a; Maximilian et al.. 1985). This reduction could be related to the frequency of head trauma (number of played matches) and therefore to the career duration. CBF reduction appears long before mental deterioration and it may be considered as the first warning in order to prevent the well known punch-drunk syndrome.

Epilepsy

A considerable CBF increase occurs in the whole brain during generalized seizures, while it remains located at the site of the focus in partial seizures (Meyer et al.. 1966; Brodersen et al. 1973; Collins et al., 1976). There is no doubt that this is due to neuronal hyperactivity during the seizure , which results in enhancement of metabolism (Sakai et al., 1978; Ingvar, 1981).

Concerning the inter-ictal period, rCBF data are controversial. Decreased blood flow in the epileptogenic focus was found by Lavy et al. (1976), Ingvar (1981), and Touchon et al. (1983), while Hougaard et al. (1976) and Meyer et al. (1978b), found increased rCBF values. Oikawa & Kanaya (1983) found (using the intra-arterial method) focal hyperemia only in 5 patients out of 20 with focal epilepsy, while the others presented focal ischemia. It was also observed that rCBF increased in the epileptogenic focus during photic stimulation. We found normal inter-ictal rCBF values in 9 patients with generalized epilepsy, while in 24 patients with partial seizures with complex symptomatology and a single or predominant EEG focus in the left temporal lobe, a significant asymmetric reduction of flow corresponding to the epileptogenic focus was seen (Rosadini et al., 1983,1984).

Migraine

A CBF reduction occurs in the aura phase of migraine (O'Brien, 1971), while a global CBF increase was observed in early studies during the pain phase (O'Brien, 1971; Sakai & Meyer , 1978). During the intercritical phase, no clear-cut evidence was produced to demonstrate CBF modifications and most authors described normal CBF values (Simard & Paulson,1973; Skinhoj, 1973; Sakai & Meyer , 1978).

A recent report (Olesen, 1985), where more sophisticated methods were employed, suggests that a patophysiological difference exists between classic and common migraine. In the former case, reduced rCBF appeared in the occipital areas during the aura , successively spreading towards the anterior regions of the brain, resulting in low global CBF values during the headache phase (a process suggesting strong similarities with the spreading depression of cortical electrical activity described in animals by Leao). In common migraine, CBF during the headache phase was normal.

Hydrocephalus

In "normal pressure" hydrocephalus reduced rCBF values resulting from increase in ventricular size have been reported (Greitz, 1969), especially in the regions supplied by the anterior cerebral arteries (Ingvar & Schwartz, 1975). In addition, cerebrovascular autoregulation was impaired (Mathew et al., 1975). Shunting procedures produced clinical improvement in some patients, which was correlated to the degree of increase in CBF (Greitz et al., 1969; Mathew et al., 1975).

A more recent report (Hayashi et al., 1984) based on the intra-arterial method, shows that CBF progressively decreases during the different stages of development of communicating hydrocephalus after SAH. Patients with lower flows had the worst clinical gradings and patients who developed vasospasm had lower flows than those who did not.

Psychiatric pathology

In this field, the most interesting results using the 133-Xe inhalation method have been obtained in the study of dementia (reviewed in Risberg & Gustafson, 1983). This disease is accompanied by a global CBF reduction. The topographical distribution of rCBF impairment allows for differential diagnosis between clinically defined forms of dementia (Alzheimer's, Pick's and Multi-Infarct): in Alzheimer's disease rCBF is reduced especially in the post-central and parieto-temporal regions bilaterally; in Pick's dementia rCBF is reduced in the fronto-temporal regions; multiple focal abnormalities with a variable distribution and right-left asymmetries were observed in multi-infarct dementia. rCBF abnormalities related well to post-mortem anatomopathological findings. rCBF measurements are of diagnostic value in evaluating demented patients and may be important in differentiating organic dementia from depression in elderly people. Indeed, it seems that CBF is normal during depression (Gustafson et al., 1981).

In chronic schizophrenia, rCBF impairment was found in frontal regions, while high values were present in post-central areas with an inversion of the normal CBF pattern, although mean flow was normal (Ingvar, 1981; 1982). It is not clear how much of these alterations are due to the disease and how much to prolonged medical treatment. Indeed, it was reported that haloperidol treatment induced a "hypo-frontal" pattern after two weeks (Nilsson et al., 1977). A more recent study did not find hypofrontality in schizophrenics (Mathew et al., 1982), showing a normal flow pattern, in agreement with recent positron emission tomography (PET) studies (Sheppard et al., 1983).

Extracranial pathology

Many extracranial pathologies may affect CBF. Lung diseases which modify arterial pCO_2 alter CBF values (Grubb et al., 1974) and so do various hemopathies responsible for acute and chronic cerebral complications (Couch & Hassanein, 1976; Andersen & Gormsen, 1976; Tohgi et al., 1978; Otsuki et al., 1983).

Hypertension causes a modification in cerebral vascular autoregulation (Lassen & Agnoli, 1972; Strandgaard, 1976); Shaw et al. (1984) demonstrated CBF reductions in both hemispheres in chronic hypertensive patients. Meyer et al. (1985) reported that anti-hypertensive treatment was effective in increasing CBF in the first two years of follow-up, but not any more so after three years. Decreased reactivity to CO_2 activation in hypertensive patients was also reported (Traub et al., 1982).

We studied 26 untreated and 10 treated patients with mild or moderate essential hypertension. A global CBF reduction was seen in the first group, compared to normal volunteers in the same age range (Fig. 6). In the first stages of the disease rCBF impairment developed in the temporal regions (the district served by the middle cerebral artery), then spreading to the whole brain.In treated patients only slight rCBF reduction in the left temporal region was seen. This finding could be attributed to arteriosclerotic changes. But, since therapy seemed effective in reducing brain damage, we suggest here that mild hypertension may result in functional vessel damage, which is reversible if adequately treated (Rodriguez et al., in press).

A flow impairment most probably occurs also in uremic and hepatic encephalopathy, although these pathologies have not been adequately studied. The experience of our laboratory shows that at least in the latter condition (10 patients with sub-clinic cirrhotic encephalopathy) a marked, statistically significant, CBF reduction occurs affecting the whole brain. We found a significant correlation between flow impairment and elevated serum free-triptophan concentrations,as well as between low ISI values and bad performances in neuropsychological tests (Rodriguez et al., 1983b; see Fig.7). This finding is confirmed in a study on a larger population (Testa et al., in preparation).

In diabetes mellitus altered cerebrovascular reactivity was demonstrated by rCBF measurements during CO_2 activation (Dandona et al., 1978). This could have a role in predisposing diabetic patients to cerebrovascular disease, since their cerebrovascular reserve is diminished.

Moreover, it was shown that, in insulin dependant diabetics, falls in CBF occur during the day independently of blood glucose concentration (Dandona et al., 1979). It was also suggested that altered cerebrovascular reactivity is not related to the classic vascular disease of diabetes.

DRUGS and rCBF.

Since valuable information for selection and evaluation of therapy may be obtained with the 133-Xe inhalation technique, the effects on rCBF both of over-the-counter and prescription drugs have been investigated.

The well known vasoconstrictive effect of caffeine has been confirmed by 133-Xe inhalation studies. Mathew & Wilson (1985a) found reduced global flows 30 and 90 minutes after the oral administration of 250 mg. of caffeine.

Kubota et al. (1983) found significant lower flows (12%) in chronic smokers as compared to non-smokers. Smokers also had high total serum cholesterol and reduced high density lipoprotein cholesterol values suggesting that advanced atherosclerosis had developed as a consequence of chronic smoking, reducing CBF and increasing risk of cerebrovascular disease. More evidence on this line was published by Rogers et al. (1983).

Berglund & Risberg (1981) studied the effect of alcohol withdrawal in chronic alcoholics. In the first two days of withdrawal, significantly reduced global CBF values were found. Patients with clouded sensorium and extended preceding drinking period had largest CBF decreases. Those patients who had relatively high temporal and low parietal flows had the worst symptoms. CBF normalized during the following week.

Risberg (1980) also reported a case of amphetamine intoxication. The patient had normal CBF when measured in a "dry" state, but after drug abuse 23% and 33% flow increases on right and left side respectively were recorded. In some frontal regions flows doubled "baseline" values. These results were not confirmed by Mathew & Wilson (1985b) with intra-venous infusion of 15 mg. dextroamphetamine in 22 normal subjects.

Berglund et al. (1977) studied a case of bromide psychosis. The markedly reduced CBF recovered to normal after 5 days on hemodialisis treatment.

In toxic encephalopathy caused by long-term exposure to organic solvents a diffuse CBF decrease was observed (Risberg & Hagstadius, 1983).

The effect of various drugs used in the therapy of cerebrovascular disease has been also investigated. Nootropil (piracetam) was tested in 9 presenile demented patients and no effect on rCBF was seen (Gustafson et al., 1978). The same results were obtained in a pilot study with 4-aminopyridine (Risberg, 1980). Gaab et al. (1982) saw a global CBF increase of about 8% in cerebrovascular patients who assumed a calcium antagonist (Nimodipine) one hour before examination. The effect of the drug was stronger in hypoemic areas where rCBF increases reached 20%. Nimodipine was also effective in raising CBF during vasospasm following SAH. McHenry et al. (1983) observed CBF increase both in intra-venous and in oral administration of papaverine in normal subjects.

Midazolam (a benzodiazepine) was found to decrease CBF, and it should therefore be used safely in patients with intracranial hypertension. At the same time, the drug induces an increases in CBF reactivity to CO_2 which could counteract its beneficial effects on cerebrovascular tone in non-ventilated patients (Forster et al., 1983).

Finally, propanolol decreased cerebrovascular reactivity to CO2 without affecting hypocarbia-induced vasoconstriction (Aoyagi et al., 1976). Using the same compound, we found a general CBF decrease in a small group of cirrhotic patients.

CONCLUSION

The 133-Xe inhalation method offers a wide range of possible applications. In the studies on the physiology of cerebral circulation and metabolism, many relationships between regional metabolic activity of the brain and behaviour have been established. The results summarized here often confirm traditional neurophysiological knowledge and sometimes open new perspectives and give rise to new problems,outlining an extremely complex picture of brain functions.

The influence of many physiological variables (eg. sex) on rCBF has yet to be clarified. The reason for higher CBF levels in females than in males is unknown and specific rCBF studies on the role of hormones and life habits in women as compared to men are needed. However, this finding is of utmost importance since the higher perfusion rates in females could be related to the lower incidence of stroke in this sex.

In the clinical field, rCBF measurements are of diagnostic value as in the case of dementia. Also in cerebrovascular pathology an important diagnostic aid is given by rCBF, though not as unquestionable as in demented patients. Indeed, many authors agree in saying that a good correlation exists between "structural" and "perfusional" brain damage. The rCBF technique has been adopted with success in the study of cerebral metabolic and vascular modifications occuring during the follow-up of treated patients (eg. hypertensives) or in the natural history of a disease (eg. head trauma). Also the effects of drugs on cerebral circulation are an interesting, promising and relatively unexplored field of investigation. Moreover, the method proves to be useful in detecting abnormalities even in some conditions where no clinical symptoms are apparent, as in the case of boxers.

New insights and perhaps solutions to the problems posed by rCBF studies may come from the study of correlations with other techniques exploring regional brain function such as electroencephalography (EEG). The information yielded by the two methods is obviously different not only because of different physiological substrates producing the recorded phenomena, but also because of the much poorer temporal resolution of rCBF measurements (5-7 minutes) as compared to EEG, which is able to detect events lasting a few milliseconds. Nevertheless, both techniques explore the topography of brain functions, and an important aid in correlating rCBF to EEG could come from computer-assisted mapping procedures now available for both methods. Simultaneous presentation of topographic distribution of rCBF and EEG is now possible using the mapping system developed in our laboratory and partly described above, but adequate statistics for data evaluation have yet to be found.

Other more sophisticated and expensive methods for investgation of rCBF are now avaiable. Positron emission tomography (PET) yields much more detailed information on the metabolic activity of the tissue and it has the advantage of three dimensional resolution, while the inhalation method "sees" only the surface of the brain.

Nevertheless, keeping in mind the experimental results reviewed here, we are convinced that the ascertained conditions of the rCBF inhalation system -including low cost, better temporal resolution (5-7 minutes as compared to 20-60 in PET), availability of the tracer, non-invasiveness, low radiation doses to the patient, the opportunity to study large populations also of normal subjects- make it one of the most useful methods fot the study of rCBF in clinical practice.

REFERENCES

Ackerman RH, Gouliamos AD, Grotta JC, Correia JA, Chang JA, Fallick C, Taveras JM: Extracranial vascular disease and cerebral blood flow in patients with transient ischemic attacks. Acta Neurol Scand 60(Suppl.72):442-444, 1979.

Andersen LA, Gormsen J: Platelet aggregation and fibrinolytic activity in transient cerebral ischemia. Acta Neurol Scand 55: 76-82, 1976.

Aoyagi M, Deshmukh VD, Meyer JS, Kawamura Y, Tagashira Y: Effect of beta-adrenergic blockade with propranolol on cerebral blood flow, autoregulation and CO2 responsiveness. Stroke 7:291-295, 1976.

Arvigo F, Cossu M, Fazio B, Gris A, Pau A, Rodriguez G, Rosadini G,Sehrbundt Viale E, Siccardi D, Turtas S, Valsania V, Viale GL: Cerebral blood flow in minor cerebral contusion. Surg Neurol 24,211-217,1985.

Arvigo F, Cossu M, Pau A, Rodriguez G, Rosadini G, Sehrbundt Viale E, Siccardi D, Turtas S, Viale GL: Cerebral blood flow after mild head injury in adolescents. rCBF Bull 8: 167-170, 1984.

Awad I, Little JR, Furlan AJ, Weinstein M: Correlaton of clinical and angiographic findings in brain ischemia with regional cerebral blood flow measured by the Xenon inhalation technique. Neurosurg 11:1-5, 1982.

Barron SA, Jacobs L, Kinkel WR: Changes in size of normal lateral ventricles during aging determined by computerized tomography. Neurol (Minneap) 26:1011-1013, 1976.

Berglund M, Nielsen S, Risberg J: Regional cerebral blood flow in a case of bromide psychosis. Arch Psych Nervenkr 223:197-201, 1977.

Berglund M, Risberg J: Regional cerebral blood flow during alcohol withdrawal. Arch Gen Psych 38:351-355, 1981.

Blauestein UW, Halsey JS, Wilson EM, Wills EL, Risberg J: 133-xenon method, analysis of reproducibility: some of its physiological implications. Stroke 8: 92-102, 1977.

Bolmsjo M: Hemisphere cross-talk and signal overlapping in bilateral rCBF measurements using Xenon 133. Europ J Nucl Med 9: 1-5. 1984.

Brodersen P, Paulson OB, Bolwig TG, Rogon ZE, Rafaelson OJ, Lassen NA: Cerebral hyperemia in electrically induced epileptic seizures. Arch Neurol 28:334-338, 1973.

Brody H: Organization of the cerebral cortex III:a study of aging in the human cerebral cortex. J Comp Neurol 102:511-516,1955

Brown MM, Wade JPH, Marshall J: Fundamental importance of arterial oxygen content in the regulation of cerebral blood flow in man. Brain 108:81-93, 1985.

Bruce DA, Langfitt TW, Miller JD, Schutz H, Valpalahti MP, Stanck A, Goldberg HI: Regional cerebral blood flow,intracranial pressure and brain metabolism in comatose patients. J Neurosurg 38: 131-144, 1973.

Burke AM, Younkin D, Gordon J, Goldberg H, Graham T, Kushner M, Obrist W, Jaggi J, Rosen M, Reivich M:Changes in cerebral blood flow and recovery from acute stroke. Stroke 17:173-178, 1986.

Collins RC, Kennedy C, Sokoloff L, Plum F: Metabolic anatomy of focal motor seizures. Brain Res 150:536-542, 1976.

Cossu M, Cabri M, Decarli F, Montano VF, Rodriguez G, Siccardi A, Traverso R, Rosadini G: Regional cerebral blood flow: normal values in healthy volunteers obtained by a 32 probes 133-Xenon inhalation system. Boll Soc Ita Biol Sperim 43:766-772, 1982.

Couch JR, Hassanein RS: Platelet aggregation,stroke and transiet ischemic attack in middle-aged elderly patients. Neurol 26: 888-895, 1976.

Dandona P, James IM, Newbury PA, Woollard ML, Beckett AG: Cerebral blood flow in diabetes mellitus: evidence of abnormal cerebrovascular reactivity. Br Med J 2:325-326,1978.

Dandona P, James IM, Woollard ML, Newbury P, Beckett AG: Instability of cerebral blood-flow in insulin-dependent diabetics. Lancet 2:1203-1205, 1979.

Dastur DK: Cerebral blood flow and metabolism in normal human aging, pathological aging and senile dementia. J Cer Blood Flow Metab 5:1-9, 1985.

Davis SM, Ackerman RH, Correia JA, Alpert NM, Chang J, Buonanno R, Kelley RE, Rosner B, Taveras JM: Cerebral blood flow and cerebrovascular CO2 reactivity in stroke-age normal controls. Neurology, 33:391-399, 1983.

Deshmukh VD, Meyer JS: Non invasive measurements of regional cerebral blood flow in man. SP medical and scientific books, a division of Spectrum Publications Inc New York, London, 1978.

Deutsch G, Papanicolaou AC, Eisenberg HM, Loring DW, Levin HS: CBF gradient changes elicited by visual stimulation and visual memory tasks. Neuropsychologia 24:283-287, 1986.

Donley RF, Sundt TM, Anderson RE, Sharbrough FW:Blood flow measurements and the "look through" artifact in focal cerebral ischemia.Stroke, 6:121-131, 1975.

Fang HCH: Observations on aging characteristics of cerebral blood vessels, macroscopic and microscopic features. In: Terry ed.: Neurobiology of aging, vol. 3. Raven Press, New York, pp.155-166; 1976.

Fieschi C, Battistini N, Beduschi A: Regional Cerebral Blood Flow and intraventricular pressure in acute head injuries. J Neurol Neurosurg Psychiat 37: 1378-1388, 1974.

Fieschi C, DesRosier M: Cerebral blood flow measurements in stroke.In: R.Russel ed.: Cerebral Arterial Disease. Churchill Livinstone, Edimburgh, pp.85-106; 1976.

Forster A, Juge O, Morel D: Effects of Midazolam on cerebral hemodynamics and cerebral vasomotor responsiveness to carbon dioxide. J Cer Blood Flow Metab 3:246-249, 1983.

Fujishima M, Nishimaru K, Omae T: Long-term prognosis in stroke related to cerebral blood flow.Stroke 8:680-683, 1977.

Gaab MR, Brawanski A, Bockhorn J, Haubitz I, Rode CP, Maximilian VA: Calcium antagonism: a new therapeutic principle in stroke and cerebrovascular vasospasm? rCBF Bulletin 3:47-51, 1982.

Greitz TVB: Effect of brain distension on cerebral circulation. Lancet 1,863-865, 1969.

Greitz TVB, Grepe AOL, Kalmer MSF: Pre-and post-operative evaluation of cerebral blood flow in low pressure hydrocephalus. J Neurosurg 31:644-651, 1969.

Grubb RI, Raichle ME, Eichling JO, Ter-Pogossian MM: The effects of changes in PaCO2 on cerebral blood volume, blood flow and vascular mean transit time. Stroke 5:630-639, 1974.

Gur RC, Gur RE, Obrist WD, Hungerbuhler JP, Younkin D, Rosen AD, Skolnick BE, Reivich M : Sex and handedness differences in cerebral blood flow during rest and cognitive activity. Science 217: 659-661, 1982

Gur RC, Reivich M : Cognitive task effects on hemispheric blood flow in humans: evidence for individual differences in hemispheric activation. Brain & Language 9: 78-92, 1980.

Gustafson L, Risberg J, Johanson M, Fransson M, Maximilian VA: Effects of piracetam on regional cerebral blood flow and mental functions in patients with organic dementia. Psychopharmacology 56,115-117, 1978.

Gustafson L, Risberg J, Silfverskiold P: Cerebral blood flow in dementia and depression. Lancet 1:275, 1981.

Hagstadius S, Risberg J: The effects of normal aging in man on rCBF during resting and functional activation. rCBF Bull 6:116-120, 1983.

Halsey JH, Blauenstein UW, Wilson EM, Wills EL: The rCBF response to speaking in normal subjects, and the time course of alterations in patients recovering from left and right hemisphere stroke. Neurol 27:351-352, 1977.

Halsey J, Blauenstein U, Wilson E, Wills E : Regional cerebral blood flow comparison of right and left hand movement. Neurol 29: 21-28, 1979.

Halsey J, Blauenstein U, Wilson E, Wills E: Brain activation in the presence of brain damage. Brain and language 9: 47-60, 1980a.

Halsey JH, Blauenstein UW, Wilson EM, Wills E: rCBF activation in a patient with right homonymous hemianopia and alexia without agraphia. Brain & Language 9:137-140, 1980b.

Halsey JH, Nakai K, Wariyar B: Sensitivity of rCBF to focal lesions. Stroke 12:631-635, 1981.

Hartmann A: Prolonged disturbances of regional cerebral blood flow in transient ischemic attacks. Stroke 16:932-939, 1985.

Hayashi M, Kobayashi H, Kawano H, Yamamoto S, Maeda T: Cerebral blood flow and ICP patterns in patients with communicating hydrocephalus after aneurysm rupture. J Neurosurg 61:30-36, 1984.

Hazelrig JB, Katholi CR, Blauenstein UW, Halsey JH, Wilson EM, Wills EL: Total curve analysis of regional cerebral blood flow with 133-Xe inhalation: Description of method and values obtained with normal volounteers.IEEE Trans on Biomed Eng 28:609-616, 1981.

Heilbrun MP, Olesen J, Lassen NA: Regional cerebral blood flow studies in subarachnoid hemorrhage. J Neurosurg 37:36-44, 1972.

Heilbrun MP, Reichman OH, Anderson RE, Roberts TS: Regional cerebral blood flow in studies following superficial temporal middle cerebral artery anastomosis. J Neurosurg 43:706-716, 1975.

Heiss WD, Zeiler K,Havelec L, Reisner T, Bruck J:Long-term prognosis in stroke related to cerebral blood flow. Arch Neurol 34:671-676, 1977.

Hougaard K, Oikawa T, Sveinsdottir E, Skinhoj E, Ingvar DH, Lassen NA: Regional cerebral blood flow in focal cortical epilepsy. Arch Neurol 33:527-535, 1976.

Ingvar DH: "Hyperfrontal" distribution of the cerebral grey matter flow in resting wakefulness: on the anatomy of the conscious state. Acta Neurol Scand 60:21-25, 1979.

Ingvar DH: Measurements of regional cerebral blood flow and metabolism in psycopathological states. Eur Neurol 20:294-296, 1981.

Ingvar DH: Mental illness and regional brain metabolism. TINS 5:199-203, 1982.

Ingvar DH, Risberg J: Increase of regional cerebral blood flow during mental effort in normals and in patients with focal brain disorders. Exper Brain Research 3:195-211, 1967.

Ingvar DH, Schwartz MS: The cerebral blood flow in low pressure hydrocephalus. In Lundberg N, Porten U, Brock M (eds): Intracranial pressure II.New York, Springer Verlag, 153-156, 1975.

Ingvar DH, Philipson L: Distribution of cerebral blood flow in the dominant hemisphere during motor ideation and motor performance. Ann Neurol 2:230-237, 1977.

Ishihara N, Meyer JS, Deshmukh VD, Hsu MC: Non-invasive measurements of regional cerebral blood flow (rCBF) in man. Normal values, effects of age, cerebral dominance and activation. Neurol 20:401, 1977.

Jablonski T, Prohovnik I, Risberg J, Stahl KE, Maximilian VA, Sabsay E: Fourier analysis of 133-Xe inhalation curves: Accuracy and sensitivity. Acta Neurol Scand 60, Suppl.72: 216-217, 1979.

Jarret F, Polcyn R, Levin A, McCormick D: The use of hypercapnia in the study of regional cerebral blood flow abnormalities with 133-Xe. J Surg Research 32:104-109, 1982.

Jensen KB, Hoedt-Rasmussen K, Sveinsdottir E, Stewart BM, Lassen NA: Cerebral blood flow evaluated by inhalation of 133-Xenon and extracranial recording: a methodological study. Clin Sci 30:485-494, 1966.

Keating EG, Ewing J, Sheehe P: The clinical uselfulness of inhalation rCBF measurements in patients with completed stroke. In:Harper AM, Jennet WB, Miller JD, Rowan RO eds: Blood Flow and Metabolism in the Brain. Churchill Livingstone, Edimburgh, pp.823-824, 1975.

Kety SS, Schmidt CF: The nitrous oxide method for the quantitative determination of cerebral blood flow in man: Theory, procedure and normal values. J Clin Investig 27:476-483, 1948.

Knopman DS, Rubens AB, Klassen AC, Meyer MW, Niccum N: Regional cerebral blood flow patterns during verbal and non verbal auditory activation. Brain & Language 9:93-112, 1980.

Kubota K, Yamaguchi T, Abe Y, Fujiwara T, Hatazawa J, Matsuzawa T: Effects of smoking on regional cerebral blood flow in neurologically normal subjects. Stroke 14:720-724, 1983.

Lassen NA: Normal average value of cerebral blood flow in younger adults is 50/ml/100 g/min. J Cer Blood Flow Metab 3:347-349, 1985.

Lassen NA, Agnoli A: The upper limit of autoregulation of cerebral blood flow on the pathogenesis of hypertensive encephalopathy. Scand J Clin Lab Invest 30:113, 1972.

Lassen NA, Ingvar DH: Radioisotopic assessment of regional cerebral blood flow. In: Progress in nuclear medicine. Basel:Karger Vol I pp:376-409, 1972

Lavy S, Melamed E, Potnoy Z, Carmon A: Inter-ictal regional cerebral blood flow in patients with partial seizures. Neurol 26:418-422, 1976.

Lavy S, Melamed E, Cooper G, Bentin S, Rinot Y: Regional cerebral blood flow in patients with Parkinson's disease. Arch Neurol 36:344-348, 1979.

Leli DA, Hannay HJ, Falguot JC, Katholi CR, Halsey JH: Age effects on focal cerebral blood flow changes produced by a test of right-left discrimination. Neuropsychologia 21: 525-533. 1983.

Mallet BL, Veall N: The measurement of regional cerebral clearance rates in man using Xenon-133 inhalation and extracranial recording. Clin Sci 29:179-191, 1965.

Mathew NT, Meyer JS, Hartmann A, Ott EO: Abnormal cerebrospinal fluid blood flow dynamics. Implications in diagnosis, treatment, and prognosis in normal pressure hydrocephalus. Arch Neurol 32, 657-664, 1975.

Mathew RJ, Duncan GC, Weinman ML, Barr DL:Regional cerebral blood flow in schizophrenia. Arch Gen Psych 39:1121-1124, 1982.

Mathew RJ, Wilson WH: Caffeine-induced changes in cerebral circulation. Stroke 16:814-817, 1985a.

Mathew RJ, Wilson WH: Dextroamphetamine-induced changes in regional cerebral blood flow. Psycophar 87:298-302, 1985b.

Matsuda H, Maeda T, Yamada M, Gui LX, Hisada K : Age-matched normal values and topographic maps for regional cerebral blood flow measurements by 133-Xe inhalation. Stroke 15:336-342, 1984.

Maximilian VA: Cortical blood flow asymmetries during monoaural verbal stimulation. Brain & Language 15:1-11, 1982.

Maximilian VA, Prohovnik I, Risberg J, Hakansson K: Regional cerebral blood flow changes in the left cerebral hemisphere during word pair learning and recall. Brain & Language 6: 22-31, 1978.

Maximilian VA, Prohovnik I, Risberg J: Cerebral hemodynamic response to mental activation in normo and hypercapnia. Stroke 11:342-347, 1980.

Maximilian VA, Rosadini G, Rodriguez G, Montano VF, Arvigo F, Sannita WG: Impaired cerebral perfusion in asymptomatic boxers. J Cer Blood Flow Metabol 5, Suppl. 1, 27-28, 1985.

McHenry LC, Merory J, Bass E, Stump DA, Williams R, Witcofski R, Howard G, Toole JF: Xenon-133 inhalation method for regional cerebral blood flow measurements:normal values and test-retest results. Stroke 9,396-399, 1978.

McHenry LC, Stump DA, Howard G, Novack TT, Bivins DH, Nelson AO: Comparison of the effects of intravenous papaverine hydrochloride and oral pavabid hp capsulets on regional cerebral blood flow in normal individuals. J Cer Blood Flow Metabol 3:442-447, 1983.

Melamed E, Larsen B: Regional cerebral blood flow during voluntary conjugate eye movements in man. Acta Neurol Scand 56,Suppl. 64:530-531, 1977.

Melamed E, Lavy S, Bentin S, Cooper G, Rinot Y: Reduction in regional cerebral blood flow during normal aging in man. Stroke 11:31-35, 1980.

Meric P, Luft A, Seylaz J, Mamo H: Analysis of reproducibility and sensitivity of atraumatic measurements of regional cerebral blood flow in cerebrovascular diseases. Stroke 14:82-87, 1983.

Meyer JS, Gotoh F, Favale E: Cerebral metabolism during epileptic seizures in man. Electroenc Clin Neurophysiol 21:10-22, 1966.

Meyer JS, Ishihara N, Deshmukh VD, Naritomi H, Sakai F, Hsu M, Pollack P: Improved method for noninvasive measurement of regional cerebral blood flow by 133-Xenon inhalation. Part I: Description of the method and normal values obtained in healthy volunteers. Stroke 9:195-205, 1978a.

Meyer JS, Rogers RL, Mortel KF: Prospective analysis of long term control of mild hypertension on cerebral blood flow. Stroke 16:985-990, 1985.

Meyer JS, Sakai F, Naritomi H, Grant P: Normal and abnormal patterns of cerebrovascular reserve tested by 133-Xe inhalation. Arch Neurol 35:350-359, 1978b.

Meyer JS, Sakai F, Yamaguchi F, Yamamoto M, Shaw T: Regional changes in cerebral blood flow during standard behavioral activation in patients with disorders of speech and mentation compared to normal volunteers. Brain & Language 9:61-77, 1980.

Mies G, Auer LM, Ebhardt G, Traupe H, Heiss WD: Flow and neuronal density in tissue surrounding chronic infarction. Stroke 14:22-27, 1983.

Mosmans PCM,Veering MM, Jonkman EJ: ISI values and interhemispheric differences in patients with ischemic cerebrovascular disease; correlations with clinical and angiographic findings. Stroke 17:58-64, 1986.

Naritomi H, Meyer JS, Deshmukh VD, Pollack P: Non-invasive measurements of regional cerebral blood flow in TIA's and stroke due to carotid and vertebrobasilar disease. Acta Neurol Scand 56(Suppl 64):2514-2515, 1977.

Naritomi H, Meyer JS, Sakai F, Yamaguchi F, Shaw T: Effects of advancing age on regional cerebral blood flow. Arch Neurol 36:410-416, 1979.

Nilsson A, Risberg J, Johanson M, Gustafson L: Regional changes of cerebral blood flow during haloperidol therapy in patients with paranoid symptoms. Acta Neurol Scand 56 (SUppl 64):478-479, 1977.

Nordstrom CH, Sjesjo BK: Regulation of brain energy metabolism under normoxic and hypoxic conditions. Brain Heart Infarct, Berlin, Springer Verlag, pp 33-40, 1977.

Norrving B, Nilsson B, Risberg J: rCBF in patients with carotid occlusion: resting and hypercapnic flow related to collateral pattern. Stroke 13, 155-162, 1982.

O'Brien MD:Cerebral blood changes in migraine. Headache 10, 139-143, 1971.

Obrist WD, Dolinskas CA, Gennarelli TA, Zimmerman RA: Relation of cerebral blood flow to CT scan in acute head injury. In : Popp AJ, Bourke RS, Nelson LR, Kimerberg HK (eds): Neural Trauma. New York, Raven Press, 41-50, 1979.

Obrist WD, Silver D, Wilkinson WE, Harel D, Heyman A, Wang HS: The 133 Xenon inhalation method: Assessment of rCBF in carotid endarterectomy. In: Langfitt TW, McHenry MR, Wollman H (eds) Cerebral circulation and metabolism. New York, Springer Verlag, 398-401, 1975a.

Obrist WD, Thompson HK, King CH, Wang HS: Determination of regional cerebral blood flow by inhalation of 133-Xenon. Circ Res 20 : 124-135, 1967.

Obrist WD, Thompons HK, Wang HS, Wilkinson WE:Regional cerebral blood flow estimated by Xenon 133 inhalation. Stroke 6:245-256, 1975b.

Obrist WD, Wilkinson WE: The non invasive 133-Xenon method : evaluation of CBF indices. In : Moosy J, Reinmuth Om (eds),Cerebrovascular disease 12th Princeton Conference, Raven Press, New York, p 119-124, 1981.

Oikawa T, Kanaya H: Regional cerebral blood flow activated with photic stimulation in focal cortical epilepsy. In: Cerebral blood flow, metabolism and epilepsy. Baldy-Moulinier M, Ingvar DH, Meldrum BS (eds).John Libbey, London, pp 26-32, 1983.

Olesen J: Controlateral focal increase of cerebral blood flow in man during arm work. Brain 94:635-646, 1971.

Olesen J: Cerebral blood flow methods for measurement regulation, effects of drugs and changes in disease. Copenhagen:Fadls Forlag, pp:11-18, 1974.

Olesen J: Migraine and regional cerebral blood flow. TINS 8:318-321, 1985.

Otsuki Y, Kondo T, Shio H Kameyama M, Koyama T: Platelet aggregability in cerebral trombosis analysed for vessel stenosis. Stroke 14:368-371, 1983.

Overgaard J: The distribution of cerebral blood flow values in traumatic coma.In : Grossman RG, Gildemberg PL (eds): Head Injury: basic and clinical aspects. New York, Raven Press, 239-245, 1982.

Overgaard J, Tweed WA: Cerebral circulation after head injury. Part 1 : cerebral blood flow and its regulation after closed head injury with emphasis on clinical correlations. J Neurosurg 41: 531-541, 1974.

Overgaard, Tweed WA: Cerebral circulation after head injury. Part 2: Effects of traumatic brain edema. J Neurosurg 45:292-300, 1976.

Overgaard J, Tweed WA: Cerebral circulation after head injury. Part 4 : Functional anatomy and boundary-zone flow deprivation in the first week of traumatic coma. J Neurosurg 59:439-446, 1983.

Potchen EJ, Davis DO, Wharton T: Regional cerebral blood flow in man. I. A study of the Xenon 133 washout method. Arch Neurol 20:378-383, 1969.

Prohovnik I, Hakansson K, Risberg J : Observations on the functional significance of regional cerebral blood flow in resting normal subjects. Neuropsychologia 18:203-217, 1980.

Prohovnik I, Knudsen E, Risberg J: Accuracy of models and algorithms for determination of fast compartment flow by noninvasive 133-Xe clearance. In: Magistretti P (ed): Functional radionuclide imaging of the brain, New York,Raven press,pp 87-116, 1983a.

Prohovnik I, Risberg J, Hagstadius S,: Temporal lobe activation by verbal memorization studied with 133-Xe inhalation rCBF. J Cer Blood Flow Metab 3: (Suppl 1):276-277, 1983b.

Raichle ME, Grubb RB, Gado MH, Eichling JO, Ter-Pogossian MM: Correlation between regional cerebral blood flow and oxidative metabolism. Arch Neurol 33:523-526, 1976.

Rao NS, Ali ZA, Omar HM, Halsey JH: regional cerebral blood flow in acute stroke:preliminary experience with the 133-Xenon inhalation method. Stroke 5:8-12, 1974.

Reivich M, Obrist WD, Slater R, Greenbeerg J, Goldberg HI: A comparison of the Xenon-133 intracarotid injection and inhalation technique for measuring regional cerebral blood flow. In : Harper AM, Jennet WB, Miller JD, Rowan JO (eds) : Blood flow and metabolism in the brain. Edimburgh , Churchill Livingstone, 1975.

Risberg J: Regional cerebral blood flow measurements by 133-Xe inhalation methodology and applications in neuropsychology and psychiatry. Brain & Language 9:9-34, 1980.

Risberg J: Regional cerebral blood flow in neuropsycology. Neuropsycologia 24:135-140, 1986.

Risberg J, Ali Z, Wilson EM, Wills EL, Halsey JH: Regional cerebral blood flow by 133-Xenon inhalation. Preliminary evaluation of an initial slope index in patients with unstable flow compartments. Stroke 6:142-148, 1975a.

Risberg J, Gustafson L: 133-Xe cerebral blood flow in dementia and in neuropsychiatry research. In: Magistretti P (ed): Functional radionuclide imaging of the brain, New York, Raven Press, pp.151-160, 1983.

Risberg J, Hagstadius S: Effects on the regional cerebral blood flow of long-term exposure to organic solvents. Acta Psychiat Scand 67(Suppl. 303):92-99, 1983.

Risberg J, Halsey JH, Wills EL, Wilson EM: Hemispheric specialisation in normal man studied by bilateral measurements of the regional cerebral blood flow. A study with the 133-Xenon inhalation technique. Brain 98:511-524, 1975b.

Risberg J, Maximilian AV, Prohovnik I: Changes of cortical activity patterns during habituation to a reasoning test. Neuropsychologia 15:793-798, 1977a.

Risberg J, Prohovnik I: rCBF measurements by 133-Xenon inhalation : Recent methodological advances. Prog in Nucl Med 7:70-81, 1981.

Risberg J, Prohovnik I: Cortical processing of visual and tactile stimuli studied by non-invasive rCBF measurements. Human Neurobiol 2:5-10, 1983.

Risberg J, Uzzell BP, Obrist WD: Spectrum subtraction technique for minimizing extracranial influence on cerebral blood flow measurements by 133-Xenon inhalation. Stroke 8:380-382, 1977b.

Rodriguez G, Arvigo F, Marenco S, Nobili F, Romano P, Sandini G, Rosadini G: Regional cerebral blood flow in essential hypertension: data evaluation by a mapping system. Stroke, in press.

Rodriguez G, Ferrillo F, Montano VF, Rosadini G, Sannita WG: Regional cerebral blood flow in boxers. Lancet 2(8354):858, 1983a.

Rodriguez G, Rosadini G: Flussimetria ematica cerebrale regionale con metodo non invasivo. In : Moglia A, Arrigo A (eds): Compromissione e ripresa della motilita': correlati neurofisiologici. EMI-RAS Pavia pp:219-240, 1984.

Rodriguez G, Testa R, Rosadini G, Arvigo F, Celle G, Gris A, Sannita WG, Sukkar GS, Traverso R: Reduction of regional cerebral blood flow in subclinical hepatic encephalophaty. IRCS Medical Science 11:763, 1983b.

Rogers RL, Meyer JS, Shaw TG, Mortel KF, Hardenberg JP, Zaid RR: Cigarette smoking decreases cerebral blood flow suggesting increased risk for stroke. JAMA 250:2796-2800, 1983.

Roland PE: Cortical organization of voluntary behaviour in man. Human Neurobiol 4:155-167, 1985.

Roland PE, Larsen B: Focal increase of cerebral blood flow during stereognostic testing in man. Arch Neurol 33:551-558, 1976.

Roland PE, Larsen B, Lassen NA, Skinhoj E: Supplementary motor area and other cortical areas in organization of voluntary movements in man. J Neurophysiol 43: 118-136, 1980a.

Roland PE, Skinhoj E, Lassen NA, Larsen B: Different cortical areas in man in organization of voluntary movements in extrapersonal space. J Neurophysiol 43:137-150, 1980b.

Rosadini G, Ferrillo F, Rodriguez G, Sannita WG, Arvigo F: rCBF and quantitative EEG correlations in epileptic patients. In: Cerebral blood flow,metabolism and epilepsy. Baldy-Moulinier M,Ingvar DH, Meldrum BS (eds).John Libbey, London, pp 26-32, 1983.

Rosadini G, Rodriguez G, Sandini G, Arvigo F, De Carli F, Romano P: A rCBF statistical-mapping system:data management in stroke patients. Acute Brain Ischemia 1985. Siena 11-14 sept. 1985.

Rosadini G, Rodriguez G, Sannita WG, Arvigo F: Correlations between rCBF and computerized EEG in temporal lobe epilepsy. Acta Neurochirurgica Suppl 33:119-122, 1984.

Rosestein J, Wang AD-J, Symon L, Suzuki M: Relationship between hemispheric cerebral blood flow,central conduction time,and clinical grade in aneurysmal subarachnoid hemorrhage. J Neurosurg 62:25-30, 1985.

Sakai F, Meyer JS: Regional cerebral hemodinamics during migraine and cluster headache measured by the 133-Xe inhalation method. Headache 18:122-132, 1978.

Sakai F, Meyer JS, Naritomi H, Hsu M: Regional cerebral blood flow and EEG in patients with epilepsy. Arch Neurol 35:648-657, 1978.

Sandini G, Rodriguez G, Romano P, Rosadini G: Topographic mapping of rCBF data: techniques and statistical comparisons. In: Pet and Nmr: new perspectives in neuroimaging and clinical neurochemistry. Battistin L (ed): Alan R Liss Inc, New York in press.

Scheinberg P, Meyer JS, Reivich M, Sundt TM, Waltz AG: XIII Cerebral circulation and metabolism in stroke. Cerebral circulation and metabolism in Stroke study group. Stroke 7:212-234, 1976.

Schmiedek P, Gratzl O, Spetzler R Steinhoff H, Enzenbach R, Brendel W, Marguth F: Selection of patients for extracranial by-pass surgery based on rCBF measurements. J Neurosurg 44:303-312, 1976.

Shaw TG, Mortel KF, Meyer JS, Rogers RL, Hardenberg J, Cutaia MM : Cerebral blood flow changes in benign aging and cerebrovascular disease. Neurology 34:855-862, 1984.

Sheppard G, Gruzelier J, Manchanda R, Hirsch SR, Wise R, Frackowiak R, Jones T: 150 positron emission tomographic scanning in predominantly never-treated acute schizophrenic patients. Lancet 2:(8365-8366):1448-1452, 1983.

Simard D, Paulson OB:Cerebral vasomotor paralysis during migraine attack. Arch Neurol 29:207-209, 1973.

Skinhoj E: Hemodinamic studies within the brain during migraine. Arch Neurol 29:95-98, 1973.

Slater R, Reivich M, Goldberg H, Banka R, Grenberg J: Diaschisis with cerebral infarction. Stroke 8:684-690, 1977.

Smith CB: Aging and changes in cerebral energy metabolism. TINS 6:203-208, 1984.

Strandgaard S: Autoregulation of regional cerebral blood flow in hypertensive patients. Circulation 53, 720-727, 1976.

Stump DA, Williams R: The noninvasive measurement of regional cerebral circulation. Brain & Language 9:35-46, 1980.

Takano T: A development of a soft ware system for generating a functional image of regional cerebral blood flow and its clinical application to the patients with cerebrovascular disease. Jap J Nucl Med (Tokyo) 16:201-215, 1979.

Tanahashi N, Meyer JS, Rogers RL, Kitagawa Y, Mortel KF, Kandula P, Levinthal R, Rose J: Long-term assessment of cerebral perfusion following STA-MCA by-pass in patients. Stroke 16:85-91, 1985.

Thomas M, Hennerici M, Marshall J: Cerebral blood flow after carotid occlusion and extracranial-intracranial bypass. J Neurol Neurosurg Psychiat 47:148-152, 1984.

Thomlinson BE, Blessed G, Roth M:Observations on the brains of non demented old people. J Neurol Sci 7:331-356, 1968.

Tohgi H, Yamanouchi H, Murakami M, Kameyama M: Importance of the hematocrit as a risk factor in cerebral infarction. Stroke 9:369-374, 1978.

Traub M, Shapiro AP, Dujovny M, Nelson D: cerebral blood flow changes with diuretic therapy in elderly subjects with systolic hypertension. Clin Exper Hyper A4(7):1193-1201, 1982.

Touchon J, Valmier J, Baldy-Moulinier M: Regional cerebral blood flow in temporal lobe epilepsy: inter-ictal studies.In : Cerebral blood flow Baldy-Moulinier M,Ingvar DH, Meldrum BS (eds).John Libbey, London, pp 26-32, 1983.

Tuteur P, Reivick M, Goldberg HI, Cooper ES, West JN, McHenry LC, Cherniach N: Transient responses of cerebral blood flow and ventilation to changes in PaCO2 in normal subjects and patients with cerebrovascular disease. Stroke 7:584-590, 1976.

Veall N, Mallet BL:Regional cerebral blood flow determined by 133-Xenon inhalation and external recording: the effect of arterial recirculation. Clin Sci 30:353-369, 1966.

Wada JA, Clarke R, Hamm A: Cerebral hemispheric asimmetry in humans. Arch Neurol 32:239-246, 1975.

Wang HS, Busse EW: Correlates of regional cerebral blood flow in elderly community residents. In: Harper AM, Jennett WB, Miller JD, Rowan RO (eds) : Blood flow and metabolism in the brain. Edimburgh, Churchill Livingstone, 8.17-8.18, 1975.

Wood F: Theoretical, methodological and statistical implications of the inhalation rCBF technique for the study of brain-behavior relationships. Brain & Language 1-8, 1980.

Wood F, Taylor B, Penny R, Stump D: Regional cerebral blood flow response to recognition memory versus semantic classification tasks. Brain & Language 9,113-122, 1980a.

Wood F, McHenry L, Roman-Campos G, Poser CM: Regional cerebral blood flow response in a patient with remitted global amnesia. Brain & Language 9:123-128, 1980b.

Wood F, Stump D, McKeehan A, Sheldon S, Proctor J: Patterns of regional cerebral blood flow during attempted reading aloud by stutterers both on and off haloperidol medication: evidence for inadequate left frontal activation during stuttering. Brain & Language 9:141-144, 1980c.

Yamaguchi F, Meyer JS,, Sakai F, Yamamoto M: Case reports of three dysphasic patients to illustrate rCBF responses during behavioural activation. Brain & Language 9:145-148, 1980.

Yonekura M, Austin G, Poll N, Hayward W: Evaluation of cerebral blood flow in patients with transient ischemic attacks and minor stroke. Surg Neurol 15:58-65, 1981.

POSITRON EMISSION TOMOGRAPHY

AN INTRODUCTION TO THE PHYSICS AND INSTRUMENTATION OF POSITRON EMISSION TOMOGRAPHY

G. W. Bennett

Brookhaven National Laboratory

Upton, NY 11973

INTRODUCTION

Positron emission tomography is a procedure which uses pharmaceuticals labeled with positron emitting radionuclides to examine regional functional performance in-vivo. It offers the best prospects for spatial resolution in emission tomography. The maturation of this field is due to the skills and industry of a dedicated cadre of chemists, engineers, pharmacists, physicians, and physicists.

Positron emitters may be characterized by chemical species, half-life, positron energy, and the presence of other emissions. The major cyclotron-produced nuclides used for PET are listed in Table I. They have relatively short half-lives and represent chemical species common in human body composition, making their implementation more straightforward than may be the case for other positron emitters. However, their major advantages are the practical range of lifetimes, low positron energies, and absence of other radiations. Their main drawback is that their short lifetimes, which minimize dose and permit sequential studies, require that the production accelerator be nearby.

A second class of positron emitters are the radionuclide daughters via beta decay of a longer lived parent nuclide. These generator produced isotopes have an effective shelf lifetime, for storage and transport, determined by the parent, while the eluate can have a half-life appropriate to medical imaging applications. Representative generator systems are listed in Table II (adapted from Yano, 1986).

Table I. Cyclotron Produced Radionuclides for PET

Radionuclide	Half-Life (Min)	β+ Decay %	E_{MAX} (MeV)	Daughter
Carbon-11	20.4	99.8	0.96	^{11}B, stable
Nitrogen-13	9.96	100	1.20	^{13}C, stable
Oxygen-15	2.07	99.9	1.73	^{15}N, stable
Fluorine-18	109.7	96.9	0.63	^{18}O, stable

Table II. Characteristics of Positron Emitter Generators

Parent/ Daughter	Parent $T_{1/2}$	Daughter $T_{1/2}$	Daughter Emissions		
			Mode (%)	$E_\gamma(\%)$ keV	$E_{\beta+}$ MAX MeV
^{44}Ti/^{44}Sc	47 years	3.93 hr	β^+(95) EC (5)	511 (190) 1,159 (100)	1.47
^{52}Fe/^{52m}Mn	8.28 hr	21.4 min	β^+(96) IT (2)	511 (193) 1,435 (98)	2.63
^{62}Zn/^{62}Cu	9.26 hr	9.73 min	β^+(98) EC (2)	511 (196)	1.31
^{68}Ge/^{68}Ga	288 days	68.1 min	β^+(90) EC (10)	511 (178) 1,077 (3)	1.90
^{72}Se/^{72}As	8.4 days	26.0 hr	β^+(77) EC (23)	511 (154) 834 (80)	2.50
^{82}Sr/^{82}Rb	25.0 days	1.25 min	β^+(96) EC (4)	511 (192) 776 (14)	3.35
^{118}Te/^{118}Sb	6.0 days	3.5 min	β^+(75) EC (22)	511 (150) 1.23 (3)	2.67
^{122}Xe/^{122}I	20.1 hr	3.6 min	β^+(77) EC (23)	511 (154) 564 (21)	3.12
^{128}Ba/^{128}Cs	2.43 days	3.6 min	β^+(61) EC (39)	511 (122) 443 (22)	2.89

It is worth noting that in general the maximum positron energy and the frequency of other emissions is greater for these generator produced isotopes than is the case for the "classic", cyclotron produced nuclides of Table I.

Generator Kinematics

Consider a system consisting of a parent nuclide (P) and a daughter product (D). Assuming that initially (t = 0) none of the daughter is present, the daughter activity varies with time as

$$A_D(t) = A_P(0) \frac{T_P}{T_P - T_D} \left[\exp(-t/\tau_P) - \exp(t/\tau_D) \right]$$

where τ indicates mean life time, T indicates half life.

If the lifetime of the parent is much greater than that of the daughter, and if the change in parent activity is negligible during the period of interest, then the daughter activity simplifies to

$$A_D(t) = A_P \left[1 - \exp(-t/\tau_D) \right]$$

that is, the daughter "charges up" to the same activity as the parent.

If the parent lifetime is greater than that of the daughter, and parent activity change is significant during the period of interest, the daughter activity increases from zero, attains a maximum value at

$$t_m = \frac{\tau_P \tau_D}{\tau_P - \tau_D} \ln(\tau_P/\tau_D)$$

then decreases, with an activity determined by the parent's lifetime. During this equilibrium the ratio of daughter to parent activity becomes

$$A_D/A_P = \frac{T_P}{T_P - T_D}$$

thus the product activity exceeds that of the parent!

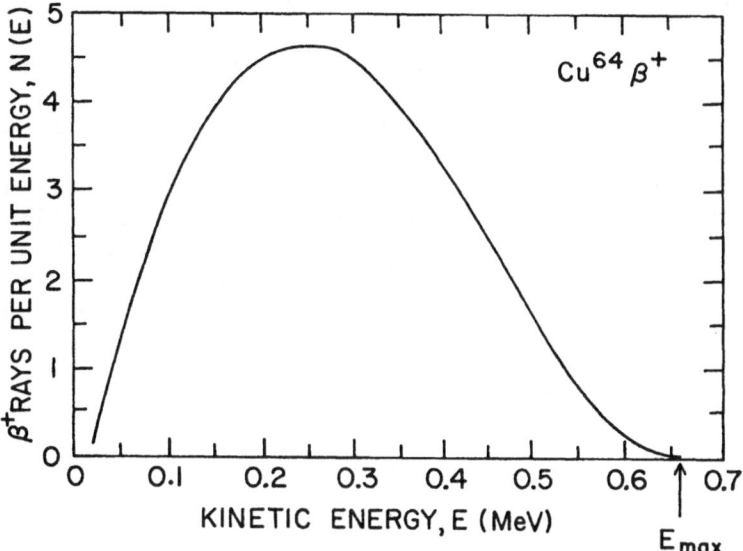

Fig. 1. Energy spectrum for positrons emitted from ^{64}Cu. (From Evans, 1955).

Positron Interactions

Note that beta decay is a three body process where the recoiling nucleus, the positron (in this case) and a neutrino are emitted. The neutrino is uncharged, has a minute or zero mass, and can carry away some of the energy available in the decay process, but is undetectable with PET instrumentation. As a consequence the energy available for the positron is variable. A representative positron energy distribution is shown in Fig. 1.

The positron, after emission, dissipates its kinetic energy by interaction with the charged particles, principally electrons, in its path, traveling a distance from its emission site depending on its initial energy and on the (electron) density of the medium.

147

The positron is an anti-electron. It is a stable particle having the same mass as the electron, 511 KeV, but is oppositely charged. The interaction of matter and antimatter, termed annihilation, results in conversion of the original particle and antiparticle into electromagnetic radiation. The result is the appearance, generally, of two gamma rays, each of 511 KeV, termed annihilation photons or quanta. Although approximately 10% of the positrons interact "in flight", the balance annihilate at rest.

The simplistic, or zero order approximation to positron annihilation is that the two gamma rays are colinear and oppositely directed, emerging from the decaying nucleus (Fig. 2). A more realistic scheme allows that the gamma rays may originate from a site millimeters away from the decaying nucleus because of the initial kinetic energy of the positron; in addition, the gamma rays are not colinear (Fig. 3). That is, the positron transfers its kinetic energy to the electrons in its path by scattering, until it can form a "positronium" system with an electron. The two particles rotate about their mutual center of mass (which is not stationary because of thermal energy) descending through quantized energy levels until annihilation. Nearly all annihilation quanta occur in pairs but a minute fraction of the annihilations produces 3 quanta instead of two.

The positrons are emitted isotropically so that the finite "range" of these particles results in a blurring of the image. One tactic used in autoradiography to reduce this effect, and proposed for PET, (Iida et al., 1986) is the introduction of an intense uniform magnetic field. The charged positrons tend to spiral about the magnetic field lines. The component of initial positron velocity parallel to the magnetic field is unaffected. The other two velocity components, in a plane perpendicular to the magnetic field, produce a force resulting in a circular orbit in that plane, where the orbit radius is smaller as the magnetic field strength is increased. The combination of linear and circular motions produces the spiral motion, reducing the blurring in all but the one dimension in the field direction.

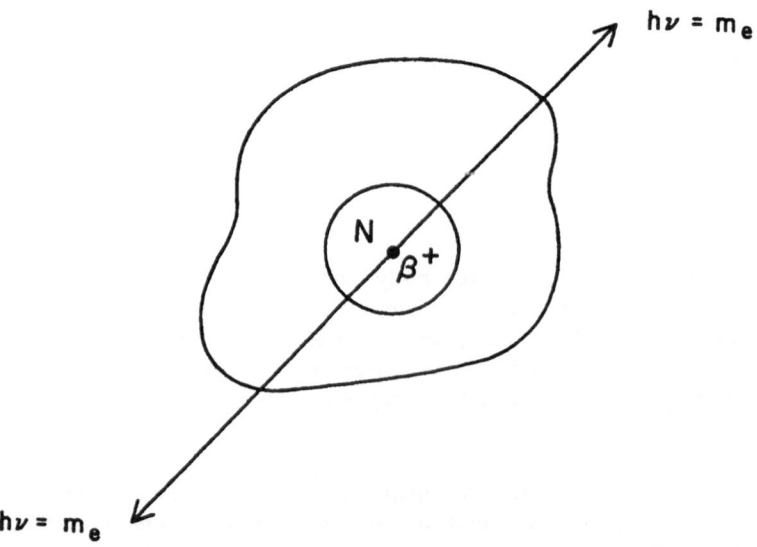

Fig. 2. Naive but workable view of positron emission and annihilation.

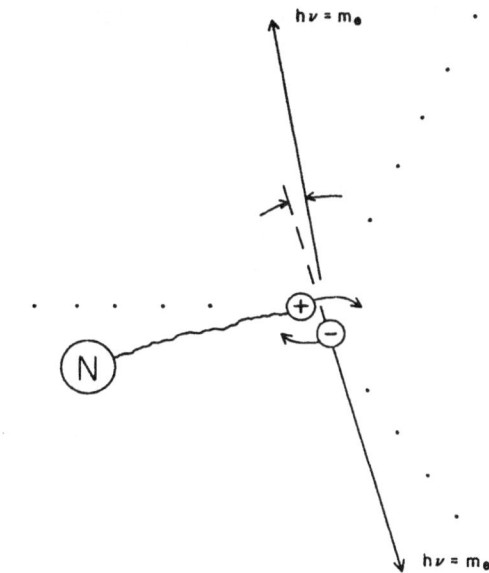

Fig. 3. A more accurate view of positron emission depicts the positron removed from the nucleus before annihilating, and the annihilation photons as not colinear.

The thermal motion of the positronium center of mass produces the non-colinearity of the annihilation photon. The departure from 180° is approximately 1/3° at room temperature.

The positron itself thus passes from consideration, replaced by a pair of annihilation photons, the particles which uniquely identify and point to a positron.

Gamma Ray Interactions

The interaction of gamma rays in matter is quite different from that of charged particles such as positrons. The positrons gradually lose their kinetic energy through a large number of small energy transfer interactions with other charged particles – atomic electrons. Neutral particles, however, interact but once, and may penetrate matter a vast number of atomic dimensions without absorption or scattering. Because of this process of "interacting at a point", a fixed fraction of a beam of photons is attenuated for each successive unit thickness of absorber traversed by the particles. The interaction is essentially probabilistic and leads to the exponential absorption law for a beam of such particles.

$$N(x) = N(0)\exp(-ux)$$

where $N(0)$ is the incident flux, $N(x)$ the flux at depth x, and u is the total linear attenuation coefficient.

Gammas with energy less than 1.02 MeV can interact only through the photo-electric effect, where the photon is absorbed and its energy transferred to an atomic electron, or through the quasi elastic collision process, Compton scattering. In this mode some of the original gamma's energy is transferred to an atomic electron, and a gamma ray of reduced energy emerges at an angle to the original trajectory. Using E_0

for the original energy, E for the final gamma energy, ϕ for the photon deflection angle, T for the electron energy, and m for the rest mass energy of the electron, the Compton Scattering relations are

$$E = E_0\left[1 + \frac{E_0}{m}(1 - \cos\phi)\right]$$

$$T = E_0 - E$$

for annihilation gammas $E_0 = m = 511$ KeV, therefore,

$$E = \frac{m}{2-\cos\phi} \, .$$

However, there is an upper limit to the energy which the electron can absorb from a photon,

$$T_{max} = E_0\left[1 + \frac{m}{2E_0}\right]$$

corresponding to $\phi = 180°$, backscattering of the gamma. For annihilation radiation T_{max} is 2/3 E_0 or nearly 341 KeV. The minimum electron energy is zero.

It is worth noting that the total attenuation coefficient is the sum of absorption (photo electric) and scattering components. The photoelectric coefficient varies as $E^m Z^n$ where Z is the atomic number of the material traversed; n ranges from 4 to 4.6, the energy exponent is usually given the value -3. The scattering coefficient varies directly as Z and decreases monotonically with photon energy. These will be useful in later discussions of detector materials, and attenuation.

Detectors

Positron emission tomography detectors operate by detecting electrons. Positrons have too little energy and range to exit the body; the annihilation gamma pairs have a high probability of exiting the body and intersecting a detector, and within the detector the electrons produced in absorption or scattering of the gamma ray are the medium for signal production.

The material used in detectors and the geometry employed are useful to characterize PET systems. The operation of the various systems are compared by means of performance parameters.

The primary parameters used to classify detectors are resolution (spatial, spectral and temporal), non-linearity (spatial and spectral) and sensitivity. Resolution is defined as the ability to distinguish nearby elements in the scene. Spatial resolution is then the minimum separation between two "points" in the field which appear distinct. In practice, the resolution is defined operationally as the full width at half maximum (FWHM) for a "point" source. Point here means much less than the resulting measured resolution. The spectral resolution is similarly measured as the FWHM for a narrow

150

(or "line") energy emission. Temporal resolution is limited by mechanical motion (the time to completely sample the field of view), or by statistics for a system which is stationary. Non-linearity is evidenced by the departure of the image spatial distribution from a precise mapping of the object distribution; straight lines in the object do not appear straight in the image. Spectral non-linearity is the apparent change in energy and energy resolution for a uniform monochromatic source viewed in different regions in the image. Sensitivity is usually defined as the detected count rate per unit activity per unit volume in a 20-cm diameter source uniformly distributed. The two components of sensitivity are geometric, relating to the solid angle of the detector elements subtended at the source point, and intrinsic, the fraction of events detected per photon incident on a detector element.

What properties are desirable in a detector for 511 keV gamma rays? Since the essential processes are electronic interactions, high density increases the probability of electron interactions. Obviously, the photo-electric interaction is preferable to Compton scattering in its ability to differentiate an annihilation photon from one that has scattered in tissue. Since the probability of photo-effect increases strongly with atomic number, the detector should have high Z.

The archetype detector is a scintillator mounted on a photomultiplier tube. In a scintillator, the electron produced by a gamma ray interaction produces photons in the visible range. These are conducted to the photomultiplier which converts the low energy photons into electrons, then amplifies these by many orders of magnitude. The overall process is linear. The energy of the initial gamma ray is proportional to the number of electrons produced and the charge output of the tube. Properties of typical scintillation materials are listed in Table III.

Table III. Scintillator Characteristics

Material	Percent Efficiency Rel. to NaI(Tl)	Decay Constant (μsec)	Wavelength (peak)	Index of Refraction	Density (g/cm^3)	Hydroscopic
NaI(Tl)	100	.23	410	1.85	3.67	Yes
CsF	5	.005	390	1.48	4.11	Yes
CsI(Tl)	45	1	565	1.8	4.51	Yes
CsI(Na)	85	.63	420	1.84	4.51	Yes
$Bi_4Ge_3O_{12}$ (BGO)	8	.3	480	2.15	7.13	No
$CdWO_4$	18	5	530	2.2	7.9	No
$CaF_2(Eu)$	50	.94	435	1.44	3.18	No
LiI(Eu)	35	1.4	485	1.96	4.08	Yes
BaF_2	10	.63*	310	1.49	4.88	No
$Ce:Gd_2SiO_5$ (GSO)	17	.06	430		6.71	No
Plastic	30	.002-.02	~400	~1.6	1.05	No

*Fast decay component at 220 nm is 700 ps. 80% of light is emitted in slow component. 20% is emitted in the fast component.

The concluding remark regarding the prototype detector is that it involves a second such detector, on the opposite side of the source, in coincidence (Fig. 4). The simultaneous detection of two annihilation photons registers the existence of a positron, and a line between the two detection sites localizes the nucleus which produced the positron. This detector pair has a sensitivity distribution measured by moving a point source through the intervening space with the detection rate at each point plotted as shown in Fig. 5. The equivalent figure for a single, collimated detector is seen in Fig. 6. It should be clear why the detector pair is said to exhibit "electronic collimation" and why PET is potentially superior to single photon tomography in resolution and sensitivity.

Positron scanners were commercially available in the early 1960s. A pair of detectors were mounted to a rectilinear scanning apparatus; the output device was a pen-plotter with the position of the pen related to the position of the detectors and the number of pen contacts proportional to the local signal intensity. A scanner of that era and a representative scan are shown in Fig. 7 (Aranow, 1962).

The standard PET geometry is a ring of scintillation detectors surrounding the source distribution (Fig. 8). Each detector is linked via a coincidence circuit to a number of detectors on the opposite side. Note that as the number of diametrically opposed detector elements in coincidence are increased, the sensitive area is increased,

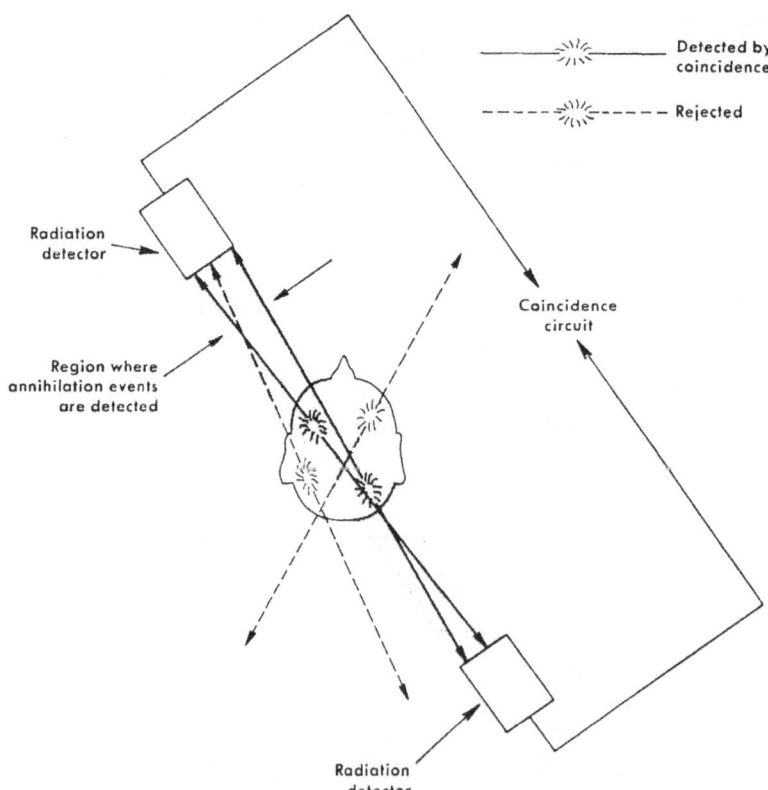

Fig. 4. A model positron detector consists of two photon detectors in coincidence (from Ter-Pogossian 1981).

152

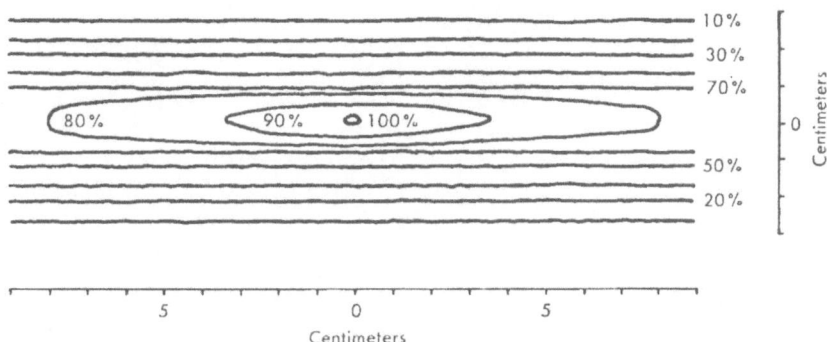

Two 5 × 5 cm diameter
NaI(Tl) detectors separated

Fig. 5. Sensistivity distribution for two 5 cm diameter NaI (Tl) detectors in coincidence separated by 60 cm (from Ter-Pogossian 1981).

but the oblique rays have a reduced efficiency and an increased rate of cross talk,
where the photons can scatter into neighboring detectors with increased detection
probability.

The variety of geometric variation in PET detectors is manifold (Fig. 9) (from
Budinger, 1978). Adjoining ring structures define a sensitive volume, often allowing
coincidence signals between different rings, thus increasing the geometric efficiency at
the penalty of non-uniform axial sensitivity. The ultimate geometric efficiency is
offered by a distribution of detector elements completely surrounding the patient. A
close approximation to this geometry (which has been proposed) is a truncated
ellipsoid which nearly encloses the patient.

Other geometries include hexagonal arrays and planar detectors. A successful
early model was the PETT III (Fig. 9i) consisting of 6 banks with 8 NaI detectors in
each. Each detector in each bank could signal a coincidence with each of the 8
detectors in the diametrally opposed bank. The banks translated in 6 steps to sample
the space between detectors, rotated one-twentieth (3 degrees) of the angular interval
between banks, repeating the translational sampling after each rotation. Only one
plane was sampled; the minimum scan duration was two minutes. The PETT III is
soon to be installed at the Smithsonian Institute. A more modern system such as the
PETT VI consists of 4 rings, each containing 72 detectors of CsF 20 mm × 24 mm ×

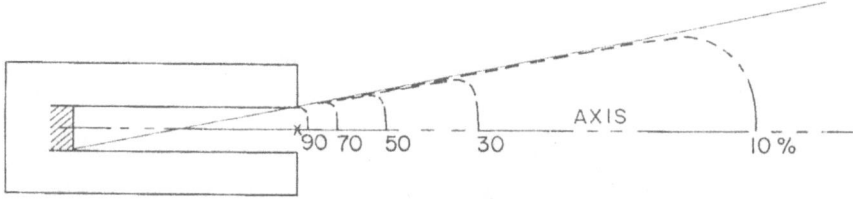

Fig. 6. Sensitivity distribution for a single scintillation detector having a cylindrical
collimator with length 5 times its diameter. 100% sensitivity is chosen at the
center of the collimator aperature.

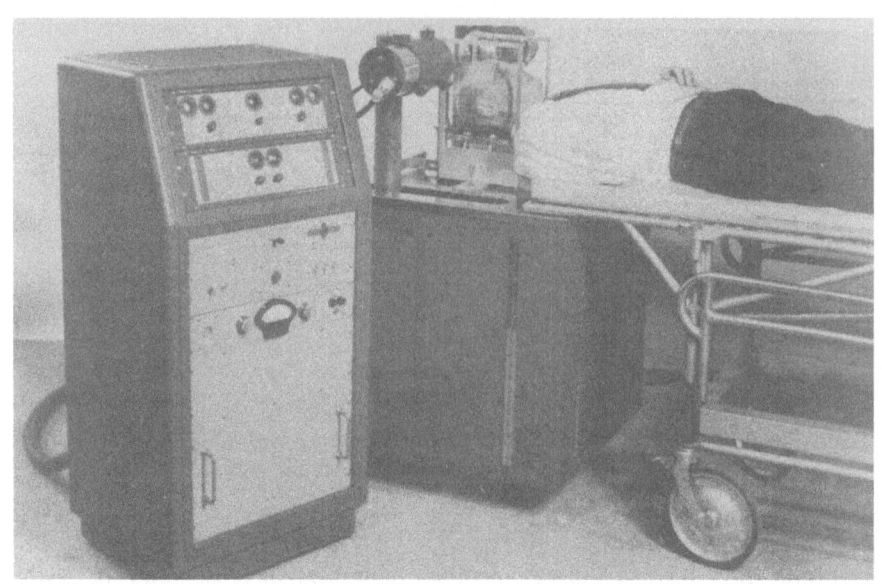

Fig. 7a. Commercial positron brain scanner unit, 1962.

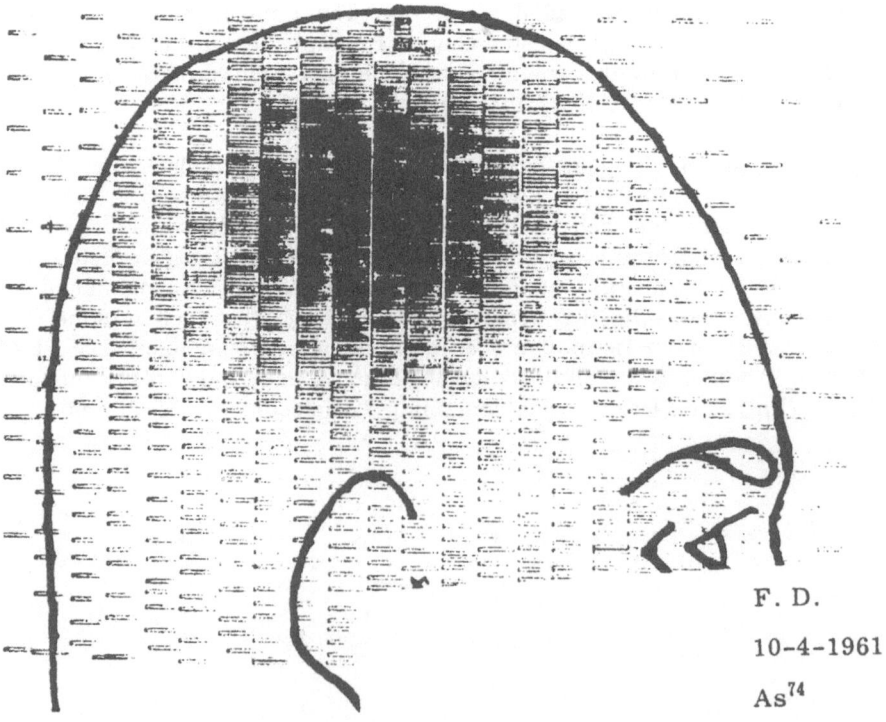

F. D.
10-4-1961
As^{74}

Fig. 7b. Representative brain scanner output.

A

Fig. 8. Model positron emission tomography ring detector geometry (from Derenzo et al., 1981).

65 mm long. Coincidences between elements in adjacent rings produces 7 effective planes. Spatial sampling is accomplished by orbiting or wobbling the axis of the device in a small circle. The time for one cycle of motion is 1 second.

A scintillation camera is a single large crystal viewed by an array of photomultiplier tubes. Signals from the tubes go to summing amplifiers via approprately weighted resistors. Position signals are produced proportional to the centroid of the light in the crystal. Planar detectors such as scintillation cameras have been used in coincident pairs for PET. Large geometric efficiency results, but overall performance is poor. Since scintillation cameras are designed for optimal performance with Technetium 99m (140 KeV gamma) the crystals are too thin for good efficiency with annihilation radiation.

The concept of scintillation camera design has been applied (Burnham, et al., 1982) in one dimension to produce a ring scanner with the characteristics of a single crystal. The standard ring system design requires one phototube per crystal, and the

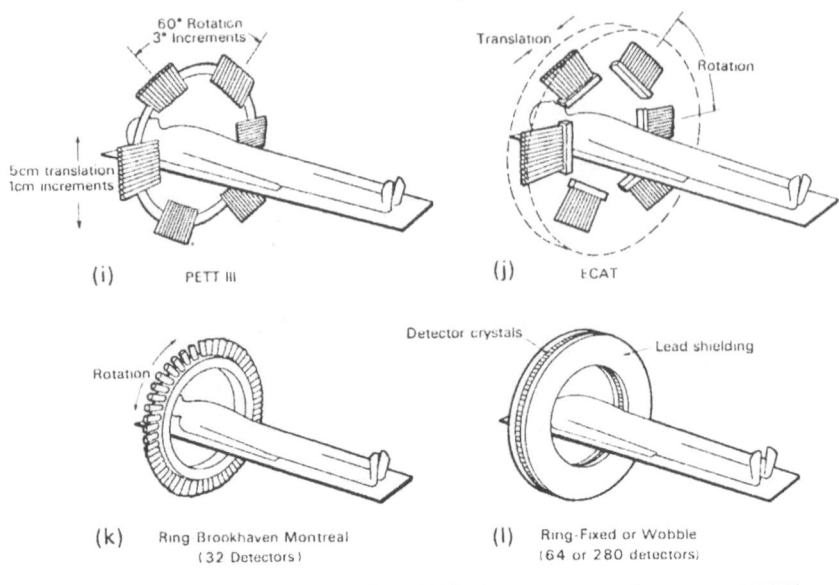

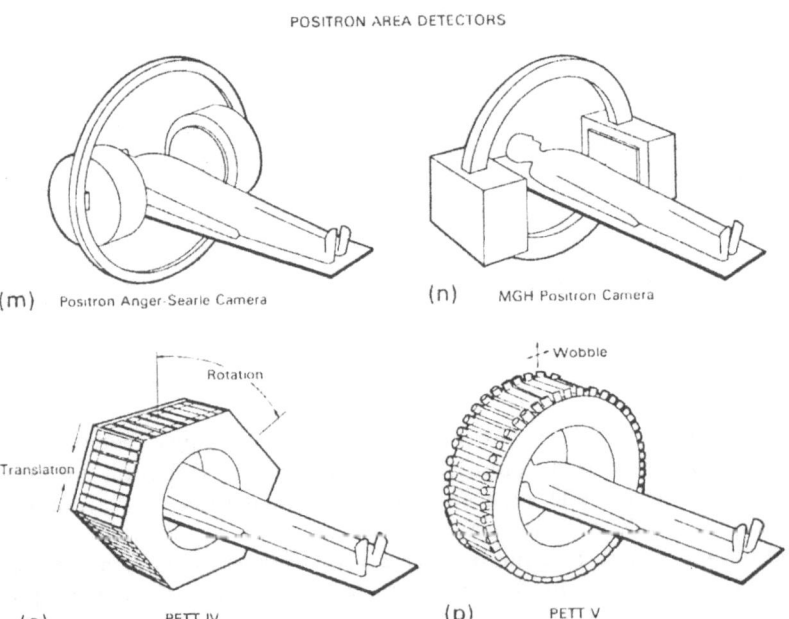

Fig. 9. Various PET geometries.

resolution is determined by the crystal size (and spacing). The application of the gamma camera principle affords a reduction in the number of photomultiplier tubes and associated circuits without degrading performance. Measured resolution in reconstructions is of the order of 5 mm for a detector with 46 cm internal diameter.

The gamma camera concept in two dimensions has been adapted for a hexagonal PET (Karp et al., 1986). Each of the six sides of the detector is a slab of NaI 110 mm wide, 470 mm long, and 25 mm thick, coupled to a two-dimensional array of 30 square phototubes (50 mm × 50 mm), the outputs of each of which are digitized directly. The spatial coordinates of the gamma ray interaction in the crystal are determined from a maximum likelihood calculation. The depth of an interaction in the crystal is inferred from the width of the light distribution. Accurate depth determination eliminates parallax errors and could provide an ultimate spatial resolution of 4.5 mm.

A current area of development is "time of flight" PET (Ter-Pogossian et al., 1981). By noting the difference in arrival times of two coincident gammas the common site of origin can be localized along the line determined by the two detection sites (Fig. 10). For this application, a very short scintillation decay time is paramount, suggesting use of cesium or barium fluoride. Coincidence resolving times of 3×10^{-10} seconds have been achieved with CSF. Since photons travel at 3×10^{10} cm/sec this allows a positron uncertainty along the event line of about 4.5 cm FWHM. Since spatial resolution of the order of 1 cm is achieved with comparable geometry but without time of flight, the main consequence is an improvement in signal to noise in the final image.

Plastic scintillators are attractive because they are inexpensive, easily fabricated, and fast. They are used in a novel 8-ring design with 1024 detectors per ring, each detector 2.4 mm wide to provide high spatial resolution (McIntyre, Spross, and Wang, 1986). The number of photomultiplier tubes per ring is only 72. This economy is achieved by the use of optical fibers in an address-coding scheme where each detector element is connected to a number of tubes, to provide radial, azimuthal and axial location of each event. Plastic scintillators have low intrinsic efficiency since their density and effective atomic number are both low, and the great radial extent required to achieve practical sensitivity necessitates radial address in each ring to avoid parallax errors.

A coding scheme was used in an early planar system, the MGH Positron Camera (Brownell et al., 1972). Each planar detector consisted of an array of 127 NaI crystals 20 mm diameter, 38 mm long. Each of 72 phototubes viewed part of four different crystals (Fig. 11). This coding simplified the coincidence logic and reduced the cost of photomultiplier tubes and associated circuitry.

Crystal scintillators are not the only materials used for positron detectors. High Density Avalanche Chambers (HIDAC) have high Z metal structures (Pb,W) which have high efficiency for photo electric interactions, but are porous enough to allow the photo electron to escape to a channel through the metal (Fig. 12). In a channel the photo electron produces secondary electrons by ionization of a gas, and the secondary electrons are accelerated by an electric field, producing electron gain. This jet of ionization electrons is directed to a multiwire proportional chamber which can efficiently measure the coordinates of the electron jet, and provide a signal to establish coincidence in a diametrically opposed HIDAC detector. The converters may be layers of lead foil, separated by insulation, pierced by a close-packed array of holes by a computer-controlled machine tool, and with the successive layers electrically connected to a resistive divider which supplies the accelerating voltage (Jeavons et al., 1981). An alternate converter structure (Lanza et al., 1986) is a mesh of tungsten wires with successive layers at higher electrical potentials. The wire diameter and pitch are chosen to optimize the conflicting probabilities of high photon interaction, and high electron escape. A third geometry uses close packed lead glass tubing with a resistive surface (Fujieda et al., 1986). The presence of an electrical potential difference along

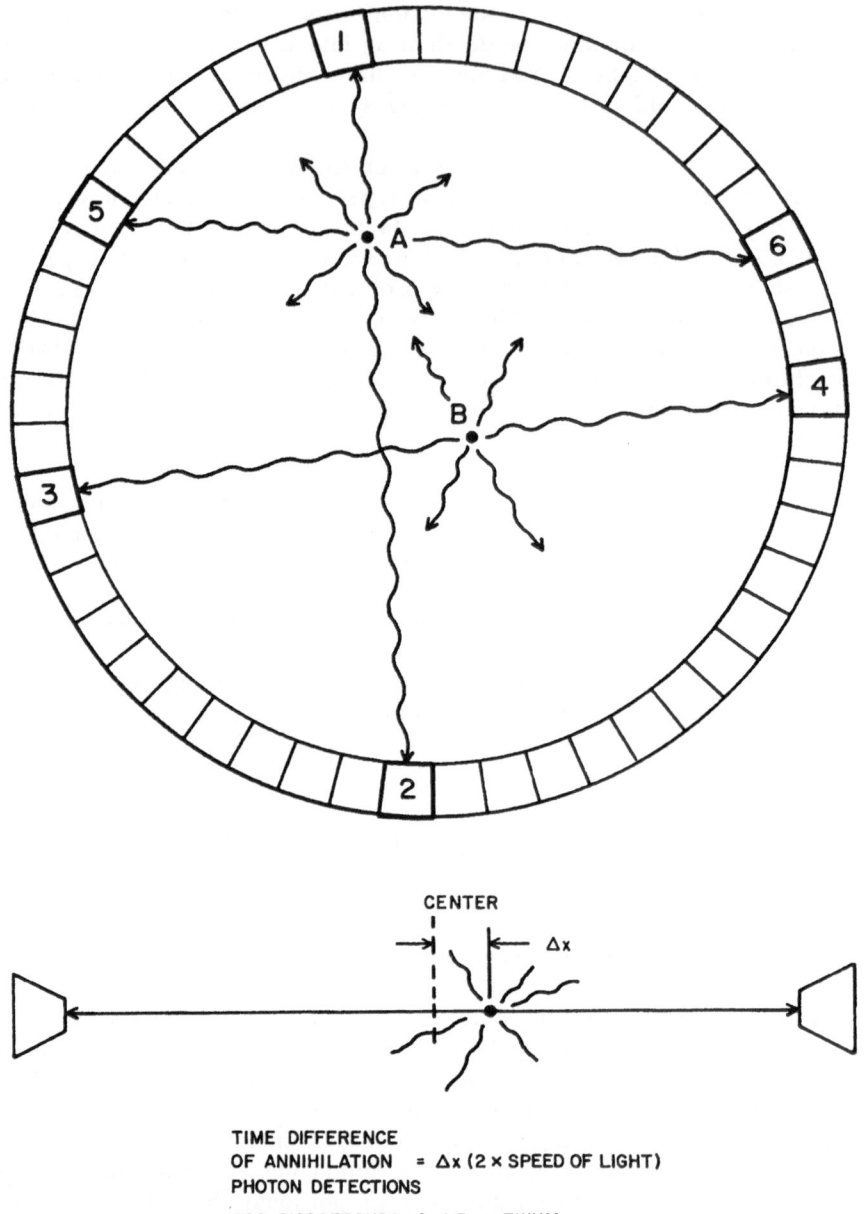

TIME DIFFERENCE
OF ANNIHILATION = Δx (2 × SPEED OF LIGHT)
PHOTON DETECTIONS

300 PICOSECONDS ⟹ 4.5 cm FWHM

Fig. 10. Time-of-flight PET uses the difference in arrival time of the annihilation
photons to limit the region along the event line where the annihilation
occurred.

the tube accelerates the ionization electrons, produced by a primary photo electron
escaping from the tubing wall, providing electron gain.

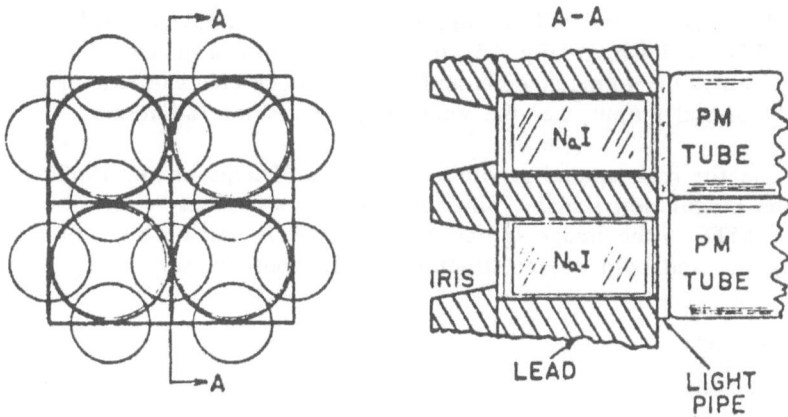

Fig. 11. Coding technique used in the MGH positron camera to couple the crystals to the phototubes in the detector.

The intrinsic efficiency for annihilation photons in a single HIDAC detector is only a few percent, small compared to that for typical scintillator systems, and abysmal when one notes that the positron detection efficiency is the square of the singles efficiency. However, a stack of detectors may be placed on each side with the net intrinsic efficiency equal to the sum of those for each unit. Each detector assembly is only a few cm thick but successive elements will be further from the test object and thus have reduced geometric efficiency. In addition, the reconstruction algorithm employed demands spatial invariance, limiting the angle between an event line and the detector to detector centerline, producing a further reduction in geometric efficiency.

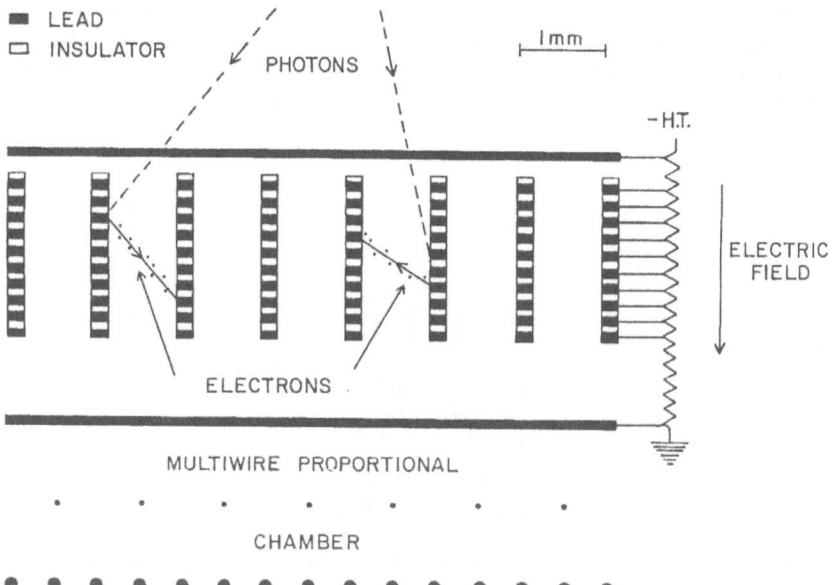

Fig. 12. The HIDAC detector consists of a convector section where gamma rays produce electrons which are amplified and accelerated into a multi-wire chamber which gives their location in space and time (Jeavons et al., 1981).

The advantages of the HIDAC approach include extremely good spatial resolution, 2 mm FWHM intrinsic (Fig. 13), low materials cost, a relatively robust structure without the sensitivity to mechanical and thermal shock characteristic of crystals, and intrinsically digital output, i.e. the wires impose discrete addressing.

The case for high spatial resolution is made (Phelps et al., 1982) in Fig. 14, showing the modulation transfer function (MTF) for systems with various intrinsic resolutions. MTF is the image (perceived) contrast for an object contrast of 100% (alternating hot and cold object structure). Here contrast is

$$C = \frac{I_{hot} - I_{cold}}{I_{hot} + I_{cold}}$$

where I is the image event density in the hot or cold areas of the object. For the 12 mm resolution system structures with spatial frequencies greater than .08 mm^{-1}, corresponding to a 12.5 mm wavelength, have an image contrast of only 5%, requiring a large number of counts to resolve an object with this spatial frequency when one considers the effect of statistical noise. The system with 2 mm resolution has only an 8% loss in image contrast at the same frequency, thus requiring far fewer counts for the same noise equivalence.

Quantitation

Ideally, emission tomography supplies the investigator with accurate maps of the quantitative distribution *in vivo* of the radionuclide injected. This requires an appreciation of the myriad processes which cause the measured distribution to deviate quantitatively and qualitatively from the true distribution. The variety of effects causing this deviation, absorption, scattering, "accidentals" are illustrated in Figure 15. Accurate attenuation correction requires a knowledge of the distribution of the attenuation coefficient throughout the volume of interest. This distribution may be obtained by performing a transmission scan of the subject before injecting the radionuclide. An external ring or sheet source is imaged alone, then data is taken with the external

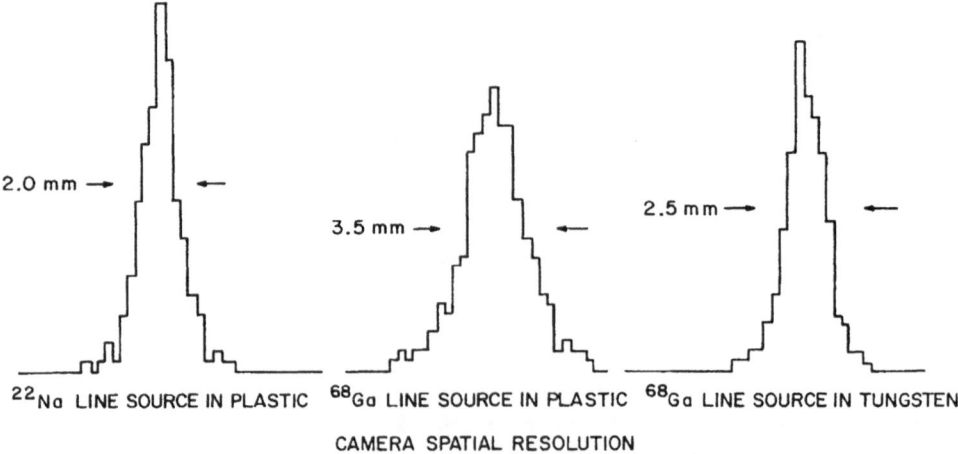

Fig. 13. Spatial resolution measured with a HIDAC camera. ^{68}Ga emits positrons with higher energy than ^{22}Na. This increased range effect is minimized by replacing the plastic with tungsten (from Jeavons et al., 1981).

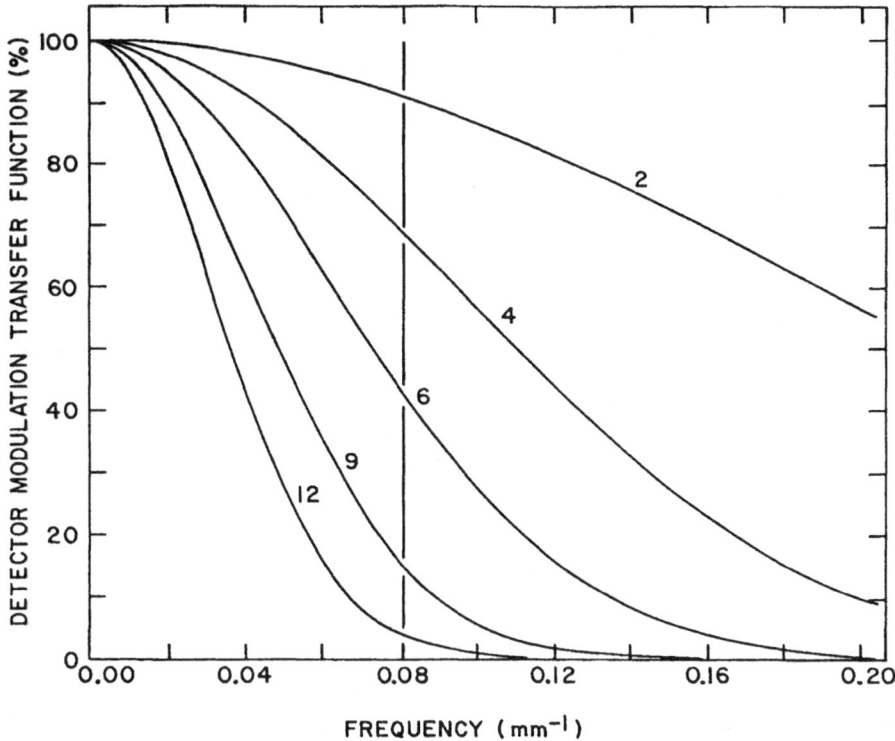

Fig. 14. Modulation transfer functions vs spatial frequency for detectors with spatial resolution indicated. Vertical line shows conventional cutoff frequency for a ramp reconstruction filter to produce an image resolution slightly worse that that from the detector of 12 mm FWHM. Note amplification of frequencies upto this cutoff as detector resolution is improved.

source and subject in place. The attenuation correction for each ray is the ratio of event rate for that ray for the source alone relative to the event rate for subject and source. The external source is then removed and the emission study can commence. The corrected image is obtained by multiplying the emission rate in each ray or channel by the ratio of target alone to target plus subject from the transmission study. The practical difficulties arising from this approach include increased procedure time, increased patient dose, increased statistical noise, and repositioning artifacts due to different positions of the subject for the two phases of the process. The last problem is often unavoidable if the time from injection to start of emission imaging is longer than a small fraction of an hour; some radiopharmaceuticals, for example, require of the order of an hour to equilibrate in the organ of interest, and to wash out of surrounding tissue and improve contrast.

An alternate approach is to determine the surface contour of the patient for the volume of interest, or more practically, to choose an ellipse "equivalent" to the surface contour and assume a uniform attenuation distribution within that contour. This approach eliminates the need for a transmission scan of good statistics. Often the emission image is sufficient to allow fitting of the surface ellipse, or a preliminary transmission scan can be used for that purpose. Obviously, the problem of registration

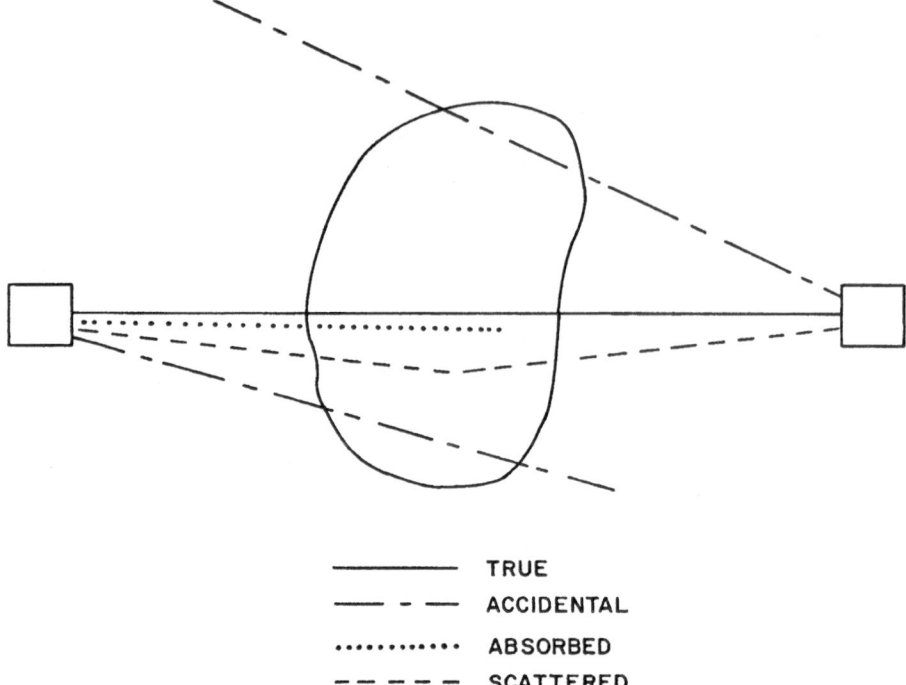

————————	TRUE
—— - ——	ACCIDENTAL
··············	ABSORBED
— — — — —	SCATTERED

Fig. 15. The processes which distort quantitation in positron emission tomography.

between the true attenuation distribution and that used for the attenuation correction is compounded further by these assumptions.

Little use has been made to date of the attenuation information from transmission x-ray (CAT) scans due to the problem of scaling dimensions from one instrument to the other. The problem of registration is magnified by uncertainty in the axial as well as the lateral dimensions, and to these must be added the problem of correcting the attenuation data from the (broad) energy of the CAT x-ray source to annihilation energy.

Attenuation reflects the two processes – absorption and scattering. The number and degree of scattered events included in the image is determined by the geometry and the spectral response of the detector system. Lowering the threshold energy of the scintillation detectors increases the observed event rate, but at the price of accepting more photons which have scattered in the body. Similarly, increasing the separation of the detectors improves scatter rejection (Atkins, 1978).

Random coincidences occur when photons from the annihilation of separate positrons generate signals within an interval less than the coincidence time "window." Correction for accidental or random coincidences also poses problems of statistical noise and positioning just as in attenuation correction. The magnitude of the random counts included in the total observed events may be estimated from the apparent event density beyond the object's boundary, or measured by a delayed coincidence circuit, or calculated from the singles rates knowing the coincidence resolving time. The delayed coincidence circuit has a delay, much greater than the resolving time, inserted in the circuit path on one side. The prompt coincidence circuit generates pulses for true and accidental events, the delayed circuit signals only the random events; the difference reflects the number of true coincident events.

The shape of the accidentals distribution is also required to make a correction to the distribuition. A low order approximation is that the accidentals are uniformly distributed. This is a workable premise if the randoms are a small fraction of the total events. In practice, the accidentals distribution is not uniform but is quite "broad;" i.e., it exhibits little high spatial frequency structure and is highly symmetric. The randoms distribution may be mapped for a test source by triggering the event position recording process only via delayed coincidences, or by imaging the distribution of a single photon emitting nuclide with appropriate energy. The single gamma emitting isotope most commonly used for this purpose is Strontium-85 (E_γ – 514 KeV, intensity = .933, $T_{1/2}$ = 64.8 days).

The primary step in PET quantitation is imaging a known distribution representative of the clinical application. A head imaging instrument is typically run each day with a cylindrical phantom containing known concentrations of Ga-68/Ge-68 in equilibrium in aqueous solutions in four or five compartments. A reconstruction of this distribution, corrected for absorption and other processes, gives a conversion scale from detected events per unit volume to activity per unit volume. This procedure also serves as a final quality assurance check on the system, when the results are compared with preceding values.

SUMMARY

Positron-emitting radionuclides permit the use of electronic collimation and thus achieve higher resolution and better sensitivity than can be obtained with gamma-emitting radiotracers. The evolution of PET imaging systems can be traced from the use of opposed collimated scanning detectors, which had all the limitations of traditional single photon imaging devices, to the present systems which surround the subject with a large volume of detector material. The improvements in system resolution now approach the theoretical limitation imposed by positron-range, and angular deviation. The use of coding permits the use of shared electronics for reading out multiple detectors, which promises to decrease the cost of PET imaging devices, at some penalty of degraded performance. Improvements in computer architectures and capabilities permit faster reconstruction of the multiple planes imaged by multi-slice imaging systems. Software for distortion correction and image processing are still under development, and the ability to map between different devices requires the coordination of efforts between different groups in the same institution. The development and validation of the mathematical models for tracer kinetic analyses will continue to occupy the attention of clinicians and scientists involved in these developments. The desire to make these devices simple and cheap enough to be used in routine patient care is occupying the attention of industry and pioneering users, but this goal has not yet been achieved.

REFERENCES

Aronow, S. 1962, Positron Brain Scanning, in: "Progress in Medical Radionuclide Scanning," R. Kniseley, G. Andrews, and C. Harris, ed., Oak Ridge Inst. of Nuclear Studies, Oak Ridge Tennessee.

Atkins, F. 1978 "Monte Carlo analysis of photon scattering in radionuclide imaging," thesis, Univ. of Chicago.

Brownell, G., Burnham, C., Hoop, B., and Kazemi, H., 1972, Positron Scintigraphy with Short-lived Cyclotron Produced Radiopharmaceuticals and a Multicrystal Positron Camera, in: "Medical Radioisotope Scintigraphy", IAEA, Vienna.

Budinger, T. 1978 "A primer on reconstruction algorithms," Lawrence Berkeley Laboratory Report LBL-8212.

Burnham, C., Bradshaw, J., Kaufman, D., Chester, D., Stearns, C., and Brownell, G., 1982 Application of a one dimensional scintillation camera in a positron ring detector, IEEE Trans. Nucl. Sci. NS-29:461.

Derenzo, S., Budinger, T., Huesman, T., Calsoon, J., and Vuletich, T. (1981) Imaging properties of a positron tomograph with 280 BGO crystals, IEEE Trans Nuc. Sci. NS-28:81.

Evans, Robley D. The atomic nucleus, McGraw Hill, New York 1955, p. 538.

Fujieda, I., Mulera, T., Perez-Mendez, V., and Del Guerra, A., 1986 "Further measurements of electron transmission and avalanche gain in narrow lead glass tubing IEEE Trans. Nuc. Sci. 33:587.

Iida, H., Kanno, I., Miura, M., Murakami, M., Takahashi, K., and Vemura, K. 1986 A simulation study of a method to reduce positron annihilation spread distribution using a strong magnetic field in positron emission tomgraphy, IEEE Trans. Nucl. Sci. NS33:597.

Jeavons, A., Schorr, B., Kull, K., Townsend, D., Frey, P., and Donath, A. 1981 A large area stationary positron camera using wire chambers in "Medical Radionuclide Imaging 1980" IAEA, Vienna.

Karp, J., Muehllehner, G., Beerbohm, D., and Markoff, D., 1986, Event localization in a continuous scintillation detector using digital processing, IEEE Trans. Nuc Sci, NS-33: 550.

Lanza, R.C., Osborne, L.S., Holman, B.L., and Zimmerman, R.E., 1986, Mesh chambers for positron emission tomography, IEEE Trans Nuc Sci, NS-33: 482.

McIntyre, J.A., Spross, R.L., and Wang, K.H., 1986, Construction of a positron emission tomograph with 2.4 mm detectors, IEEE Trans Nuc Sci, NS-33: 425.

Phelps, M., Huang, S., Hoffman, E., Plummer, D., and Carson, R. 1982 An analysis of signal amplification using small detectors in positron emission tomography J. Comp Ass Tomo, 6:551.

Ter-Pogosian, M. Physical aspects of emission computed tomography, in Radiology of the skull and brain, vol. 5, Newton, T., and Potts, D., ed. C. V. Mosbry Co. St. Louis 1981.

Ter-Pogossian, M., Mullani, N., Ficke, D., Markham, J., and Snyder, D. 1981 Photon time-of-flight assisted positron emission tomography J. Comp. Ass. Tomo. 5:227.

Yano, Y., 1986 "Essentials of a Rubidium-82 generator for nuclear medicine," Lawrence Berkeley Laboratory Report LBL-21810.

ACKNOWLEDGMENTS

The submitted manuscript has been authored under Contract No. DE-AC02-76CH00016 with the U.S. Department of Energy.

NUCLEAR MAGNETIC RESONANCE

NUCLEAR MAGNETIC RESONANCE (NMR) SPECTROSCOPY: BASIC PRINCIPLES AND SOME APPLICATIONS TO STUDIES OF CEREBRAL METABOLISM

J. Feeney

MRC Biomedical NMR Centre
National Institute for Medical Research, Mill Hill
London NW7 1AA, England

INTRODUCTION

Over the last 20 years there has been widespread application of NMR to biological problems. Initially much of the work was on isolated biological molecules and was concerned with studies of molecular structures and conformations, dynamics and interactions of complex biological molecules such as proteins, nucleic acids, phospholipids and oligosaccharides. In 1973 two developments occurred which considerably extended the range of application of NMR to biological studies. In the first it was shown that it was possible to measure ^{31}P NMR spectra of metabolites in cells and intact tissues (Moon and Richards[1] and Hoult et al.[2]) and these early experiments stimulated the use of NMR for 'in vivo' studies which have now been extended to investigations of intact organs in human subjects. The second important advance was the demonstration by Lauterbur[3] and Mansfield and Grannell[4] that NMR could be used as a proton density imaging technique further development of which resulted in the present day availability of commercial NMR clinical imaging systems and provided the basis for the blood flow studies described later in this text.

THE NMR EXPERIMENT AND SPECTRAL FEATURES

The basic requirement for an isotope to show an NMR effect is that its nucleus must possess a magnetic moment, designated μ. There are more than 100 known isotopes with magnetic moments and some of those having potential importance in biological studies are listed in Table 1. The relative sensitivity of nuclei to NMR detection at constant magnetic field strength is related to $\sim\mu^3$ (see Table 1): thus the proton (^1H) is

the most NMR sensitive naturally occurring isotope. Each nucleus is also
characterised by a spin quantum number I which takes 1/2 integral values
as indicated in Table 1.

TABLE 1

Nuclear Properties of Some Isotopes of Biological Interest

Isotope	Spin Number I	Natural Abundance %	Larmor Frequency MHz (2.355T)	Sensitivity At constant B_o
^{1}H	1/2	99.985	100	1.00
^{2}H	1	1.56×10^{-2}	15.35	9.65×10^{-3}
^{13}C	1/2	1.108	25.15	1.59×10^{-2}
^{15}N	1/2	0.365	10.14	1.04×10^{-3}
^{19}F	1/2	100	94.09	0.833
^{23}Na	3/2	100	26.45	9.25×10^{-2}
^{25}Mg	5/2	10.05	6.12	2.68×10^{-3}
^{31}P	1/2	100	40.49	6.63×10^{-2}
^{35}Cl	3/2	75.53	9.80	4.70×10^{-3}
^{39}K	3/2	93.08	4.67	5.08×10^{-4}

If we introduce a sample containing protons into a magnetic field
(B_o) then the protons can occupy ($2I + 1$) different nuclear energy
levels: thus for ^{1}H which has a value of $I = 1/2$ there are two such
energy levels. Each level is characterised by a magnetic quantum number
m which has values ($I, I-1, \ldots -I$) and thus for ^{1}H the two states have m
values of $+1/2$ and $-1/2$ as indicated in Figure 1. The energies of the
levels are $\pm \mu_Z B_o$ and thus the energy separation between the two levels is
given by $2\mu_Z B_o$ (where the components of the magnetic moment along the
direction of B_o are $\pm \mu_Z$). As in other spectroscopic methods transitions
can be induced between the different energy levels (selection rule $\Delta m =
\pm 1$) by irradiating with energy of the correct frequency to satisfy the
Bohr condition

$$\mu_Z B_o / I = h \nu_o \qquad (1)$$

where h is Planks constant. The frequency, ν_o, referred to as the Larmor
frequency, is seen to depend on the field strength B_o: for ^{1}H nuclei at
2.355 Tesla the frequency is 100 MHz (i.e. in the R.F. region of the

electromagnetic spectrum). At this field strength other magnetic nuclei

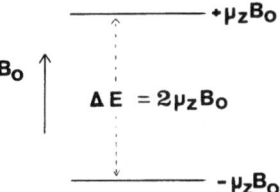

Figure 1. Nuclear energy levels for a nucleus of spin number I = 1/2 in a magnetic field B_o.

absorb radiofrequency energy at completely different Larmor frequencies (e.g. ^{13}C, 25.15 MHz; ^{31}P, 40.49 MHz).

An alternative description of the Larmor equation is

$$\omega_o = \gamma \, B_o \qquad (2)$$

where ω_o is the Larmor frequency expressed as an angular frequency ($\omega_o = 2\pi\nu_o$) and γ is the magnetogyric ratio of the nucleus ($\gamma = 2\pi\mu_Z/Ih$). The adjacent energy levels are populated according to a Boltzman distribution where

$$(N_{upper} = N_{lower} \, e^{-\mu_Z B_o/IkT}) \qquad (3)$$

and because the energy separation is small there is only a small excess population in the lower energy level. This has an important bearing on the sensitivity of the NMR experiment because only the excess population in the lower field effectively absorb energy at the Larmor frequency. The nuclei in the lower level absorb energy while those in the upper level give out energy by stimulated emission.

Chemical Shifts

Nuclei in a solid sample examined in the presence of a magnetic field experience not only the applied magnetic field B_o but also strong local fields from the dipole-dipole interactions between nuclei in fixed positions in the sample and these give rise to very broad NMR absorption signals. However for nuclei in a gas or liquid these dipole-dipole interactions are averaged to zero as a result of the rapid tumbling and rotational motions of the molecules and in such cases narrow absorption bands are observed. If one examines the NMR absorption spectrum of a sample containing molecules with nuclei in different chemical environments then each different type of nucleus can absorb energy at a separate radiofrequency for a fixed value of the applied magnetic field, B_0. Thus in Figure 2 we see that the ^{1}H NMR spectrum of acetic acid shows separate absorption frequencies for the two different types of protons (i.e. in the CH_3 and COOH groups). The nuclei in the two different groups have slightly different electronic environments which effectively shield the nucleus to different extents from the applied magnetic field B_0. The magnetic fields actually experienced by the nuclei can be described as

$$B_{CH3} = B_0 \ (1 - \sigma_{CH3})$$

$$B_{COOH} = B_0 \ (1 - \sigma_{COOH})$$

where σ_{CH3} and σ_{COOH} are the dimensionless shielding constants for these protons. Because the shielded nuclei experience slightly different

Figure 2. ^{1}H NMR spectrum of acetic acid at 60 MHz.

magnetic fields they will absorb radiofrequency energy at different resonance frequencies according to Equation (1). The frequencies at which various types of protons absorb energy are characteristic of particular types of proton. It is usual to measure the frequencies as differences from that of a reference material (tetramethylsilane, TMS) and to quote them in parts per million (ppm) by using the relationship

$$\delta = \frac{\nu_{Sample} - \nu_{Reference}}{\nu_{Reference}} \times 10^6$$

These dimensionless values are referred to as <u>chemical shifts</u> and are independent of the applied field strength. Extensive collections of chemical shift data for protons and other magnetic nuclei in all types of molecule have been compiled and these data can be used to assist in the structural analysis of molecules.

Spin-Spin Coupling Constants

For some molecules the chemically shifted absorption bands appear as multiplets when examined under high resolution conditions (see Figure 3). These splittings arise from interactions between the spins in the groups of differently shielded nuclei. The energy of a particular nuclear spin depends on the orientation of the nuclear spin of a nearby nucleus, the effect being transmitted via indirect interactions involving the valency electrons. Thus the signal for the methyl protons in acetaldehyde is a doublet, each component corresponding to the two possible orientations of spin (+1/2 and −1/2) of the interacting formyl proton. The signal for the formyl proton appears as a quartet, each component corresponding to a different value of total spin of the three interacting methyl protons (+3/2, +1/2, −1/2, −3/2). The interaction between two spins is a scalar interaction the strength of which is known as the spin-spin coupling constant, usually denoted by J and measured in Hz. Coupling constants are independent of magnetic field strength.

The multiplicities on the absorption bands are useful in assigning signals in complex molecules and the measured values of the coupling constants often prove useful in structural and conformational studies[5,6].

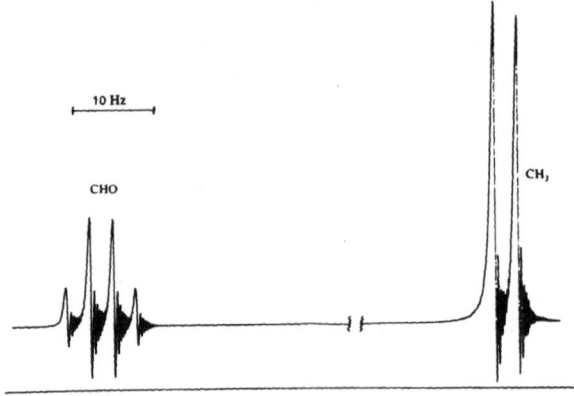

Figure 3. ^{1}H NMR spectrum of acetaldehyde at 60 MHz.

Signal Intensities

Under ideal conditions the intensity of a nuclear magnetic resonance absorption (i.e. the integrated area under the signal) is directly proportional to the number of nuclei contributing to the signal. This makes the NMR technique extremely useful for the quantitative analysis of mixtures of compounds. Relative concentrations can be determined directly from measurement of the intensity ratios of the various signals and absolute concentrations can be obtained if a suitable calibration procedure is adopted and if appropriate corrections for the different relaxation times are applied[7].

Relaxation Times

An understanding of relaxation times can best be reached by considering the classical description of the NMR experiment where we have a collection of nuclear magnetic moments precessing about the direction of B_0 (z axis) with an angular frequency ω_0 where $\omega_0 = \gamma B_0$ (Equation 2). In such a collection of nuclear moments we have a mixture of those aligned along the field direction and those antiparallel to the field direction: each of these magnetic moments can be resolved into components along the field direction and in the xy plane (Figure 4). In the xy plane the resolved components cancel out each other and thus there is no resultant magnetisation in this plane at equilibrium. However because of the excess of nuclei aligned along the field direction there is a resultant component of magnetisation along the z axis (Figure 4b): it is changes in this component which are monitored in relaxation studies. If the value of M_z is perturbed from its equilibrium value, which can be done for example by applying an rf field B_1 along the x axis, it will

170

then return to its equilibrium value by an exponential relaxation process characterised by a time T_1 known as the <u>spin lattice</u> relaxation time or the <u>longitudinal</u> relaxation time. Methods for measuring T_1 will be considered in Chapter 3.

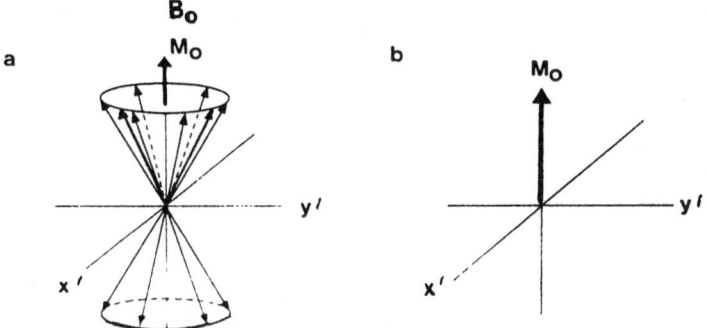

Figure 4. (a) Precessing nuclear magnetic moments in a rotating frame of coordinates (b) the resultant magnetisation vector M_0 along the z axis (the direction of B_0).

When the B_1 rf field is applied along the x axis to the system at equilibrium the nuclear magnetic moments begin to precess in phase: if we resolve the components in the xy plane we now have a resultant bulk magnetisation in this plane. When the B_1 field is removed this bulk magnetisation disappears as the magnetic moments lose their phase coherence and the rate at which this process takes place is characterised by T_2 – the <u>spin-spin</u> relaxation time or the <u>transverse</u> relaxation time.

<u>Rotating coordinates</u>. When the rotating B_1 field is applied the resulting magnetisation vector M_0 will precess in a complex manner under the combined influence of the static magnetic field and the applied rf field B_1. This motion can be considerably simplified by considering the system to be in a frame of reference rotating about the z axis at frequency $\omega_0 = \gamma B_0$. Under the condition that the frame is rotating at the Larmor frequency then the effective static field seen by the magnetisation vector is zero and we no longer need to worry about the

precession of the spins about B_o. The rotating frame coordinates are indicated by superscripts on the x and y axis as shown (see Figure 4).

When B_1 is now applied in this rotating frame and along the x' axis the M_o magnetisation rotates about the x' axis in the zy' plane. The angle θ through which it rotates in t seconds is given by $\theta = \gamma B_1 t$. A 90° pulse is one that rotates the M_o magnetisation by 90° (i.e. from the z axis down to the y' axis) and the time required for this is given by t $= \pi/(2\gamma B_1)$ secs. A 180° pulse will require twice as much time. Such pulses can be used to manipulate the resultant magnetic moment vector in the various experiments described later.

FOURIER TRANSFORM NMR

The experimental requirements for NMR measurements are a magnet of high stability and homogeneity (<1 in 10^8 for highest performance), a source of rf power, (a transmitter and transmitter coil), a receiver for detecting the signals when the rf energy is absorbed at the Larmor frequency and a computer for controlling the rf pulse sequences, and for the data collection and processing. Modern NMR spectrometers use Fourier Transform methods[30,34] for processing the data into a form where we can easily identify the frequency components in the spectra. In this method all the nuclear spins are excited simultaneously using a short intense rf pulse: the response to this is a complex signal measured as a function of time — the so called Free Induction Decay (FID) signals (see Figure 5a) which contains the magnetisation components in the x'y' plane. The time allowed for acquisition of the FID is called the acquisition time, τ_A. This defines the maximum resolution achievable in the experiment which is given by $1/\tau_A$.

This time domain FID is collected and digitised for storage in the computer memory before being transformed into the frequency domain using the mathematical process of Fourier Transformation (FT) (see Figure 5b). Because all the nuclei are excited simultaneously this method is very efficient for collecting the NMR data. The sensitivity is increased by accumulating a large number of FID signals prior to Fourier transformation: if n FID signals are acquired and added together then the improvement in signal to noise ratio is given by $\sqrt{n}$.

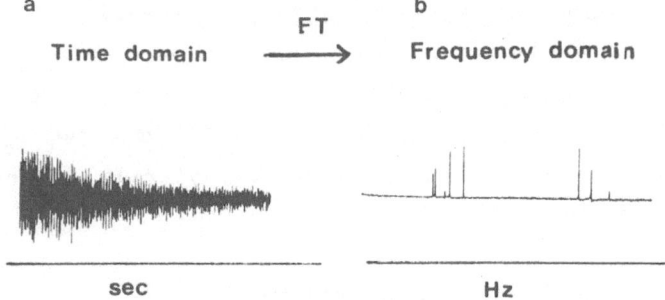

Figure 5.(a) Time-domain free induction decay (FID) spectrum. (b) Corresponding frequency domain spectrum after Fourier transformation of this FID (Cooper[31]).

In a Fourier Transform NMR spectrometer, the rf frequency pulse is generated by an rf transmitter and after power amplification and gating (the timing for this is provided by the so-called pulse programmer) the rf pulse is applied to the sample in the probe. The resulting signal response from the sample is amplified and detected in the receiver which is set to detect signals in the x'y' plane. The receiver is only switched on during the period of data acquisition and not during the time the rf pulse is being applied since this would overload the receiver.

In order to carry out a successful Fourier transform NMR experiment one requires a rf transmitter of sufficient power to provide a 90° pulse in a very short time (usually less than 10 μs): such a B_1 field will have a distribution of frequencies sufficient to excite simultaneously all the different nuclei of the same isotope in the molecule.

Most commercial high-resolution high-field instruments operate with the rf B_1 field set at a constant fixed amplitude, and the required pulse angle is obtained by varying the pulse time t.

If we give the resultant magnetisation vector a 90° pulse along x', then it is rotated into the x'y' plane to lie along the y' axis. In the simplest FT experiment it is the magnetisation components along the y' axis which are detected. If the resonance frequency of these nuclei are exactly at the carrier rf frequency (rotating frame frequency) then the magnetisation $M_{x'y'}$ would remain at a constant value along the y' axis in the rotating frame. If however there is a difference in frequency between the observed nuclei and the carrier $(\omega_i - \omega_c)$ then the transverse magnetisation $M_{x'y'}$ magnetisation will rotate in the x'y' plane as shown in Figure 6: the angle ϕ changes with time $\dot{\phi} = (\omega_i - \omega_c)t$ and the projection of the $M_{x'y'}$ component on the y' axis varies as a cos relationship given by

$$M_{y'}(t) = M_{x'y'} \cos (\omega_i - \omega_c)t.$$

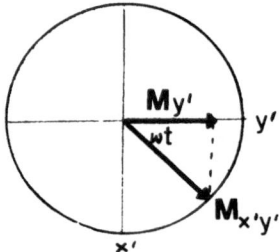

Figure 6. $M_{x'y'}$ component of magnetisation rotating in the x'y' plane and its projection $M_{y'}$ on the y' axis.

The $M_{y'}$ component is modified further by the T_2 relaxation process which causes it to decrease exponentially such that

$$M_{y'}(t) = M_{x'y'} \exp(-t/T_2) \cos (\omega_i - \omega_c)t$$

and the FID signal for this is shown in Figure 7a. This is the signal

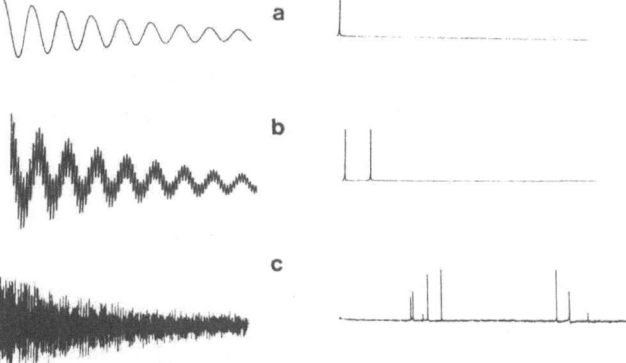

Figure 7. (a) FID from single line spectrum offset from carrier frequency.
(b) FID from two line spectrum. (c) FID from multiline spectrum
(Cooper[31]).

obtained from a single line offset from the carrier frequency. Figure 7b
shows the time domain and frequency domain spectra for two lines offset
from the carrier and Figure 7c shows these spectra for a multiline
spectrum[31].

We have already seen that the time domain spectrum can be converted
into the frequency domain spectrum by means of a Fourier transformation
and this is expressed as

$$f(\omega) = \int_{-\infty}^{+\infty} f(t) \exp (-i\omega t)\ dt$$

In this expression $f(t)$ can be replaced by $M_{y'}(t)$. The term i is $\sqrt{-1}$ and
this results in the frequency domain spectrum following transformation
having two parts — a REAL part and and IMAGINARY part. These can be
calculated separately in the Fourier transformation as a cosine transform
(REAL) and a sine transform (IMAGINARY). These correspond to the
absorption mode spectrum and the other to the dispersion mode spectrum.
They differ simply in phase: one is in phase with the frequency of
excitation (dispersion) and the other is 90° out of phase with it
(absorption).

Sometimes, due to instrument imperfections, the calculated REAL part contains distortions in the phase. These can be effectively removed by adding in components from the dispersion signal to give a 'pure' absorption signal. Such phasing routines are part of the conventional computer software.

To summarise then, the REAL part is given by:

$$M_i(\omega) = \int_0^{+\infty} M_{x'y'} \exp(-t/T_2) \cos(\omega_i - \omega_c) \cos\omega t . dt.$$

where the time domain spectrum includes the T_2 contribution and the cos term appropriate for the difference in frequency between a particular line (ω_i) and the carrier (ω_c).

APPLICATION OF HIGH RESOLUTION NMR TO STUDIES OF THE BRAIN

There have been numerous NMR spectroscopic studies of the brain ranging from 'in vitro' studies of superfused brain slices[8] to 'in vivo' studies on brains of animals and humans (see references 9,10,11) and these results have been reviewed in the literature (see for example Avison et al.[12]). Most of the investigations have used the ^{31}P nucleus[9-12,15] but some studies have also used ^1H[13,14,33] and ^{13}C nuclei[16]. Using these methods it has been possible to study metabolic processes not only in normal tissue but also in diseased state resulting from metabolic stress caused by hypoxia, anoxia, ischaemia, hypoglycaemia or seizure[12].

Brain Studies using ^{31}P NMR

Bachelard and coworkers[8,17] have studied changes in energy metabolism by examining ^{31}P spectra of superfused brain slices from guinea pig. A typical spectrum from their work is shown in Figure 8 and the assignments of signals to detectable metabolites (those present at concentrations > 1 mM) are indicated on the spectrum. These include signals from nucleoside phosphates NTP, NDP (mainly ATP), phosphocreatine (PCr), inorganic phosphate (P_i) and phosphomonoesters and —diesters: by measuring the intensities of these signals one can monitor their levels in a quantitative and non-invasive manner. In the spectrum shown in Figure 8 the P_i signal is much larger than in spectra from 'in vivo' brains because P_i is used as a component of the superfusate medium.

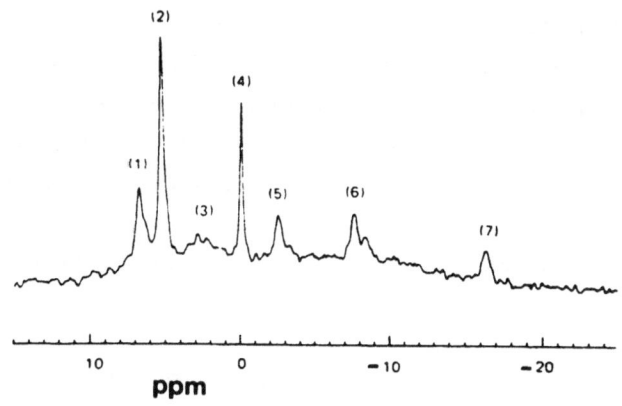

Figure 8. ^{31}P NMR spectrum of guinea-pig cerebral cortex slices superfused in vitro with bicarbonate buffered medium containing glucose and gassed continuously with O_2/CO_2. Signal assignments are (1) Phosphomonoesters; (2) P_i (used in medium); (3) phosphodiesters; (4) phosphocreatine; (5) γ-NTP, β-NDP; (6) α-NTP, α-NDP; (7) β-NTP (Bachelard et al.[8]).

Comparisons between the amounts of ADP and P_i measured by NMR in 'in vivo' measurements and by chemical methods (following biopsy and freeze clamping) indicate that only part of these pools are visible in the NMR experiment probably because some of these metabolites exist in bound states involving immobilised proteins and other biological macromolecules and thus have signals which are too broad to detect in high resolution NMR experiments.

The chemical shift of the P_i signal has been shown to provide a sensitive measure of the intracellular pH[1,2] and can give reliable pH values ($\pm$ 0.1 pH) if a suitable calibration curve is used[6]. Thus for example a value of pH = 7.1 $\pm$ 0.1 has been determined 'in vivo' for rat and rabbit brains[18].

Shoubridge et al.[19] have shown that it is possible to measure the rates of some enzymic reactions (ATP synthetase and creatine kinase) by observing the transfer of saturation between ^{31}P signals from exchanging species. This involves selective irradiation at the frequency of an exchanging nucleus (such as γ-P of ATP) and then measuring the decrease in intensity of another ^{31}P nucleus involved in the exchange (such as PCr) as a function of irradiation time. Such results can be analysed to provide exchange rates[19,20]. Several groups have used this approach to see if the creatine kinase reaction is in equilibrium at the steady state.

$$PCr + ADP \longrightarrow Cr + ATP \quad Forward$$
$$Cr + ATP \longrightarrow PCr + ADP \quad Reverse$$

Shoubridge et al.[19] has observed different rates for the forward and reverse reactions in rat brains (1.46 and 0.68 μmol s^{-1}/g wet weight). One suggested explanation of the different rates is based on considerations of multisite exchange between different pools of ATP and PCr (see for example Brindle's review[20]). However, Balaram[21] has found the rates in this system to be equal (2 μmol s^{-1}/g wet weight) and Bachelard and coworkers[8,22], in their studies on superfused brain slices from guinea pig, also measured similar rates for the forward and reverse reactions (0.68 and 0.78 μmol s^{-1}/g wet weight respectively). The discrepancy in rate measurements could reflect the difficulties in making accurate measurements involving species with short T_1 values and low intensity signals such as the γ-ATP peak. The interpretations of these measurements are thus difficult and are the cause of much dispute in the literature.

Brain studies using ^{1}H NMR

'In vivo' ^{1}H NMR studies of animal brains are technically difficult to carry out not only because the large H_2O and lipid CH_2 signals require suppression but also because the measured spectra are ill-resolved as a result of many overlapping signals resonating within a relatively small chemical shift range. Notwithstanding these problems, Shulman and his coworkers[13,14] have shown that it is possible to detect resonances from phosphocholine, phosphocreatine, creatine, aspartate, glutamate, N-acetylaspartate, γ-amino butyric acid (GABA), alanine and lactate in excised rat brain tissue and extracts and that several of these signals could also be detected in the spectrum of rat brain 'in vivo' although this spectrum was somewhat broader (see Figure 9).

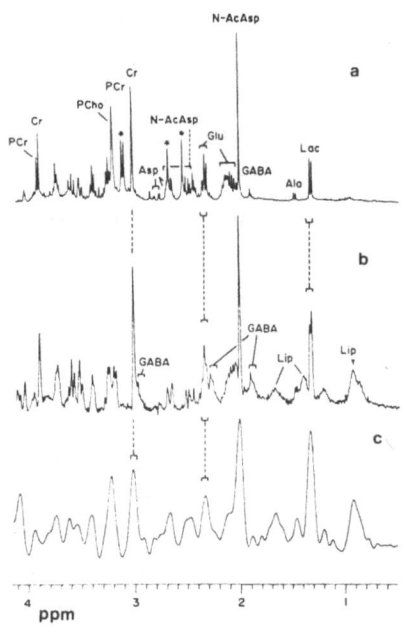

Figure 9. ^{1}H NMR spectra at 360 MHz of (a) acid-extract of rat-brain (b) excised rat brain tissue (c) posthypoxic rat-brain (spectrum obtained 'in vivo' using a surface coil). (Behar and coworkers[13]).

Several workers have used multipulse sequences which select out specific multiplets in the ^{1}H spectra: for example, Behar and coworkers[13] have isolated the methyl resonance from lactate in the complex ^{1}H spectra from rat brain and shown how this can be monitored directly to provide the concentration of lactate.

Studies of Metabolic Stress and Diseased Brain Tissues

Behar[33] has used ^{1}H NMR studies to measure changes in lactate concentration in anoxic rat brains and he has been able to measure the buffering capacity of rat brain by combining the ^{1}H data on lactate measurements with ^{31}P data used to measure cerebral acidification as indicated by the pH determination from the $^{31}P_i$ chemical shift (see Reference 12). In some studies, pH and lactate concentration have been measured simultaneously and in some cases have been found not to change together. For example, Petroff and coworkers[23] in their 'in vivo' studies on rat brains recovering from mild seizures found that the pH returned to its equilibrium value at a faster rate than did the level of lactate. Thulborn and coworkers[9] in their studies of brain ischaemia in gerbils found a decrease in PCr and ATP, a rise in P_i and a fall in pH and other workers have found similar results for cerebral hypoxia in dogs[18] and rabbits[25].

Hypoglycaemia, hypoxia, anoxia, ischaemia and seizure have all been investigated in 'in vivo' animal brain studies[12] and, more recently diseased states in human brain have become amenable to study [11,32]. For example 'in vivo' brain studies on rabbits[10] and rats[28] with insulin induced hypoglycaemia have shown that PCr and ATP both decrease while P_i increases as does the pH: these changes were found to be reversed by glucose infusion.

Gadian and coworkers[26,27] have used NMR to measure local pH and levels of lactate and phosphorus containing metabolites in gerbil brain whilst simultaneously measuring cerebral blood flow with a hydrogen clearance technique. They were thus able to define the time course of the metabolic changes which follow from regional ischaemia or from reperfusion and could define threshold flow values associated with various metabolic changes.

While most of the reported studies have been 'in vivo' studies on animal brains, 'in vitro' studies using superfused brain slices do offer some advantages for studying metabolic stress processes. Such systems allow one to manipulate the metabolic rate in the tissue by making controlled changes to the perfusate. An encouraging aspect of the studies on brain slices carried out by Bachelard and coworkers[8] was that ^{31}P spectra obtained on such preparations examined over long periods of time were very similar to those obtained earlier by Shoubridge and coworkers[19] in their 'in vivo' studies on rat brain. Using a superfused brain slice preparation Bachelard and coworkers[8,16] have shown that it is necessary to reduce the glucose level of the perfusate to below 0.5 mM before the amounts of phosphocreatine and ATP start to fall significantly and that when they do fall they do so to the same extent. A reduction in O_2 from 60 KPa to 16KPa while reducing the phosphocreatine (by 70%) had no effect on ATP showing an interesting difference between hypoglycaemic and   hypoxic insults.

The availability of large bore instruments is now opening up the possibility of carrying out NMR spectroscopy on human brains 'in vivo'. Pioneering experiments on neonatal brains have been made in the laboratory of Reynolds, Wilkie and their coworkers[11,,32] where the technique is being developed for the early detection of cerebral hypoxia and other brain-damage conditions in new born infants.

Figure 10 shows two ^{31}P spectra from neonatal brains: most of the signals are assigned as in Figure 8. However, one difference in the neonatal brain ^{31}P spectra is the large phosphomonoester signal seen in the low field region: and this has now been assigned to phosphorylethanolamine[25,29,32]. Figure 10A shows the ^{31}P brain spectrum of a 17-day infant with profound hypotoxia (probably caused by congenital muscular dystrophy). In this case the PCr/P$_i$ ratio was measured as 1.7 and the pH = 7.0. The authors believe this spectrum to be representative of normal neonatal brain. Figure 10B shows the ^{31}P brain spectrum of a 6 day old infant with severe birth asphyxia: here the PCr/P$_i$ ratio was only 0.5 and pH = 7.2. The PCr/P$_i$ was noted to increase as the clinical condition of such infants improves.

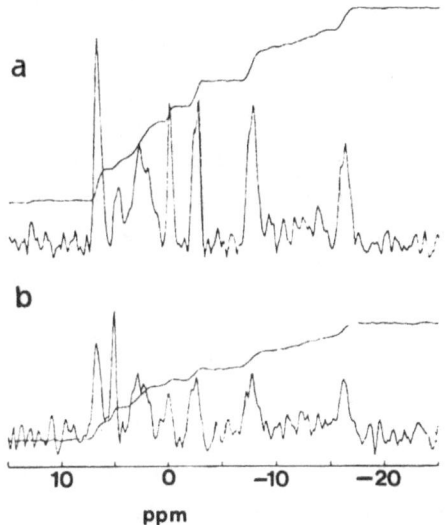

Figure 10. ^{31}P NMR spectra at 32 MHz of neonatal human brains examined 'in vivo'. (a) From a 17-day old infant with profound hypotoxia. (b) From a 6-day old infant with severe birth asphyxia. (Cady and coworkers[11].)

Clearly the NMR technique has much to contribute to the study and diagnosis of diseased states in human brains. Furthermore, the increasing availability of localisation methods for allowing spectra to be obtained from specific regions of deeply situated organs (see Chapter 2) will have a large impact on the future studies of the human brain in its normal and diseased states.

REFERENCES

1. R.B. Moon and J.H. Richards, J. Biol. Chem., 248:7276 (1973).

2. D.I. Hoult, S.J.W. Busby, D.G. Gadian, G.K. Radda, R.E. Richards and P.J. Seeley, Nature (London) 252:285 (1974).

3. P.C. Lauterbur, Nature (London) 242:190 (1973).

4. P. Mansfield and P.K. Grannell, J. Phys. Chem. 6:L422 (1973).

5. J.W. Emsley, J. Feeney and L.H. Sutcliffe, "High Resolution Nuclear Magnetic Resonance Spectroscopy", Vols. I and II Pergamon Press (1965).

6. O. Jardetzky and G.C.K. Roberts, "NMR in Molecular Biology", Academic Press, New York (1981).

7. D.G. Gadian, "Nuclear Magnetic Resonance and its Applications to Living Systems", Clarendon Press, Oxford (1982).

8. H.S. Bachelard, D.W.G. Cox, J. Feeney and P.G. Morris, Biochem. Soc. Trans., 13:835 (1986).

9. K. Thurlborn, G.H. du Boulay, L.W. Duchen and G.K. Radda, J. Cerebral Blood Flow, 2:299 (1982).

10. J.W. Prichard, J.R. Alger, K.L. Behar, O.A.C. Petroff and R.G. Shulman, Proc. Natl. Acad. Sci. USA, 80:2748 (1983).

11. E.B. Cady, M.J. Dawson, P.L. Hope, P.S. Tofts, A.M. de L. Costello, D.T. Delpy, E.O.R. Reynolds and D.R. Wilkie, Lancet 1:1059 (1983).

12. M.J. Avison, H.P. Hetherington and R.G. Shulman, "Applications of NMR to Studies of Tissue Metabolism", Ann. Rev. Biophys. and Biophys. Chem., 15:377 (1986).

13. K.L. Behar, J.A. den Hollander, M.E. Stromski, T. Ogino, R.G. Shulman, O.A.C. Petroff and J.W. Prichard, Proc. Natl. Acad. Sci., USA 80:4945 (1983).

14. K.L. Behar, D.L. Rothman, R.G. Shulman, O.A.C. Petroff and J.W. Prichard, Proc. Natl. Acad. Sci. USA 81:2517 (1984).

15. B. Chance, Y. Nakase, M. Bond, J.S. Leigh Jr. and G. McDonald, Proc. Natl. Acad. Sci. USA 75:4925 (1978).

16. D.L. Rothman, K.L. Behar, H.P. Hetherington, J.A. van Hollander, M.R. Bendall, O.A.C. Petroff and R.G. Shulman, Proc. Natl. Acad. Sci. USA 82:1633 (1985).

17. D.W.G. Cox, P.G. Morris, J. Feeney and H.S. Bachelard, Biochem. J. 212:365 (1983).

18. O.A.C. Petroff, J.W. Prichard, K.L. Behar, J.R. Alger, J.A. den Hollander and R.G. Shulman, Neurology 35:781 (1985).

19. E.A. Shoubridge, R.W. Briggs and G.K. Radda, FEBS Lett. 140:288 (1982).

20. K. Brindle, Prog. Nucl. Mag. Spectros. In press.

21. R.S. Balaban, D.G. Gadian and G.K. Radda, Kidney Int., 20:575 (1981).

22. P.G. Morris, J. Feeney, D.W.G. Cox and H.S. Bachelard, Biochem. J., 212:777 (1985).

23. O.A.C. Petroff, J.W. Prichard, T. Ogino, M.J. Avison, J.R. Alger and R.G. Shulman, Ann. Neurol. In press (1986).

24. M. Hilberman, V.H. Subramanian, J. Haslegrove, J.B. Cone, J.W. Egan, L. Gyulai and B. Chance, J. Cereb. Blood Flow Metab. 4:366 (1984).

25. L. Gyulai, L. Bolinger, J.S. Leigh, C. Barlow and B. Chance, FEBS Lett. 178 (1): 137 (1984).

26. D.G. Gadian, R.S.J. Frackowiak, H.A. Crockard, E. Proctor, K. Allen, S.R. Williams and R.W.R. Russell, J. Cereb. Blood Flow Metab. In press.

27. H.A. Crockard, D.G. Gadian, R.S.J. Frackowiak, E. Proctor, K. Allen, S.R. Williams and R.W.R. Russell, J. Cereb. Blood Flow Metab. In press.

28. K.L. Behar, J.A. den Hollander, O.A.C. Petroff, H.P. Hetherington, J.W. Prichard and R.G. Shulman, J. Neurochem., 44:1045 (1985).

29. S.J. Kopp, J. Krieglstein, A. Freidank, A. Rachman, A. Siebert and S.M. Cohen, J. Neurochem. 43:1716 (1984).

30. R.R. Ernst and W.A. Anderson, Rev. Sci. Instruments, 37: 93 (1966).

31. J.W. Cooper, "Topics in Carbon-^{13}C NMR Spectroscopy" Vol. 2: 406 (1976) Editor G.C. Levy, John Wiley and Sons.

32. P.L. Hope, E.B. Cady, P.S. Tofts, P.H. Hamilton, A.M. Costello, Lancet, 2:366 (1984).

33. K.L. Behar, Ph.D. Thesis. Yale University (1985) – results discussed in Reference 14.

34. D. Shaw, "Fourier Transform NMR", Elsevier, Amsterdam (1976).

NUCLEAR MAGNETIC RESONANCE IMAGING

Laurance D. Hall and Steven C.R. Williams

Laboratory for Medicinal Chemistry
Addenbrooke's Hospital
Cambridge, U.K.
CB2 2QQ

The purpose of this chapter is to attempt to span the ground between the introduction to spectroscopy given in the previous chapter and the more sophisticated discussion of perfusion and flow given in later chapters. As such we shall provide a brief historical introduction to the different forms of "magnetic resonance imaging". In particular we will end this chapter with a summary of a method which is currently the most widely used, and shall take this opportunity to give, effectively, a glossary of the terms which are encountered when one reads papers in clinical MRI.

HISTORICAL INTRODUCTION

In a letter to Nature in 1973, Lauterbur[1] presented the first nuclear magnetic resonance (NMR) image of a heterogeneous object, namely two tubes of water. He noted that if a linear variation in the strength of the static magnetic field (known as a linear field gradient) is applied to a structured sample, then each nucleus within that sample responds with its own NMR frequency (ω) determined by its position; the resultant NMR response is thus a one-dimensional projection of the nuclear density of the sample along the direction of the applied gradient.

Thus, in a homogeneous magnetic field ,

$$\omega_o = \gamma.B_o \, ,$$

and in the presence of a gradient the frequency term becomes:

$$\omega_x = \gamma.B_o + \gamma x.G_x$$

where γ is a proportionality constant specific for each nucleus, x is the distance along the direction of the applied linear field gradient G_x in a static magnetic field B_o.

When two gradients were applied simultaneously in orthogonal directions, a single effective gradient was produced and hence by varying their relative magnitude a series of projections was obtained. A method of image reconstruction, analogous to x-ray CT known as "back-projection" was used to generate a two-dimensional proton NMR image. A diagrammatic representation of the basic principle of this technique is shown in Figure 1.

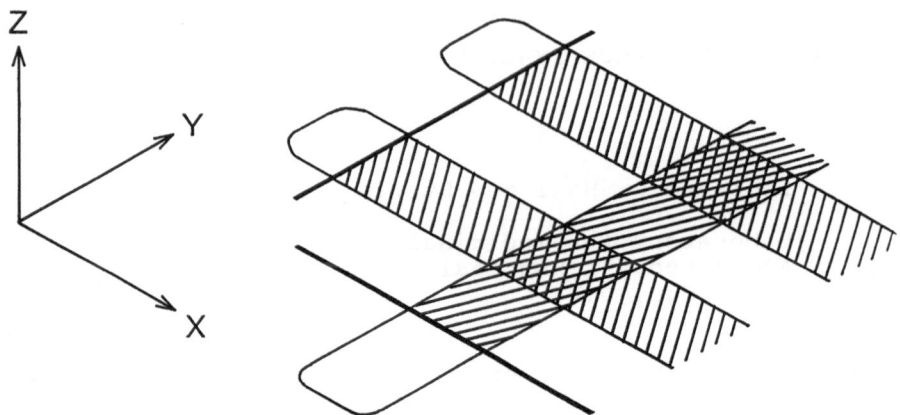

Fig. 1. A diagrammatic representation of the back-projection method. When a linear field gradient is applied in the y direction to a two compartment sample, the NMR spectrum shows two separate projections. In the profile along the x-axis, the signals from the two compartments coincide.

A field gradient can be regarded as essentially a one-dimensional probe and during the past decade a wide variety of techniques have been advanced for manipulating combinations of gradients to secure two-, three- and even four-dimensional information. Some of the techniques most widely used for encoding spatial information in the clinical environment, particularly in observing cerebral anatomy, will be mentioned later in this chapter.

Prior to this use of magnetic gradients for imaging, Gabillard[2,3] had investigated one-dimensional NMR projections with simple glass and liquid structures as early as 1951. Damadian (who had reported that proton relaxation times were significantly higher for cancerous tissue than corresponding normal tissue[4]), filed a somewhat prophetic patent in 1972 [5] that proposed, without detail, that the human body might be scanned for clinical purposes by NMR. The first NMR human image soon followed and that of a living finger was reported in 1976 [6]. The first image of the head was reported by Young's group in 1978 [7].

As a method of medical imaging we must consider some of the advantages that this technique offers. NMR-imaging, or Magnetic Resonance Imaging (MRI) as it is more commonly termed nowadays, does not use ionising radiation; and if utilised properly, it is non-invasive and without known hazard. Hence, it is a much safer means of imaging than modalities which use X-rays, γ-rays, positrons or heavy ions. We can therefore infer that the use of MRI for long term studies following various forms of therapy is an ethically sound proposition. In contrast to ultrasound, the radiofrequency radiation penetrates bony structures without attenuation, and in addition, MRI is immensely versatile, allowing transverse, coronal and sagittal slices or even slices of arbitrary orientation to be imaged.

Furthermore, it is feasible to image, and to quantitate the movement of molecules. Hence, one can envisage its use to study self-diffusion, perfusion or flow; the last two of these are discussed in greater detail elsewhere.

The discussion of MRI can be sub-divided into several aspects. The actual techniques by which the spatial distribution of the nuclei is mapped will be dealt with in some detail here. Other aspects which may be regarded as more exciting and important in the long term involve the type of information obtained from each location in the image. It is worth mentioning at this point that the image appearance depends on at least four different tissue parameters : the concentration of the nuclei under observation (normally ^{1}H), the spin-lattice relaxation time (T_1), the spin-spin relaxation time (T_2) and macroscopic motion such as flow. Imaging techniques have been designed so as to highlight or to eliminate the contributions of any of these. The best technique for each particular pathology can only be derived by clinical trials.

IMAGING - METHODOLOGY

The flexibility of MRI means that it is possible to image points, lines, slices (planes) or even entire volumes of the object of interest. The first two of these possibilities have, until recently only been regarded with historical interest since their signal-to-noise ratios are relatively low and scanning times prohibitively long for practical use as imaging modalities. However, they now provide the basis for methods of volume-selective spectroscopy.

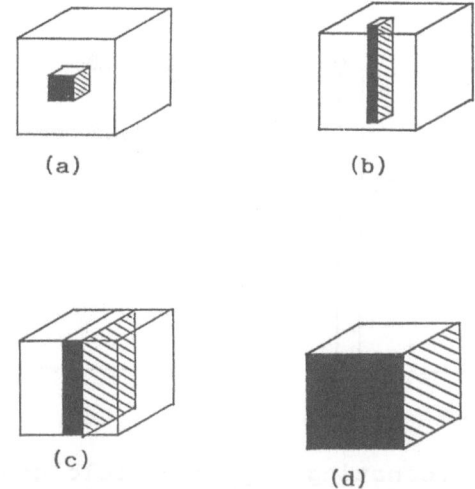

(a)

(b)

(c)

(d)

Fig. 2. Classification of various imaging techniques
a) sensitive point b) sequential line c) sequential slice and d) simultaneous three-dimensional element.

Imaging methods can be separated into two broad categories:-
(a) Mapping methods which build the image from point or line
scans, or are calculated directly with a multidimensional
"Fourier transform" (FT). A Fourier transform is a mathematical
process for changing the description of a function by giving its
value in terms of its frequency components instead of its spatial
or temporal coordinates.

(b) Reconstruction methods which require an algorithm, such as
back-projection, to calculate the image from a series of
projections which may be either line or slice integrals.
Sensitivity, resolution and computation time of this wide array
of techniques are vastly different and a definitive comparison
that encompasses all of these methodologies has yet to be
published.

(i) Point Scanning

Probably the simplest MRI method devised by Hinshaw[8] is that
of point scanning. This technique is based on the application of
sinusoidal varying field gradients in the x, y and z directions
with the object of defining a single point where the magnetic
field is time invariant. All other points will experience time
varying field gradients.

Consider, for example, a linear x-gradient which is
periodically alternated as shown in Figure 3. This alternating
magnetic field gradient causes a time dependence of the magnetic
response in the sample which is averaged to zero, except at the
crossing-point of the gradient. This defines a time-invariant
field, or "zero-field" plane in the sample, termed the sensitive
plane. If a second time-dependent gradient is likewise applied
along the y-axis, perpendicular to the first, then a sensitive
line is defined within the sample. A third, orthogonal
alternating gradient along the z-axis then defines a sensitive
point. By changing the magnitude of the gradients and the
frequency, it is possible to scan any plane, point-by-point.

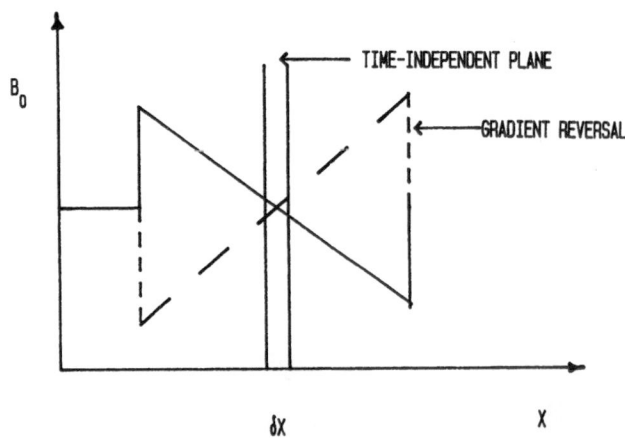

Fig. 3. An alternating magnetic field gradient
which defines a sensitive plane (dx).

In order to reduce the scanning time to a reasonable level,

the radiofrequency excitation is performed under conditions of steady state free precession (SSFP), where 90° radiofrequency pulses are applied sufficiently rapidly for an equilibrium magnetisation to be set up[9]. The signal-to-noise ratio of the method is however, poor since the NMR signal is only received from a single point at a time and no averaging is undertaken during the process of image formation.

(ii) Line Scanning

The methods of line scanning that have been proposed by Mansfield[10] and Maudsley[11] rely on the technique of selectively exciting only the nuclei in a slice of the object through the application of a linear field gradient during the excitation of the nuclei with a radiofrequency pulse of limited excitation bandwidth. This technique can be interpreted simplistically as shown in Figure 4.

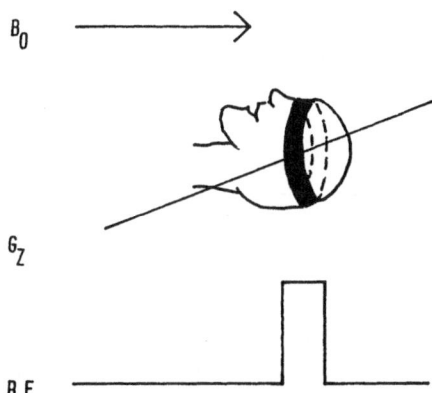

Fig. 4. Slice-selection by radiofrequency excitation in the presence of a linear field gradient in the z direction

Only those nuclei within the desired volume which experience the magnetic field appropriate for NMR frequencies and also within the bandwidth of the radiofrequency pulse are excited. The location and dimensions of the slice is determined by the magnitude and direction of the gradient, and by the shape of the radiofrequency pulse. The position of the slice may be altered by shifting the frequency of the RF pulse, and the thicknesss of the slice is controlled by the amplitude of the field gradient and by the frequency-bandwidth of the radiofrequency pulse. It can be shown mathematically that an optimum shape for this envelope in the time domain is a weighted (typically Gaussian) sinc function since this generates a rectangular envelope in the frequency domain.

This technique is not restricted to line scanning but has also been extended to methods that obtain slice images of an object. Maudsley incorporated Hahn's spin-echo technique[12] i.e. a 90°-τ- 180° pulse train, into an imaging strategy in which a

90° pulse was applied together with a gradient along the z direction (G_z) in order to select a slice perpendicular to the z-axis. After a short, specified time (τ), a 180° pulse together with a gradient along the y direction (G_y) was then performed in order to select a further slice, perpendicular to the first. The intersection of these two slices defined a line of voxels (volume slices) from which a spin-echo was received after the time interval (2τ) as shown in the Figure 5. A third gradient (G_x) applied during the evolution of the spin-echo resulted in a signal, which following Fourier transformation, yielded a distribution of the concentration of nuclei along the selected line. Again, the signal-to-noise ratio is not optimal since the signal is only received from one line of voxels at one time.

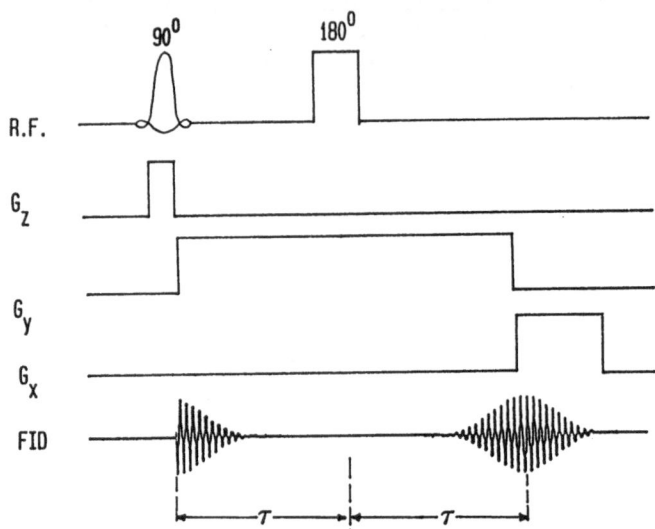

Fig. 5. Line scanning using selective excitation of two planes and incorporating the spin-echo pulse sequence.

(iii) Planar Scanning

Three, well established techniques exist to study the spin magnetisation in a plane of interest within a specimen. The first, which was mentioned briefly in the introduction, is projection reconstruction (PR) imaging[1]. The second technique, devised by Ernst's group in 1975, has been termed Fourier zeugmatography, but is more commonly known as two-dimensional Fourier transformation (2DFT) imaging[13]. The third and fastest method, namely Echo planar (EP) imaging which is presently undergoing extensive research will be reported in some detail in a later chapter[14].

a) Projection Reconstruction

Any NMR imaging technique that requires a reconstruction of the image from projection data can be implemented in either a two- or three-dimensional mode. In a typical two-dimensional plane scanning procedure a gradient G_z say, is applied along with the frequency selective pulse such that the nuclei in a single slice of the object is selectively excited. The other two gradients, G_x and G_y, are then applied during the evolution of the resulting signal known as the "free induction decay". On a single Fourier transformation of this signal, a line integral projection is produced. The series of gradient combinations used effectively "rotate" the line integral projections around the object to yield a large series of views. This is analogous to X-ray computerised tomography[15], and the image of the object is obtained by filtering the Fourier transformed signal and then reconstructing the projection data by one of several commonly used mathematical techniques[16-18].

It should be remembered that this, and the other techniques are not restricted to the examination of single slices since excitation and data collection of successive slices can be interlaced in a single scanning sequence to make more efficient use of the scanning time. Thus, the magnetisation of the nuclei in each slice is still allowed several spin-lattice relaxation (T_1) periods to enable the net magnetization to return to zero between separate excitations; but now, during this relaxation period several slices can be excited and observed (see Figure 6). This multiple slice scanning technique is now commonly used and typically fifteen slices can be simultaneously recorded.

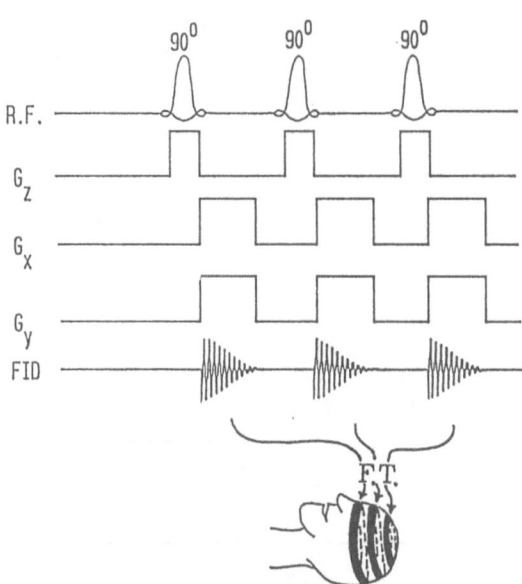

Fig. 6. Multiple slice scanning with reconstruction from line integrals.

Another advantage of this technique is the improved signal-to-noise ratio obtained. Particularly in the reconstruction process when many projections are used to reconstruct a typically 128 pixel image, we find that the image, on averaging, is over-determined. Note also that after a short time to allow for settling of the gradients, the signal is collected simultaneously from an entire slice almost immediately after excitation i.e. when the free induction decay signal is at its largest. One major disadvantage is that image quality is very quickly degraded through imperfections of either or both the main static, or gradient fields. Imperfections in the definition of the slice profile tends to degrade the discrimination between contiguous slices.

b) Direct Fourier Methods

A second, and more commonly used method for obtaining an NMR image of a plane is two-dimensional FT imaging. Firstly, a frequency selective excitation is used to define the plane of interest, and then a projection is obtained along, for example, the x-axis, as in the projection reconstruction method. All subsequent projections are also obtained along the x-axis, but a y gradient of increasing duration is used to increment the line scan of the perpendicular direction. The nuclei along the centre of the x,y plane see no gradient at all, and those farther away see progressively stronger gradients. If the y gradient is very short, the loss of coherence is very gradual along the y-axis of the plane, and cancellation of signal occurs between wide bands in the object. This process encodes spatial distribution along the y-axis in a form that is mathematically comparable to the form in which the time-domain signal encodes x-axis information.

Just as with the x-axis, information along the y-axis is decoded by a Fourier transform. The FT has to be performed twice; first say along the x-axis, and then along the y-axis. One must remember that this technique termed 2DFT had been used for several years by the chemist prior to its application for imaging[13]. Figure 7 illustrates the basic formulation of this method.

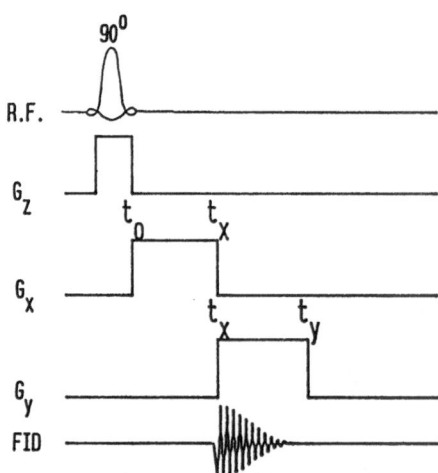

Fig. 7. Two-dimensional slice scanning with reconstruction by the direct Fourier technique.

A disadvantage of this method is that the magnetisation is not sampled until after a time $(t_x + t_y)$ so that its amplitude will have significantly decreased. In fact, if the decay of the free induction signal in the presence of a magnetic field (T_2^*) is short as is the case in most biological media, then only a very weak signal is observed. Hence, signal-to-noise may be slightly lower when compared with other techniques such as projection reconstruction, for which the signal is sampled shortly after excitation.

. These disadvantages are minimised in the spin-warp method devised by Edelstein et al[19], where the phase-encoding gradient is kept on for a constant duration, but its amplitude is incremented in successive excitations (see Figure 8). This has an effect equivalent to that produced by varying the duration of the gradient, but leads to improved signal-to-noise, greater flexibility and reduced dependence on field homogeneity.

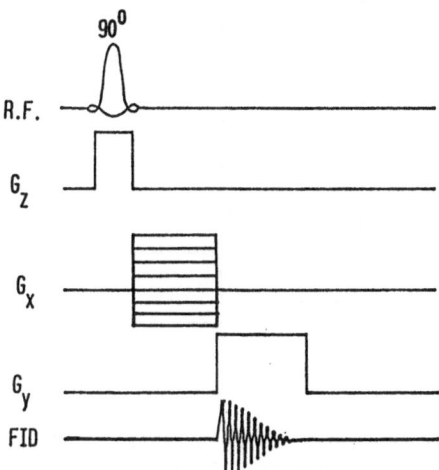

Fig. 8. Spin warp imaging. An alternative method of 2D slice scanning based on a variation of the amplitude of the phase encoding gradient rather than its duration.

The method does however remain vulnerable to any instability of the system during the period of data acquisiton e.g. the motion artifacts from breathing, peristalsis and heart beat. Gating techniques, which cause an imaging sequence to be repeated at some specific time in a biological cycle have now become routinely feasible and have eradicated most of these instabilities. The basic method is however well suited to studies of the head as in this case motion does not pose such a serious problem.

(iv) <u>Volume Imaging</u>

Three-dimensional Fourier imaging is a direct extension of the two-dimensional imaging modalities mentioned above. Basically, the 2DFT process is repeated n times to produce n sections each with a different incremented value of the z-gradient applied before readout of the FT data. This process encodes z-axis information in exactly the same way as it is encoded along the y-axis. Fourier transformation is then applied sequentially three times[20-22] as shown in Figure 9.

During the period following a radiofrequency pulse, the magnetization precesses at the frequencies determined by the magnetic field gradient acting at the site of the sample. The frequency determines the location of the resonating spins but, in this case, their coordinates are not measured directly as we do not detect the free induction decay during the time that the gradient is on. The spins have, however accumulated unique phase angles which are related to their spatial coordinates. This "phase-encoding" allows spatial information to be determined by analysis of the initial phase of the resulting signal. Both x and y dimensions can be phase-encoded by independently stepping the gradients.

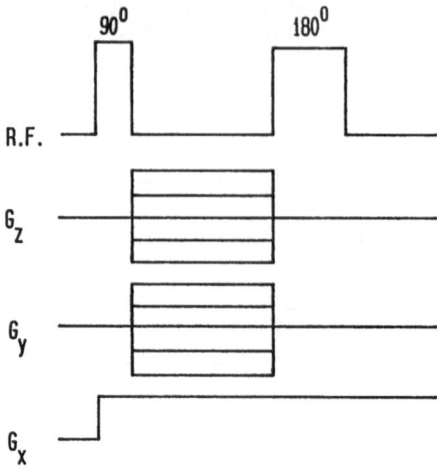

Fig. 9. Three dimensional imaging utilising
two phase encoding gradients.

The spatial resolution of the final image can be made isotropic by choosing equivalent gradient strengths, duration and incremented steps for each of the three dimensions. This natural extension of 2DFT to 3DFT need not stop at this point. An "extra" dimension can be created if the system is allowed to evolve due to some other "Hamiltonian" which causes a linear phase shift with time. Such an interaction can then be isolated by means of a multi-dimensional FT. For example, if all phase-encoding is undertaken prior to data acquisition, which therefore takes place in a homogeneous field, then chemical shift information is retained[23,24] as shown in Figure 10.

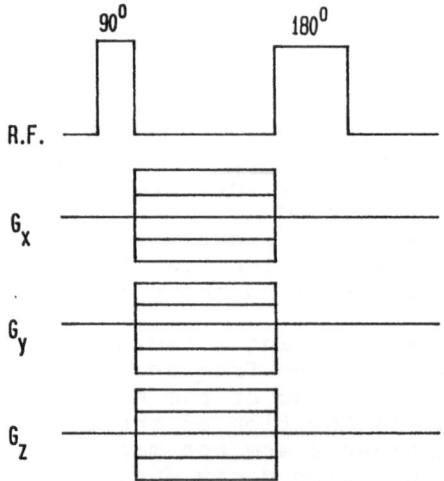

Fig. 10. Four dimensional imaging with retention
of chemical shift information using
three phase encoding gradients.

Various projection reconstruction algorithms have also been utilised to generate three-dimensional volume images. One direct method for the 3D projection is based on an extension of the filtered back-projection method described earlier. This method termed "Fourier reconstruction zeugmatography" (F.r.z) involves, initially, a Fourier transformation of the measured projections then multiplication by a squared wavenumber function before an inverse FT and finally back-projection to·yield the three dimensional image. This technique has been reported in extensive mathematical detail in the literature[25,26]. An alternative application of two succesive filtered back projection processes has also been reported[27]. Both methods, irrespective of their excellent signal-to-noise ratio, require long collection times. This may prove to be a prohibitive factor in their clinical utility.

Despite several problems that have been encountered such as those of magnetic field homogeneity and dealing with large arrays of data, use of three dimensional imaging in human subjects was reported as early as 1981[28,29]. This technique may prove to be useful for pinpointing areas of variable local cerebral perfusion, flow and diffusion. Armed with this knowledge the clinician may be able to predict the potential of cerebral ischaemia and stroke.

Another novel imaging methodology is that devised by Hoult in 1979 [30], and termed "rotating-frame Fourier zeugmatography". He proposed that instead of phase-encoding in the fashion with which we are most acquainted i.e. by variation of the magnetic field gradient, a similar effect could be induced via a radiofrequency field gradient. He designed transmitter coils whose field in the rotating frame corresponded to:

$$B_1 = B_{1O} + B_{1X}X$$

If a radiofrequency pulse is applied for a time   , then magnetisation at a position $P(x,y)$ is tipped through an angle θ ' given by

$$\theta ' = +\gamma\tau (B_{1O} + B_{1X}X)$$

Thus, the angle of flip θ ' is linearly proportional to the spatial position x, provided that B_1 is very much greater than any residual fields in $B_O(\triangle B_O)$, caused by chemical shift or gradients in the relevant direction. If a 90° pulse with a 90° phase shift (as shown in Figure 11) is then applied using say, the receiver coil, this causes the magnetisation to be placed in the x'y' plane with a phase of $\pi/2 - \theta$ ' . Likewise, a 90° pulse with -90° phase shift causes an equivalent flip but now produces a phase of $\theta ' - \pi/2$. By incrementing the pulse length τ , the rate of change of θ ' with τ provides information concerning the spin concentration along x'. Information in the y' direction is obtained by applying the appropriate static gradient in B_O.

This method can, in principle, be extended to form a three dimensional image by applying another similar radiofrequency field gradient corresponding to:

$$B_1 = B_{1O} + B_{1Y}Y$$

for various or corresponding values of τ to give an angle of nutation θ'' equivalent to

$$\theta'' = + \gamma\tau(B_{10} + B_{1Y}Y)$$

Now the total angle of nutation experienced by the magnetisation after application of both radiofrequency field gradients O is given by

$$\theta = \theta' + \theta'' = +2\gamma\tau(B_{10}) + \gamma\tau(B_{1X}X + B_{1Y}Y)$$

On application of a 90° pulse, the phase of the magnetisation will correspond to $\pi/2 - \theta$, and so on. Therefore, on incrementing τ before application of a static magnetic field gradient, produces a matrix of free induction decays. Fourier transformation of these resultant signals yields a three dimensional image of the spin intensity.

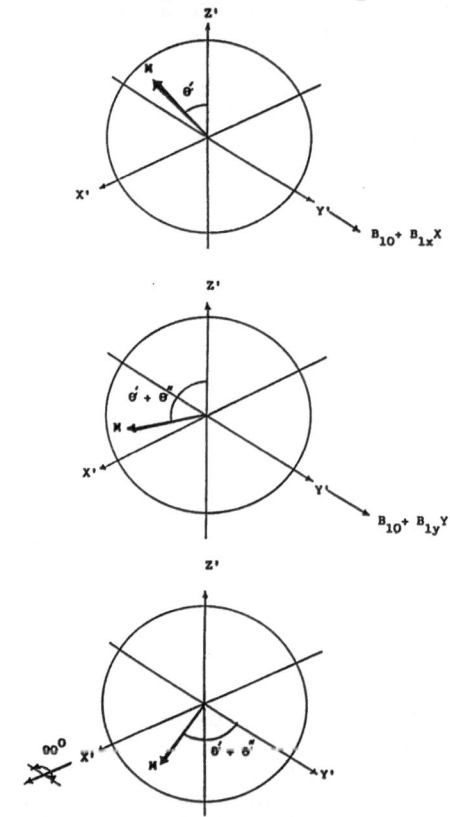

Fig. 11. A diagrammatic representaion of
Rotating-frame Fourier zeugmatography.

The advantages of this method include the fact that both magnetic field inhomogeneity and chemical shift effects can be ignored. Also, the relevant relaxation time is approximately $2T_2$ rather than T_2, and hence longer pulses with reduced radiofrequency power can be used. It should, however be noted that it is not easy to implement this technique for the 3D version of this measurement because of the stringent hardware requirements.

Conclusion

In concluding this chapter we shall describe, in simple terms, a common imaging experiment. This account, at the risk of re-iteration will include definitions of the terms most widely encountered.

If we take, for example, the spin-echo pulse sequence (90°-t-180°) and incorporate it into the two-dimensional FT, slice-selective imaging technique then a sequence of timings, gradients and radiofrequency pulses are combined as shown in Figure 12. This format is commonly encountered when reading most documentation relating to the field of MRI.

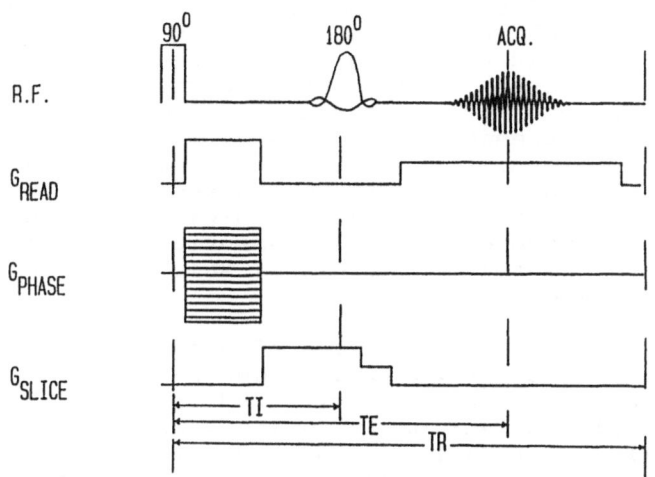

Fig. 12. An example of a typical experiment utilised in medical NMR. A spin scho imaging pulse sequence utilising two dimensional Fourier transformation.

This widely used method begins with a "hard 90° pulse" which excites the whole volume of interest in a non-selective fashion by applying a short burst of radiofrequency for a period of time (μs) long enough to cause a flip of the magnetisation through an angle of ninety degrees relative to its position at equilibrium. This is then followed by a variable time delay "TI" known as the "interpulse spacing". After TI, a "soft 180° pulse" is applied which only excites a slice within the volume of interest; this is a longer, modulated radiofrequency pulse which causes the magnetisation to be refocused as a spin echo after another equivalent time TI. Hence, the whole time from initialisation of the first pulse to the observation of the maximum echo intensity is 2*TI = TE, where "TE" is known as the "echo time". The time for repetition, "TR" of the whole sequence in order to undertake signal averaging etc is usually much longer than TE in order to allow the system to relax to equilibrium before further excitation. These time intervals can be varied and hence, contrast between nuclei with different relaxation characteristics can be optimised.

G_{READ}, corresponds to a series of events which occur to the "read gradient" simultaneously during the pulse sequence. The linear magnetic field gradient initially dephases all the nuclei that have been excited by the 90° pulse but, then rephases them at a time after the 180° pulse where the area "under the gradient" is equivalent. At this time a "gradient echo" is observed. In practice, for optimal signal-to-noise, the gradient echo and spin echo are adjusted so that they coincide. During the collection of the resulting signal, the linear gradient which was reapplied after the 180° pulse encodes a specific frequency for each nucleus along the direction of the gradient. Hence, the read gradient causes "frequency-encoding" along one dimension of the final image.

G_{PHASE}, corresponds to a series of events which occur to the "phase gradient", applied orthogonal to the read gradient and again simultaneously during the pulse sequence as shown in Figure 12. As mentioned earlier, a brief variation of the magnetic field causes a dramatic alteration of the phase of the echo. The amount of this alteration is directly proportional to the field variation, multiplied by its duration. If the magnetic field change is linearly dependent on distance, then on application of a brief linear gradient, the phase also becomes dependent on position. Hence, spatial information is encoded in the phase. If the gradient is incremented in a step-wise fashion as shown then we have, on a second Fourier transformation "phase-encoded" information along the second dimension, which produces the final two-dimensional image.

G_{SLICE}, corresponds to a series of events which occur to the "slice gradient", which is applied orthogonal to the previously mentioned gradients. Note that this gradient is only applied in the case where slice selection is desired, as is the case in this example. The linear gradient is imposed directly before the shaped 180° pulse in order to define a specific plane of interest and is then switched off before the acquisition of the signal.

Although a comparison of different imaging modalities mentioned may initially seem an insurmountable task, a convenient visualisation and comparison is possible through the k-space trajectory formulation. Concise reports by Ljunggren[31] and Tweig[32] are present in the literature. For more elaborate documentation on MRI and related areas recommended texts include Foster[33], Partain et al[34] and Morris[35].

Acknowledgements

We would like to thank Dr. Herchel Smith for an endowment (LDH) and a research scholarship (SCRW). SERC for their financial support which has enabled us to purchase a spectrometer at Cambridge and finally Dr. Boicelli and his staff for the warm hospitality they showed during the course of a most enjoyable and stimulating conference.

REFERENCES

1.   P.C. Lauterbur, Nature, 242 : 190-191 (1973).
2.   R. Gabillard, C.R.Acad.Sci.Paris, 232 : 1551-1553 (1951).
3.   R. Gabillard, Phys.Rev., 85 : 694-695 (1952).
4.   R.V. Damadian, Science, 171 : 1151-1153 (1971).
5.   R.V. Damadian, US Patent 378 9832, filed 17th March 1972.
6.   P. Mansfield and A.A. Maudsley, Proc. XIXth Congress Ampere, Heidelberg : 247-252 (1976).

7. H. Clow and I.R. Young, New Scientist, 80 : 588 (1978).
8. W. Hinshaw, J.Appl.Phys., 47 : 3709 (1979) .
9. H.Y. Carr, Phys.Rev., 112 : 1643 (1958).
10. P. Mansfield, A.A. Maudsley and T. Baines, J.Phys.E:Sci.Instrum., 9 : 271-278 (1976).
11. A.A. Maudsley, J.Magn.Reson., 41 : 112-126 (1980).
12. E.L. Hahn, J.Geophys.Res., 65 : 776 (1960).
13. A. Kumar, D. Welti and R.R. Ernst, J.Magn.Reson., 18 : 69-83 (1975).
14. P. Mansfield, J.Phys.C, 10 : L55 (1977).
15. G.N. Hounsfield, Br.J.Radiol., 46 : 1016-1022 (1973).
16. R.A. Brooks and G. DiChiro, Phys.Med.Biol., 21 : 689-732 (1976).
17. A. Klug and R.A. Crowther, Nature, 238 ; 435-440 (1972).
18. P.F.C. Gilbert, Proc.Roy.Soc.B, 182 ; 89-102 (1972).
19. W.A. Edelstein, J.M.S. Hutchinson, G. Johnson and T.Redpath, Phys.Med.Biol., 25 : 751-756 (1980).
20. A.A. Maudsley, S.K. Hilal, W.H. Perman and H.E. Simon, J.Mag.Reson., 51 : 147 (1983).
21. I.L. Pykett and B.R. Rosen, Radiology, 146 : 197 (1983).
22. L.D. Hall and S. Sukumar, J.Mag.Reson., 46 : 326 (1984).
23. L.D. Hall, S. Sukumar and L. Talagala, J.Mag.Reson., 56 :275 (1984).
24. L.D. Hall and S. Sukumar, J.Mag.Reson., 56 : 314 (1984).
25. C.M. Lai, J.Appl.Phys., 52 : 1141 (1981).
26. L.A. Shepp, J.Comp.Asst.Tomog., 4 : 94 (1980).
27. C.M. Lai and P.C. Lauterbur, J.Phys.E:Sci.Instrum., 13 :747 (1980).
28. P.C. Lauterbur, J.Comp.Asst.Tomog., 5 : 285 (1981).
29. I.L. Pykett, Sci.Am., 246 : 54 (1982).
30. D.I. Hoult, J.Mag.Reson., 33 : 183-197 (1979).
31. S. Ljunggren, J.Mag.Reson. , 54 ; 338-343 (1983).
32. D.B. Tweig, Med.Phys., 10 ; 610-621 (1983).
33. M.A. Foster, Magnetic Resonance in Medicine and Biology Pergammon Press (1986).
34. C.L. Partain, A.E. James, F.D. Rollo and R.R Price Nuclear Magnetic Resonance (NMR) Imaging, Saunders (1983).
35. P.G. Morris, Nuclear Magnetic Resonance Imaging in Medicine and Biology, Clarendon Press (1986).

MEASUREMENT AND INTERPRETATION OF T_1 AND T_2.

C.A. Boicelli[*], A.M. Baldassarri[*], A.M. Giuliani[°] and M. Giomini[§]

[*]NMR Lab. IRCCS San Raffaele - 20132 Milano - Italy
[°]ITSE-CNR - 00010 Montelibretti (Roma) - Italy
[§]Istituto di Chimica Generale dell'Università - 00185 Roma-Italy

1- Definition of the relaxation concept

The nuclear magnetic relaxation process is an ensemble of phenomena by which an assembly of nuclei returns to the equilibrium distribution appropriate to B_0 after the r.f. field B_1 has been turned off. The same phenomena are responsible for the establishment of the equilibrium magnetization when a sample (assembly of nuclei) is first placed in a static magnetic field B_0.

The relaxation process is described by the Bloch equations, which, for nuclei placed in a static homogeneous magnetic field B_0 and to which is simultaneously applied a r.f. field B_1 rotating at the Larmor frequency about the x axis, take the form:

$$\frac{dM_x}{dt} = \gamma \left[M_y B_0 + M_z B_1 \sin 2\pi\nu' t \right] - \frac{M_x}{T_2} \tag{1a}$$

$$\frac{dM_y}{dt} = -\gamma \left[M_x B_0 - M_x B_1 \cos 2\pi\nu' t \right] - \frac{M_y}{T_2} \tag{1b}$$

$$\frac{dM_z}{dt} = -\gamma \left[M_x B_1 \sin 2\pi\nu' t + M_y B_1 \cos 2\pi\nu' t \right] - \frac{M_z - M_0}{T_1} \tag{1c}$$

In these equations T_1 is the longitudinal or spin-lattice relaxation time, and T_2 is the transversal or spin-spin relaxation time. The Bloch formulation of the relaxation process is based on several assumptions:

1 - The macroscopic magnetization M behaves as the individual magnetic moment of each nucleus, μ. This implies that the nuclear moments are unaffected by any interaction. The time dependence of M in the homogeneous magnetic field B is then given by

$$\frac{dM}{dt} = M \times \gamma B \tag{2}$$

2 - The magnetization components parallel (M_z) and perpendicular ($M_{x,y}$)

to a static magnetic field B_O evolve towards equilibrium according to:

$$\frac{dM_z}{dt} = -\frac{M_z - M_o}{T_1} \quad \text{and} \quad \frac{dM_{x,y}}{dt} = -\frac{M_{x,y}}{T_2} \tag{3a,b}$$

These assumptions imply that each nucleus relax independently from the others.

When the Bloch equations hold, relaxation is described by an exponential function and the NMR line is Lorentzian. This is however an ideal condition, which in practice is met only for very dilute spin systems and in the extreme narrowing limit ($\tau_c \omega_o \ll 1; T_1 = T_2$).

Very often, and in particular for biological systems, which are heterogeneous, the time evolution of M is multiexponential, since it is the resultant of more than one decay process. Non-exponentiality of the time evolution of M can arise from the non-exponentiality of the correlation function associated with random processes with a probability distribution of non-Markovian type. Many mechanisms can contribute to the nuclear magnetic relaxation, which are related to the mobility of the molecules in the sample. The different motions are characterized each by its correlation time, τ_c. Motional components with long τ_c values, cannot be handled by means of the perturbation theory, as required by the Bloch treatment of the spin-lattice relaxation. These components arise either from slow motions or from very fast, infrequent motions. The spin Hamiltonian, in such cases, is fluctuating at random in time, the NMR transition is very broad and can be measured only by application of a rather strong r.f. field B_1, non-negligible relative to B_O. The effective field acting on the nuclei at resonance is no more $B_{eff} = B_1 i$, (as in the case when Bloch equations hold), the equilibrium magnetization is time dependent, i.e. $M(t) = \chi_o (B_O + B_1(t))$, and the time constants T_1 and T_2 can no more be defined through the equations (1).

A generalization, due to Robertson[1] , of the Bloch equations can be used to describe the approach to equilibrium of the magnetization $M(t)$:

$$d\mathbf{M} = \gamma \mathbf{M} \times \mathbf{B} - \int_0^t \underline{K}(t,t') \left\{ \mathbf{M}(t') - \chi_o \mathbf{B}(t') \right\} dt' \tag{5}$$

In eq.5 the first term on the right is the precession of M about the B direction with the Larmor frequency $\nu = \gamma B \simeq \gamma B_o$; the second term represents the relaxation, normally non-exponential, of M which is characterized through the kernel $\underline{K}$, typical for every sample.
When $\underline{K}$ takes the simple form

$$\underline{K}(t,t') = \underline{R}\,\delta(t,t') \tag{6}$$

with

$$\underline{R} = \begin{Bmatrix} T_2^{-1} & 0 & 0 \\ 0 & T_2^{-1} & 0 \\ 0 & 0 & T_1^{-1} \end{Bmatrix}$$

and δ the Dirac's δ function, the longitudinal and transversal components of the magnetization, $M_{//}$ (t) and $M_{\perp}$ (t), respectively, are expressed by simple exponential relaxation functions in terms of the associated time constants $T_{//} \equiv T_1$ and $T_{\perp} \equiv T_2$, i.e.

$$M_{//} (t) - M_O = \left[M_{//}(0) - M_O \right] \exp(-t/T_1)$$

$$\tag{7}$$

$$M_{\perp} (t) = M_{\perp} (0)\exp(-2t/T_2)$$

In this case T_1 and T_2 are relaxation times by definition.

The value of **K**, and the related relaxation times T_1 and T_2, for every sample depend critically on molecular properties such as distances, energetic inter actions and motions, which reflect the local magnetic field and on bulk organization and properties.

2 - Determination of the relaxation times : experimental methods [2,3]

The nuclear relaxation parameters are a valuable source of information about nuclear dynamics and stereochemistry. Their determination is obtained by application of suitable sequences of 90° and 180° pulses, separated by appropriate time intervals.

The most frequently used sequences for the determination of T_1 and T_2 will be briefly described and their relative advantages and disadvantages discussed. The functions describing the time evolution of the magnetization are given for a mono-exponential approach to equilibrium obeying the Bloch equations.

T_1 Measurements

Inversion-recovery method (IR)

The pulse sequence it employs is:

$$(180° - \tau - 90° - AQ - T_D)_n$$

where AQ is the acquisition time.

The 180° pulse aligns **M$_O$** in the - z' direction ($M_z = - M_O$) , then magnetization is allowed to recover towards equilibrium for a time τ and is tipped onto the x' axis by the 90° pulse and the free induction decay is acquired. Before a second pulse sequence is applied, the magnetization should be restored to its equilibrium value along z' and a time T_D is required, which must be at least five times the longest T_1. The function describing the magnetization behaviour is:

$$M_\tau = M_O \left[1 - 2 \exp (- \tau/T_1) \right]$$

$$\tag{8}$$

The advantages of the IR sequence are its simplicity, its general applicability, its dynamic range $(- M_O$ to $+ M_O)$ which is twice that of the other techniques and its relative insensitivity to pulse angle missetting. The procedure is however rather time consuming and this is its main disadvantage.

Progressive saturation method (PS)

The pulse sequence is:

$$(90° - AQ - T_D)_{3 \text{ or } 4} - (90° - AQ - T_D)_n ; \quad AQ + T_D \equiv \tau$$

The first 3 or 4 sequences are dummy scans not sampled, necessary to allow the attainment of a stationary state where an equilibrium is reached between excitation and relaxation.

The value of the stationary magnetization which is sampled depends on the pulse interval τ and the function which describes the dependence of M_z from τ is:

$$M_\tau = M_o \left[1 - \exp(-\tau/T_1) \right] \tag{9}$$

The method is simple and relatively rapid, but these advantages are paid for with several limitations. The method can be used only for the measure of relatively long relaxation times, since τ must be longer than the acquisition time AQ. Moreover reliable results are obtained only when T_2^* is much shorter than T_1, so that no transverse magnetization exists at the time when the acquisition pulse is applied. Other disadvantages are the inherent lower dynamic range (from 0 to $+M_o$) and the need for a precise setting of the pulse angle.

Saturation recovery method (SR)

The pulse sequence which is used is :

$$(90° - (HSP) - \tau - 90° - AQ - (HSP))_n \tag{10}$$

The first 90° pulse rotates the magnetization vector into the xy plane reducing to zero the z component. Then a homogeneity spoiling pulse (HSP) is applied to dephase the spins and destroy the transverse magnetization. The system of spins is then allowed to relax for a time τ (variable) before the acquisition pulse is applied. Once the FID is acquired a second pulsed field gradient is applied to destroy again the transverse magnetization and allow a rapid repetition of the sequence. The time recovery of the magnetization obeys eq. 9.

The method has the advantage of the rapidity and is applicable also for the measurement of short T_1's; its dynamic range however is lower than in the IR sequence and a non-negligible time is required to restore a good homogeneity after a HSP (up to 100 ms).

It is convenient to point out that several modifications exist for each of these methods, which are not described here; the interested reader is referred to specialized publications [2,3].

Another point which needs to be stressed is that the quantity one measures in practice is an intensity (voltage in a coil) I, which is proportional to the magnetization and not the magnetization itself.

T_2 Measurements

The simplest method to obtain T_2 is, in principle, to measure the linewidth at half height, which should be :

$$\Delta\nu_{1/2} = 1/\pi T_2$$

However this is true only for a perfectly homogeneous magnetic field. The measured linewidth depends on the inhomogeneities of the B_o field and is governed by the effective relaxation time T_2^* defined by :

204

$$\frac{1}{T_2^*} = \frac{1}{T_2} + \frac{1}{T_2^{inhom}} \tag{11}$$

The second term on the right represents the instrumental broadening which might overweight the effects of true relaxation.

Hahn or Spin-Echo method (SE)

The pulse sequence is:

$90° - \tau - 180° - 2\tau$ - echo

The 90° pulse rotates the equilibrium magnetization into the xy plane; the spins are allowed to dephase for a time τ then a 180° pulse is applied which rotates each spin vector by 180° about the x' axis. At time 2τ all the spins refocus and a maximum transverse magnetization is obtained. The measured value of the magnetization is a function of τ given by:

$$M_{2\tau} = M_o \exp(-2\tau/T_2) \tag{12}$$

A separate experiment is needed for each value and this renders the method time consuming; moreover the method is sensitive to diffusion and to missetting of the 180° pulse. The SE method on the other hand is very simple and errors of the pulse angle are non-cumulative.

The Carr-Purcell method (CP)

The method is an improvement of the SE sequence which makes the experiment faster and drastically reduces the effect of diffusion on the measured T_2 values. The pulse sequence is:

$(90° - \tau - 180° - 2\tau - 180° - 2\tau ....)$

All the 180° pulses are applied along the x' axis. At times 6τ , 8τ etc.. echoes are collected which are alternatively positive and negative.

The time dependence of the magnetization is described by the relationship (12). The first advantage of the CP sequence is a saving of time, since a single sequence yields a train of echoes. The second advantage is the possibility to virtually eliminate the effect of diffusion by making τ short, since the diffusion is effective only during the time 2τ . The main disadvantage of the method is the introduction of cumulative errors when there is a missetting of the pulse angle; the consequence is an imperfect refocussing of the spins for higher order echoes.

The Meiboom-Gill method (CPMG)

A modification of the Carr Purcell method due to Meiboom and Gill over comes the problem of cumulative errors due to missetting of the pulse angle; the pulse sequence to be used is:

$90°_{x'} - \tau - 180°_{y'} - 2\tau - 180°_{y'} - 2\tau ....$

Equation 12 describes the time evolution of the magnetization.

The CPMG method essentially eliminates the problem of diffusion and pulse adjustment, compensates for inhomogeneities of the B_1 field and for off-resonance effects; however, the results obtained with this method can be seriously affected by field/frequency drift.

Moreover, good field homogeneity is advisable when large values of τ are needed to minimize the contribution of diffusion, and extra hardware (phase shifter) is necessary.

The plot of ln $(M_0 - M_\tau)$ (or $M_{2\tau}$) vs τ (or 2τ) is a straight line and the value of T_1 (or T_2) is obtained directly from the slope.

Determination of the diffusion coefficients [4,5]

The SE technique is sensitive to the effects of diffusion. Diffusion of the spins from one part to another of a not perfectly homogeneous field causes a reduction of the echo intensity. The measured value of the magnetization at time 2τ depends on the value of τ, the diffusion coefficient D and the magnetic field gradient g_0 according to :

$$M_{2\tau}^g = M_0 \exp\left[-\frac{2\tau}{T_2} - \frac{2}{3}\gamma^2 g_0^2 D \tau^3 \right] \tag{13}$$

This relationship can be exploited to measure the self-diffusion coefficients. A field gradient

$$g_0 = \frac{\partial H}{\partial r} \qquad (r = x, y, \text{ or } z)$$

is applied to the spin system and the amplitude of the echo $M_{2\tau}^g$, obeying eq. 13, is measured. In the absence of the gradient g_0, the amplitude of the echo is simply given by equation 12. The ratio of the two measured echo amplitudes is:

$$R_D = \frac{M^g (2\tau)}{M^0 (2\tau)} \exp\left[-\frac{2}{3}\gamma^2 g_0^2 D \tau^3 \right] \tag{14}$$

from which the value of D can be obtained if the gradient g_0 is precisely known. The value of g_0 is proportional to the current I through the coils producing the gradient and a plot of ln R_D vs I^2 is a straight line, whose slope yields the value of D. However, when a steady strong gradient is applied, as is necessary for the determination of small diffusion coefficients, the echo is distorted and behaves as a first order Bessel function.

To overcome the problem a train of pulsed gradients is applied, of duration δ (μ s) and of distance Δ (μ s), starting at time t $= \tau/2$. Under these conditions the gradient is not present during the observation of the NMR signal and echoes of the correct shape can be collected even when very high gradients are used. The relationship between the R_D and the characteristics of the pulsed gradient train is:

$$R_D = \exp\left[-\gamma^2 D \delta^2 g_0^2 (\Delta - \frac{1}{3}\delta) \right] \tag{15}$$

from which the diffusion coefficient is deduced as described for the steady gradient method.

3 - Ex vivo applications

The simple equations outlined above for the determination of T_1 and T_2 are generally inadequate to describe the magnetization behaviour in biological specimen. The living matter is in fact heterogeneous and very often more than one compartment is present in an apparently uniform sample. Here, compartment is not defined as a part of a specimen physically separated from the others by boundaries, but as a population of molecules with particular properties, e.g. relaxation time [6]. Thus, when relaxation is measured on a biological sample one should always consider that more than one component contributes to the total magnetization of the sample and a multiexponential behaviour is expected, since the various components might have very different relaxation characteristics [7,8]. With such systems, eq.8 must be replaced by:

$$M_\tau = \sum_{i=1}^{n} M_{\tau i} = \sum_{i=1}^{n} M_{oi}(1 - 2e^{-\tau/(T_{1i})})$$

and

$$M_o = \sum_{i=1}^{n} M_{oi}$$

(16)

where n is the number of different compartments. Equations 9 and 12 should be modified in a similar way.

Analysis of the experimental data obtained in a relaxation experiment of these systems generally requires the use of a computer for a complete deconvolution. The different values of T_1 and T_2 can be obtained directly from the plots lnM(t) vs t only when they are formed by a sequence of straight segments, whose slopes give the various relaxation times.

It is generally impossible to assign a characteristic value of T_1 or T_2 to a specific tissue; it is however possible in certain cases to correlate the histological properties of a tissue with its NMR response or to obtain significant information from the NMR study of in-vivo or in-vitro samples.

A case where the hope to find a correlation between the NMR relaxation parameters and the histological data fails, at least on individual bases is that of cardiac hypertrophy induced by isoproterenol [9]. The proton relaxation times of the myocardium (left ventricle with apex) of normal Wistar rats and of isoproterenol treated (3 and 80 mg/Kg of body weight twice a day) animals were measured with a low resolution spectrometer operating at 20 MHz. More than one relaxation time has been measured in each case (table 1). For all the normal (untreated) animals the T_1 value between 0.50 and 0.60 s and the T_2 value between 0.040 and 0.045 s have been observed: these are the characteristic values for the healthy tissue. However, not all the treated animals show this component of the total magnetization, though they were all vital at the time of sacrifice. On the other hand, the slow relaxing component of the magnetization is not observed at all in the T_1 determinations and only in the 10% of cases in the T_2 determinations, in the case of the controls: isoproterenol administration and the subsequent cardiac hypertrophy induce the presence of additional components of the magnetization, as reported in table 1.

TABLE 1

Fractions ($\pm$ 5%) of observed specimen of rat myocardium presenting the reported relaxation time.

	Untreated Animals	Isoproterenol Treated Animals	
		3 mg*	80 mg*
T_1 (s)			
0.30 - 0.45	10	40	50
0.50 - 0.60	100	75	80
> 0.7	0	45	55
T_2 (s)			
0.01 - 0.02	60	55	10
0.040- 0.045	100	35	10
> 0.05	10	85	80

* dose twice a day per Kg of body weight.

A useful correlation between NMR parameters and histology has been obtained for human pancreatic tissue[8] . Six sequential slices of the pancre as of a patient suffering of the Werner-Morrison syndrome have been prepared, going from the tail with the tumour (sample 1) to the head of the organ which was still healthy(sample 6). The ^{1}H relaxation times of the different samples are collected in table 2.

The tumour can be differentiated from the healthy tissue on the basis of the relaxation behaviour. Both T_1 and T_2 are much longer for the tumour; moreover sample 1 shows the presence of a second type of tissue, characterized by a shorter T_1, which is histologically similar to the infiltrating tissue (T_1 = 0.324 s) of sample 4.

TABLE 2

n° of slice	T_1 /s ($\pm$ 0.08 s)	T_2/s ($\pm$ 0.003 s)
1 (Tumour, tail of pancreas)	0.475 0.350	0.070
2	0.185	0.041
3	0.113	0.042
4	0.324 0.113	0.038
5	0.196 0.133	0.039
6 (Head of pancreas, healthy)	0.162	0.035

An interesting correlation has been observed between the different phases of liver regeneration after partial hepatectomy and the proton relaxation times of the tissue [10]. Fresh tissue has been obtained by biopsy from rat liver at successive stages of the regenerative process and the ^{1}H relaxation times determination has been completed within 45 minutes from resection. Sham-operated or unoperated animals were used as controls. The measured magnetization is essentially due to the tissutal water protons. The T_1 values are reported in fig.1 as a function of time after surgery. The increasingly slower relaxation rate observed between 6 and 24 hours may be connected with the expansion of the intercellular spaces and the increased cell hydration which accompany the first stages of liver regeneration and which lead to an increase of the water mobility. After the 24th hour, i.e. after the first massive DNA duplication, the percentage of cells where DNA duplication occurs, decreases and the phenomena causing the increase of T_1 are progressively reduced.

Equations, describing the regeneration as the resultant of a proliferative and an inhibitive process, have been deduced from the relaxation data: the results are shown in fig. 2.

A certain correspondence between histological findings and NMR relaxation data have been found in the case of mitral-aortic pathologies [9]. Human papillary muscles of patients have been examined by NMR and their histological characteristics determined. The ^{1}H relaxation data are reported in table 3.

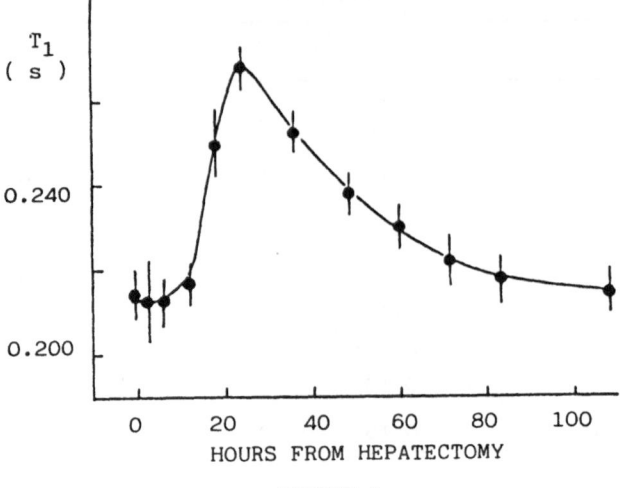

FIGURE 1

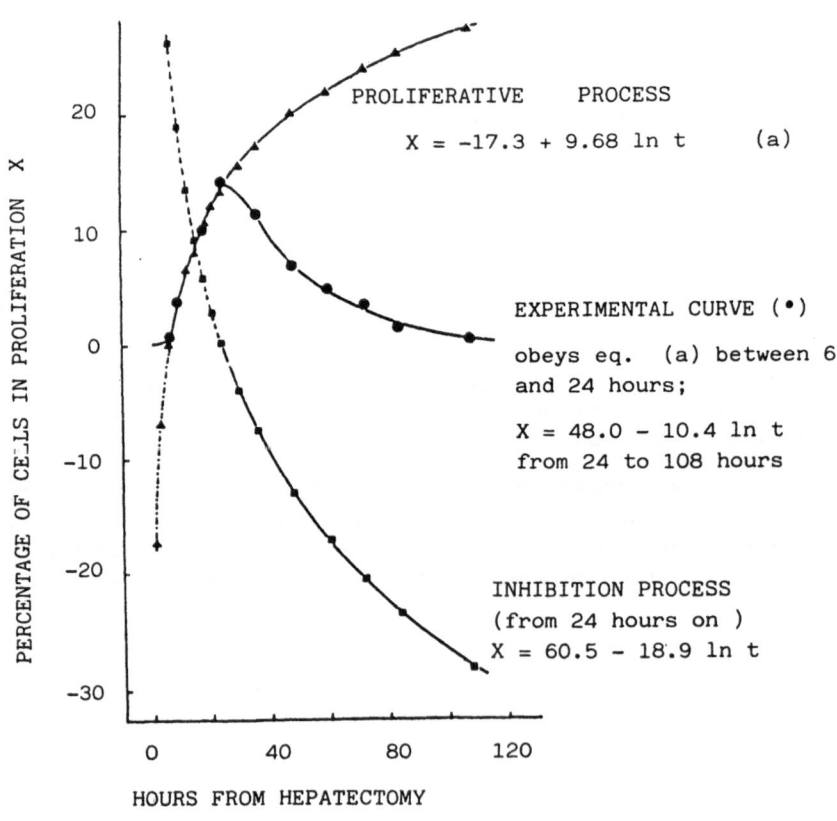

FIGURE 2

TABLE 3

Proton relaxation times of human pathological papillary muscle samples (B_o = 0.47 T; T = 37 $\pm$ 1 °C).

sample	T_1 (s)	T_2 (s)
1. Mitral-Aortic Pathology	0.31 0.51	0.076
2. Mitral-Aortic Pathology	0.31 0.50	0.063
3. Fallot Tetralogy	0.34 0.51	0.073
4. Mitral-Aortic Pathology	0.48 0.60	0.060
5. Ventricular Aneurysm after Infarct	0.80	0.017 0.071
6. Mitral-Aortic Pathology	0.67	0.077

The first three samples have common histological features (hypertrophy of myocardial cells, interstitial oedema and intracellular accumulation of lipoperoxides) though the pathology of 3 is different, and have common relaxation behaviour. The other specimen present different histological characteristics and their ^{1}H-NMR relaxation behaviour is different. The longer T_1 value observed for these samples has been related to the presence of abundant connective tissue and of ischemic areas, which is a common feature.

These findings show the possibility, in certain cases, to establish a gross correlation between histological structures and NMR relaxation data in a tissue.

NMR has been exploited to differentiate various human endocrine pancreatic tumours [9]. The precise diagnosis of these tumours is intrinsically difficult since they are all histologically very similar and the only possibility is to resort to immunochemical tests. The proton relaxation data (table 4) suggest the possibility (though only a limited number of cases has been examined because of the rarity of the pathology) to discriminate between vipoma and insulinoma, while the differentiation of gastrinoma B and insulinoma is doubtful. The NMR response of gastrinoma A and B is different. When the histological and immunochemical characteristics of the different tumours are considered and their bearing on the NMR results are examined, it appears that the ultrastuctural features do not influence the NMR behaviour of the tissue, while an important role is played by the relative amounts of the gross histological structures.

TABLE 4

[1]H relaxation times of human endocrine pancreatic tumours. Controls are the fractions of pancreas that appear histologically non-infiltrated by neoplastic cells. (B_o = 0.47 T; T = 37 $\pm$ 1°C).

Sample	T_1 (s)	T_2 (s)
Control	0.12	0.040
	0.18	
Gastrinoma A	0.60	0.040
	0.68	0.064
Gastrinoma B	0.54	0.040
	0.65	0.070
	0.88	0.110
Insulinoma	0.53	0.030
	0.68	0.080
	0.85	0.090
	1.04	0.110
Vipoma	0.35	0.070
	0.48	

3 - Concluding remarks

The presence of a field gradient modifies the time dependance of the measured magnetization value in a T_2 determination experiment, with a term in r^3 (eq. 13) which causes a non-exponential appearance of the M_{2r} vs $2r$ curve.

A similar effect is observed with biological specimen (ex-vivo or in-vivo) without applying any field gradient.

This behaviour is caused by the existence of several compartments, each of them characterized by its own value of nuclear magnetic susceptibility, χ_{oi}.

The magnetization of a homogeneous sample is given by

$$\mathbf{M} = \chi_o \, \mathbf{B}_o \tag{16}$$

and therefore, in a sample with different compartments, an intrinsic gradient of susceptibility is present, the effect of which on the measured values of the magnetization is similar to that of an applied field gradient [11].

In these conditions, the measured magnetization at each time r, is smaller than the theoretical value for a purely exponential decay and the lineshape is non-lorentzian.

It should therefore be kept in mind that the application to biological samples of any of the sequence for the determination of either T_2 or T_1, involving the measurement of the magnetization values, lead to an incorrect evaluation of the relaxation times, unless an appropriate numerical analysis of the experimental magnetization decay data is applied.

BIBLIOGRAPHY

1. F. Noach, "Nuclear Magnetic Relaxation Spectroscopy" in NMR Basic Principles and Progress (P.Diehl, E. Fluck and R.Kosfeld eds.) vol.3, Springer-Verlag, Berlin (1971) 83.

2. M.L.Martin,J.-J.Delpuech and G.J.Martin "Practical NMR Spectroscopy", Heyden, London (1980).

3. T.C. Farrar and E.D.Becker "Pulse and Fourier Transform NMR", Academic Press, New York (1971).

4. Bruker Report n°1 (1981) 8-9.

5. E.D.Stejskal and J.E.Tanner, J.Chem.Phys. $\underline{42}$ (1965) 288.

6. P.S.Belton and R.G.Ratcliffe, Progr. NMR Spectry $\underline{17}$ (1985) 241.

7. I.D.Weisman, L.H.Bennett, L.R.Maxwell,Sr. and D.E.Henson, "Cancer Detection by NMR in the Living Animal " in NMR in Medicine (R.Damadian ed.) Springer-Verlag, Berlin (1981)

8. C.A.Boicelli and A.M.Baldassarri, "Pratical Aspects of in vitro and in vivo T_1 and T_2 Measurements" in NMR in Living Systems (T.Axenrod and G. Ceccarelli eds.) Reidel Publ. Corp., Dordrecht (1985) 119.

9. C.A.Boicelli, A.M.Baldassarri, A.Bondi, A.M.Giuliani and R.Toni, "Towards NMR spectroscopy in vivo : II Relationship Between NMR Parameters and Histology" in NMR in the Life Sciences (E.M.Bradbury and E.Nicolini eds.), Plenum Publ.Corp., New York (1986) 179.

10. C.A.Boicelli, A.M.Baldassarri, M.Giomini and A.M.Giuliani, "Towards NMR Spectroscopy in vivo: the Use of Models" in NMR in the Life Sciences, (E.M.Bradbury and C.Nicolini eds.) Plenum Publ. Corp., New York (1986) 165.

11. F.F.Brown, J. Magn. Res.: $\underline{54}$, (1983) 385

FLOW STUDIES

Edgar Müller

Siemens AG, UB Med
Henkestraße 127
Erlangen, West-Germany

1. CONCEPTS OF FLOW NMR

Magnetic Resonance Imaging (MRI) sequences with selective
radio frequency pulses and magnetic field gradients combine
many well known 'classical' non-imaging experimental methods. Early attempts to quantify flow were made using
Stejskal's observation (1), that the phase of the complex
NMR signal will be shifted due to movement of the spin system in a magnetic field gradient. Subsequent development
of this concept lead to Packer's pulse sequence (2), which
succeeded in measuring 'slow coherent motion' in the range
of μm/s. Packer combined the advantage of small RF-pulse intervals in multiple echo experiments (to reduce diffusion
effects) with the accumulation of flow sensitive phase by
incorporating a series of additional magnetic gradients.

There are other applications today in which the accumulation
of flow phase is not desired. Carr and Purcell demonstrated
in 1954 (3) that phase dispersion resulting from spatial
flow distributions reduces signal intensity in all odd numbered spin echos in multiple echo experiments. For non-pulsa-
tile flow, however, there is almost complete rephasing in
all even numbered echos and therefore an increased signal
intensity in this case. This alternating echo amplitude
behavior is known as the 'odd/even echo effect'. The even
echo rephasing which occurs with this effect may be used to
highlight constant flow in MR imaging.

Another characteristic of flow may also be used to obtain
quantitative information. Because only those spins in the
region where the conditions necessary for nuclear magnetic
resonance exist will contribute to the received signal, the
motion of spins through the NMR probe should significantly
alter this received signal intensity. In NMR spectroscopy,
the physical dimensions of this region are limited by the
dimensions of the RF-coil. However, these dimensions may be
further reduced to those of a single slice through the use
of selective excitation techniques in MR imaging.

The inflow effect (unsaturated fluid enters the NMR-probe and replaces those spins which were partially saturated by the preceding RF-pulses) for example, is routinely used to enhance signal intensity in chemical NMR studies (4). The outflow effect results in a reduction in signal intensity. In outflow experiments, excited spins leave the NMR probe during the course of the measurement and therefore cannot contribute to the NMR signal. In Hahn spin echo MRI experiments, signal intensity is predominantly reduced by spins passing through thin slices during the time interval between two RF-pulses. The experimental conditions in such an experiment often allow a description of inflow/outflow effects based on apparent relaxation rates, 1/T1f and 1/T2f. These apparent relaxation rates are roughly proportional to the flow velocity.

A third possibility for measuring flow in NMR experiments is the "time of flight" principle. This idea was first proposed by Singer (5), who used spatially separated transmitter and receiver coils. The RF-pulses from the transmitter coil label the moving spins as they pass through the coil. When these moving spins then pass through the receiver coil downstream, their signal is detected.

All of these principles may be used in slice selective MRI as well. A variety of special pulse sequences (sensitive to motion via phase transport effects) and evaluation algorithms have been proposed in the past few years (6-12) to take advantage of these principles. Although it is now possible to visualize larger vessels and quantify flow in phantoms, much work remains to be done in this field. One of the topics of this book should be the detection and quantification of microcirculation of human blood by an MR imaging system, but this work was not even possible until at least 1986. However, it will be shown in this chapter that the measurement of flow velocities as slow as 0.1 mm/s is possible in phantoms with a commercial MR imaging system.

2. EXPERIMENTAL METHOD AND ANALYSIS

2.1 MATERIALS AND METHODS

Phantom measurements were performed using 0.5 Tesla and 1.0 Tesla whole body imagers (MAGNETOM). The basic spin echo sequence which was used is shown in figure 1. The selective RF-pulses in this sequence are normally SINC-shaped with a Hamming filter applied. A spoiler gradient applied after the last spin echo destroys the residual component of the transverse magnetization. The multi-echo sequence is a modification of the Carr-Purcell multi-echo sequence (3) according to Gill and Meiboom (13). Details of the sequence are reported elsewhere (14). The basic gradient echo sequence is shown in figure 2. The spin echo is achieved by inversion of gradients (gradient recalled echo) rather than by the application of a 180 degree RF-pulse as in the Hahn spin echo sequence.

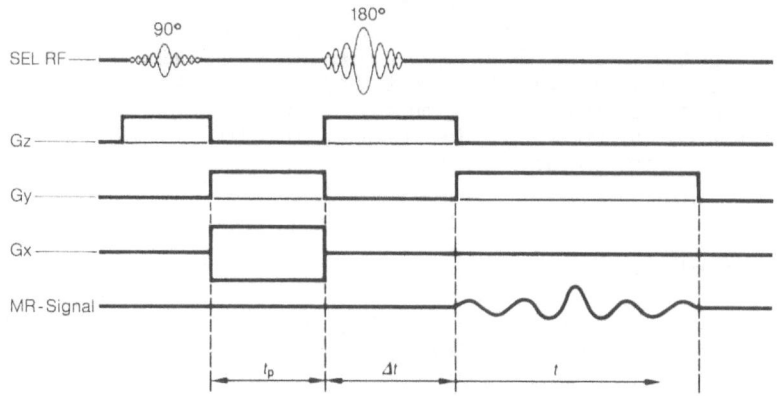

Fig. 1. Timing diagram of a Hahn spin echo sequence

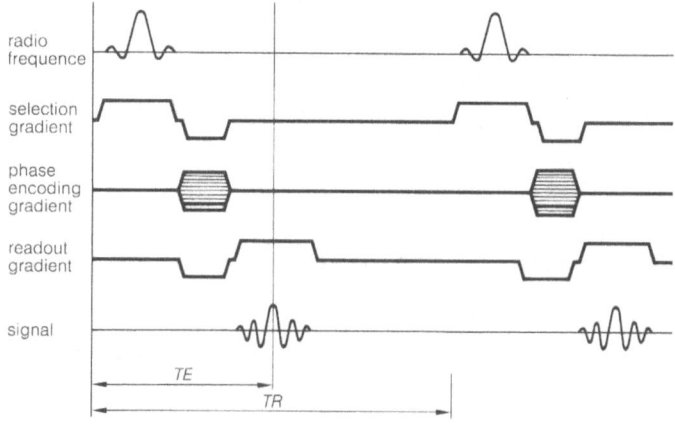

Fig. 2. Timing diagram of a gradient echo sequence

To experimentally investigate the effects of laminar flow, a phantom with cylindrical tubes of various diameters (ranging from 4 mm to 16 mm in diameter) was used. The tubes in the phantom are surrounded by a cylinder (210 mm diameter) filled with a signal producing fluid as a non-flowing reference. The viscosity of the fluid in the phantom and the tubes was the same and was chosen such that laminar flow could be maintained at flow velocities up to 50 cm/s. This viscosity was achieved by adding glycerine to the water. The T2 relaxation time of the fluid was adjusted to approximately 450 ms by adding copper nitrate to the solution.

The tubes in the phantom were connected to a pump which maintained a constant flow velocity. The volume flow rate through each tube could be adjusted individually and was regulated with calibrated flow meters.

To investigate plug flow, a phantom containing two water-filled cylinders (5 cm diameter, 40 cm length) was constructed. One of these cylinders remained stationary as a reference. The second cylinder was driven by a pneumatic system (15) which maintains a constant velocity with a time stability better than 0.1 mm/s. Because of its finite length, the moving cylinder was forced to oscillate by end switches which reversed its motion and constrained the cylinder to move with a stroke length of 10 cm. MRI measurements were gated to one direction of motion.

2.2 THEORY

a) Inflow/Outflow

The signal intensity, I, resulting from a Hahn spin echo sequence can be written roughly as:

$$I = \rho \; f(v) \; (\exp(-TE/T2))(1-\exp(-TR/T1)) \tag{1}$$

where ρ denotes the spin density within a picture element (voxel), TE and TR are the echo time and repetition time respectively, and $f(v)$ is a term which reflects the influence of flow effects. A simple example is shown in figure 3, where all spins are flowing with the same constant velocity (plug flow) through a rectangular slice. When a long TR is used (TR > SL/v), $f(v)$ should yield a linear decrease in signal as described by equation (2):

$$f(v) = 1 - v \; (TE/SL) \; (n - 1/2) \tag{2}$$

where SL is the slice thickness, v is the flow velocity and n denotes the number of the observed spin echo. For a fixed set of parameters (SL, TE, TR), the signal should vanish at the so-called cutoff velocity (11).

For non-ideal slice profiles, a better approximation of real experiments would be the correlation of the 90 degree

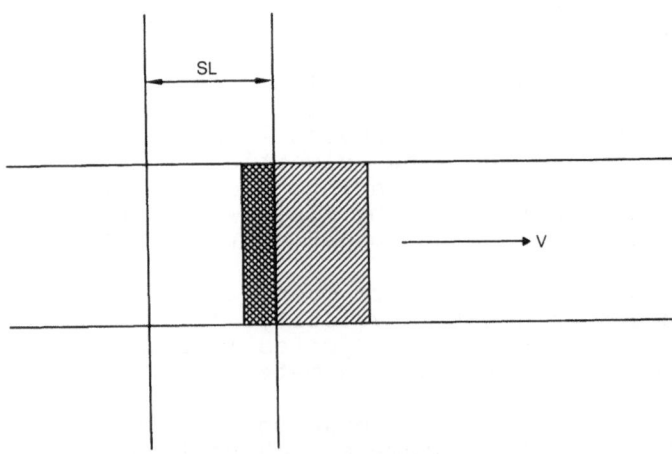

Fig. 3. Excited spins (shaded) moving out of slice (vertical lines)

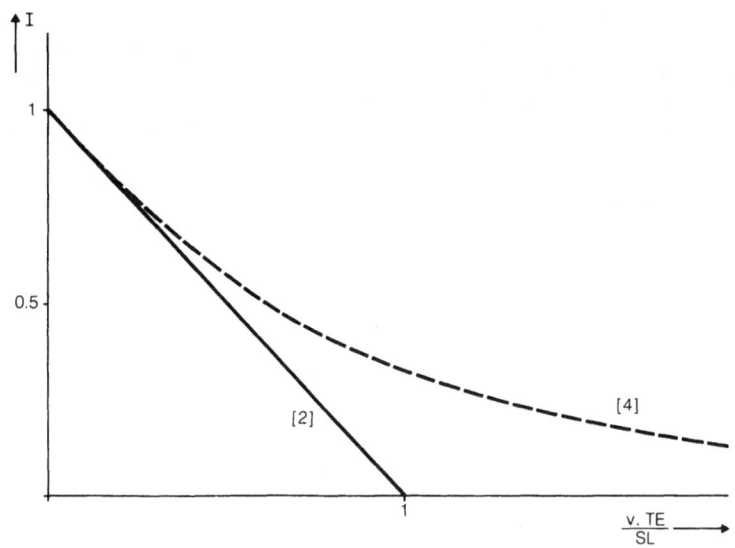

Fig. 4. Received signal intensity as spins move out of slice

excitation pulse profile with the 180 degree excitation pulse profile:

$$f(v) = (shape\ 90)*(spape\ 180) \tag{3}$$

In multi-echo experiments, we have observed a signal decay, according to equation (4), at velocities up to four times the cutoff velocity for a wide range of parameters (SL, TE, v). This is demonstrated in the next section.

$$f(v) = exp\ (-v(TE/SL)) \tag{4}$$

The lack of a cutoff velocity in real experiments clearly demonstrates that treating flow separately from selective excitation is too simplified an approach for a true flow quantification. Moreover, a simple analytical understanding of flow effects in MRI cannot be expected because of the complex interactions between flow and selective excitation techniques. However, equation (4) shows that a simple mathematical description (not an analytical understanding) of certain experiments is possible.

b) Time of Flight

The loss of signal intensity caused by outflow effects (figs. 3, 4) can be circumvented by bolus tracking. If a slice is excited by, for example, a 90 degree selective pulse, one could then apply a second RF-pulse (a 180 degree pulse in fig. 5) to the corresponding slice downstream. The advantage of MRI, compared to Singer's early experiments (previously described), is that the slice in which one tags the bolus may be located arbitrarily using electronic slice shift techniques. This capability makes it possible to sample unknown spatial velocity profiles by varying the spatial gap between the location of RF excitation pulses in a series of spin echo measurements. The signal behavior one

can expect after applying this technique is shown in figure
6. When rectangular slice profiles are assumed, this signal
behavior can be approximated by equations (5) and (6) under
laminar flow and plug flow conditions respectively. An
approximation which more closely resembles actual experimental results can be obtained by taking the actual slice
profiles into account in the calculations as shown in the
results section.

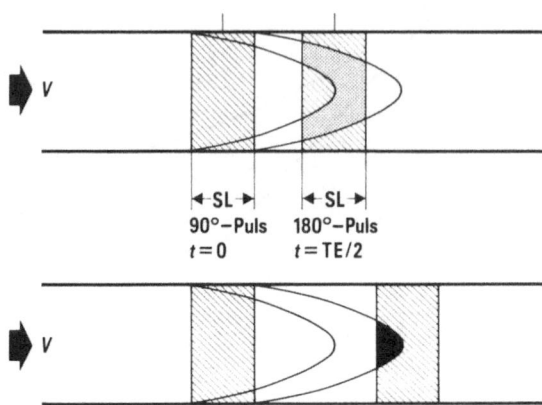

Fig. 5. Excited spins in the regions excited by the 90
degree and 180 degree pulses. Changing the distance, d, between the two RF pulses changes the
number of spins capable of emitting a signal

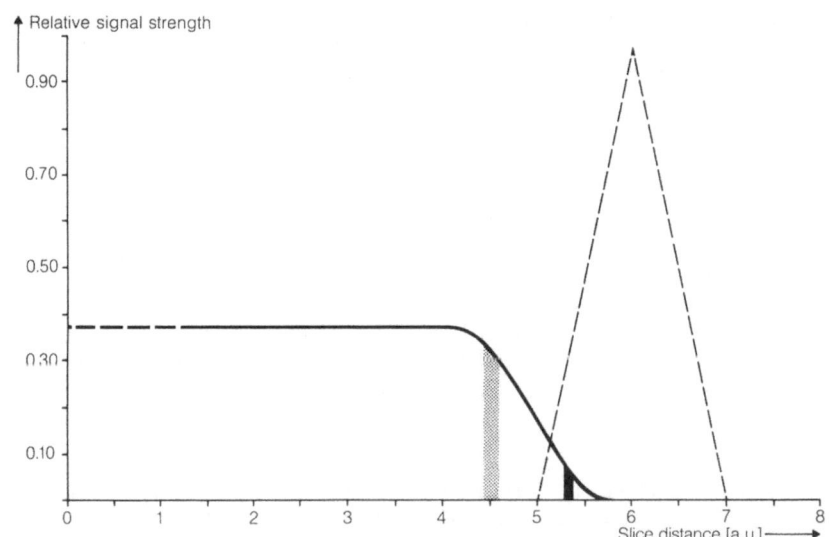

Fig. 6. Solid line – Signal received after the applications of a 180 degree pulse as a function of
slice distance, d, under laminar flow conditions.
Dashed line – Signal received after the application of a 180 degree pulse as a function of
slice distance, d, under plug flow conditions.

The following equations describe the normalized signal intensities with a variance of the slice gap (1) for plug flow (eq. (6)) and laminar flow (eq. (5)) respectively.

Laminar Flow $\hspace{6cm}$ (5)

$$I/I_o = 0 \quad ; \qquad \text{for} \quad d > v.TE + SL$$

$$I/I_o = \frac{v.TE}{2.SL} + \frac{SL}{2.v.TE} + d - (\frac{d}{SL} + \frac{d}{v.TE}).d + \frac{d^2}{2.SL.v.TE}$$
$$\text{for} \quad v.TE + SL > d > v.TE$$

$$I/I_o = \frac{SL}{2.v.TE} - \frac{v.TE}{2.SL} + d + (\frac{d}{SL} - \frac{d}{v.TE}).d - \frac{d^2}{2.SL.v.TE}$$
$$\text{for} \quad v.TE > d > v.TE - SL$$

$$I/I_o = SL / (v.TE) \qquad \text{for} \quad v.TE - SL > d$$

Plug Flow $\hspace{6cm}$ (6)

$$I/I_o = (d - SL - v.TE) / SL ; \quad \text{for} \quad v.TE - SL < d < v.TE$$

$$I/I_o = (v.TE + SL - d) / SL ; \quad \text{for} \quad v.TE < d < v.TE + SL$$

$$I/I_o = 0 \qquad ; \quad \text{for} \quad d < v.TE - SL ; \ d > v.TE + SL$$

c) Using Phase Information

Phase encoding is a fundamental principle in MRI today. The spatial location, X, of a spin ensemble is related to the phase, PHI, by the scalar product of X with the imaging gradients, G (equation (7)):

$$PHI(t) = g \int G(t) \, X \, dt \qquad (7)$$

If X is described as a function of time (for example, in fluid flow), then X must be replaced in (7) by X(t). Normally X(t) can be described as a Taylor expansion:

$$X(t) = X + vt + 1/2 \, at^2 + \dots \qquad (8)$$

In the case of constant flow (higher order motion terms are equal to zero), a combination of (7) and (8) with the pulse sequence shown in figure 1 leads to a first order approximation of a flow dependent phase shift in the MR images (16).

$$PHI(v) = constant.v \qquad (9)$$

For the Hahn spin echo sequence in figure 1, the constant in equation (9) can be expressed as:

$$constant = g \cdot Gy \, (tp^2 + tp.\Delta t) \qquad :\text{Flow parallel to y direction}$$
$$constant = 0 \qquad :\text{Flow parallel to x direction}$$

Assuming a spatial velocity distribution within a voxel, the different phase values destructively interfere (figure 7) with each other causing a reduction in signal intensity as shown in figure 8 (for laminar flow). Under laminar flow conditions, a complete cancellation of signal can be calculated (and experimentally observed) when the phase difference between the stationary spins near the surface of a cylindrical tube and the rapidly flowing spins at the center of the tube approaches 2π. This flow effect is calculated for a second echo in figure 8 and is negated by even echo rephasing phenomenon. This effect can be used to enhance the signal intensity from flowing blood.

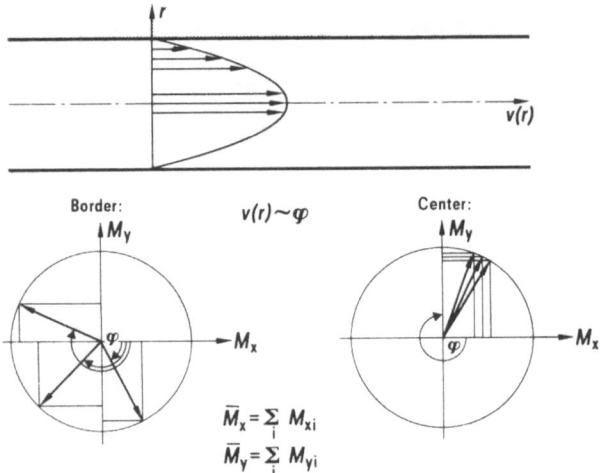

Fig. 7. Spins within a slice dephasing as a result of a local velocity distribution

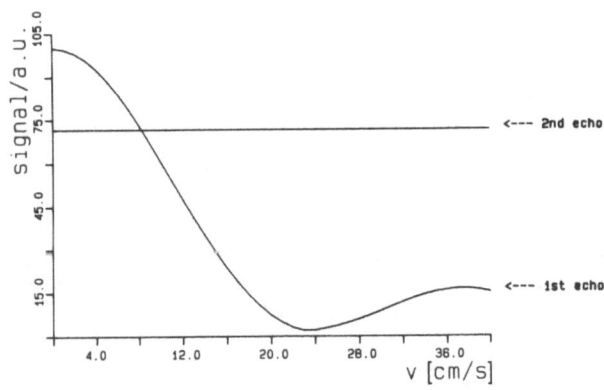

Fig. 8. Received signal as a function of flow velocity under laminar flow conditions for both first and second echos

It is obvious that pulse sequences can be tailored to enhance or suppress different flow effects. For example,

G(t) in equation (5) can be chosen so that PHI(TE) = 0 or
PHI(TE) is increased. Even echo rephasing is not the only
way to make PHI(TE) = 0. Moran (6) incorporated two negative gradient pulses symmetrically about the 180 degree RF
pulse in figure 1. These pulses do not affect stationary
tissue and can be timed so that flowing spins rephase at
the time of the first spin echo. This so-called bipolar
gradient technique is, in fact, the most promising means of
visualizing major blood vessels today.

2.3 PHANTOM MEASUREMENTS

a) Inflow/Outflow

Figure 9 shows the normal signal behaviour in 2D Hahn spin
echo experiments when a fluid is flowing through a slice.
We observe a signal increase at slow velocities due to the
inflow of unsaturated spins. This increase was first observed by Crooks (11), who called the effect "paradoxical enhancement". The effect vanishes when the repetition time,
TR, is long compared to the fluid's longitudinal relaxation
time, T1. Spins which leave the slice between the application of the two RF pulses should not contribute to the spin
echo signal and therefore will cause a decrease in signal
intensity at high velocities (outflow). The phantom experiment shown here refers to laminar flow through a 16 mm diameter tube (figure 9).

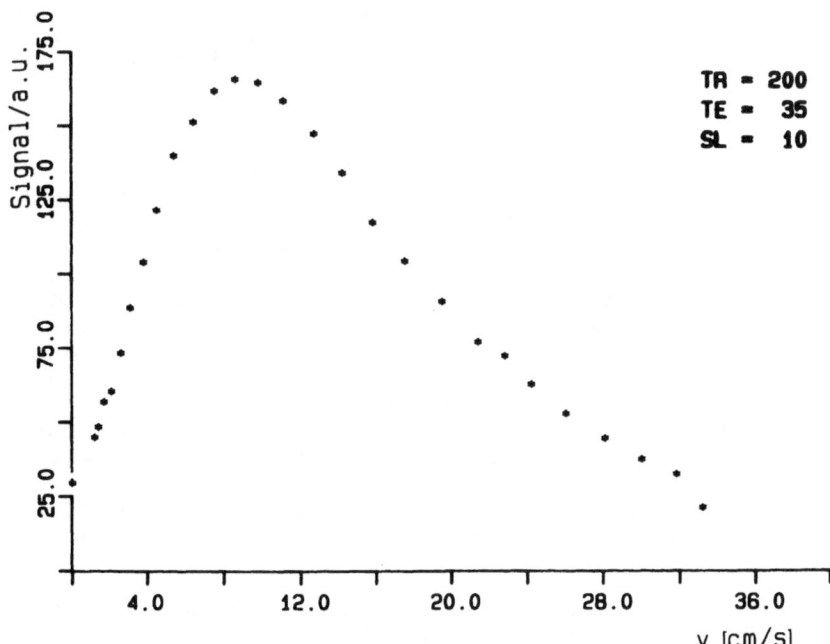

Fig. 9. Plot of received signal as a function of flow
velocity demonstrating "paradoxical enhancement"

Figure 10 shows a typical measurement of outflow by applying a CPMG multi-echo sequence to a slice oriented perpendicular to the flow direction. When flow is present, we observe an additional decrease in signal (v=0.8cm/s,
v=3.5cm/s, d=5mm) compared to the stationary fluid (v=0).
the experimental data can be fitted if one assumes an exponential f(v) as shown in equation (4).

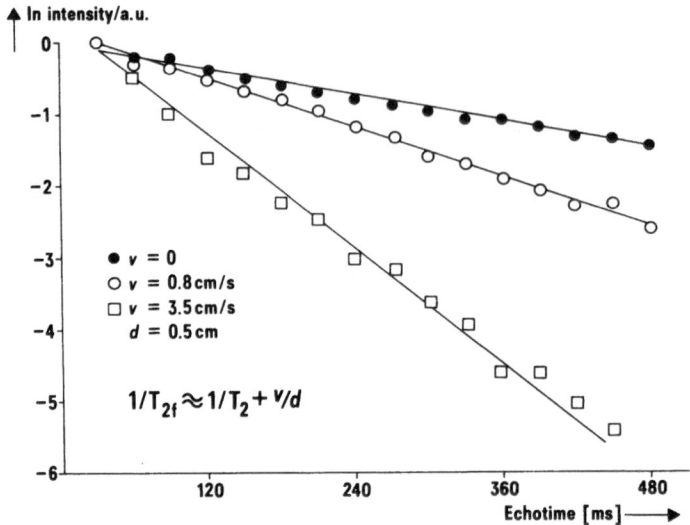

Fig. 10. Plot of received signal as a function of echo
time for a CPMG sequence showing outflow effects. Plots are fit with a line using equation (4), laminar flow

To verify that the lack of a cutoff velocity is not due to
the spatial velocity distribution within the tube, we performed plug flow experiments by moving a fluid-filled cylinder through the magnet at a constant velocity (see
MATERIALS AND METHODS). The experimental results (figure
11) confirm the lack of a cutoff velocity. Again, the data
can be fit using the assumption that equation (4) describes
the outflow of fluids for the CPMG sequence. In principle,
each two spin echos could be used to calculate T2f and
therefore the mean velocity within the pixel. This fact was
used to attempt a quantification of pulsatile flow during
one CPMG echo train (see below). It is clear, however, that
the transfer of such simple phantom studies to the complex
behavior fo pulsatile blood flow in elastic arteries will
lead to large uncertainties in the quantities being
measured.

Figure 12 demonstrates no outflow effect (compare the high
velocity branch to figure 9) in the more recently developed
gradient echo sequence, FLASH (17), but the increase in
signal can be adjusted by choosing appropriate sequence
parameters such as slice thickness. The selection of thick
slices makes this sequence sensitive to high velocities
whereas it is more sensitive to slow velocities when thin
slices are used.

224

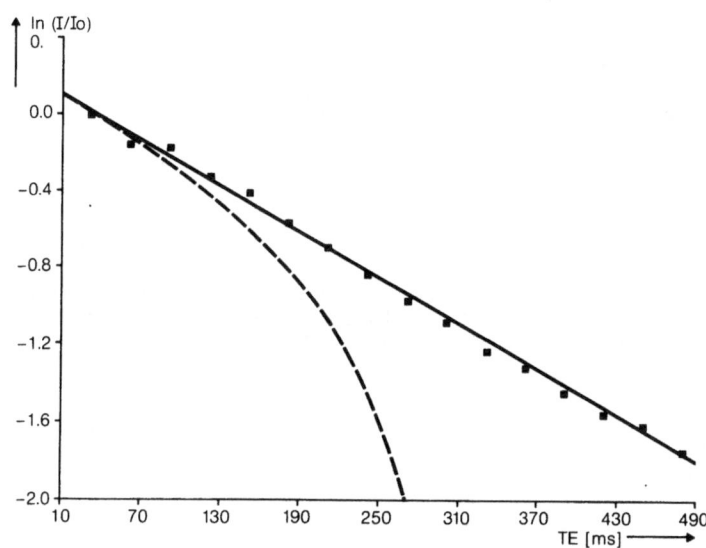

Fig. 11. Plot of the natual logarithm of I/Io as a
function of echo time for a CPMG sequence.
No cutoff velocity is seen. The dotted curve
was generated using equation (2) while the
solid line was generated using equation (4)
and provides a better approximation of the
actual data.

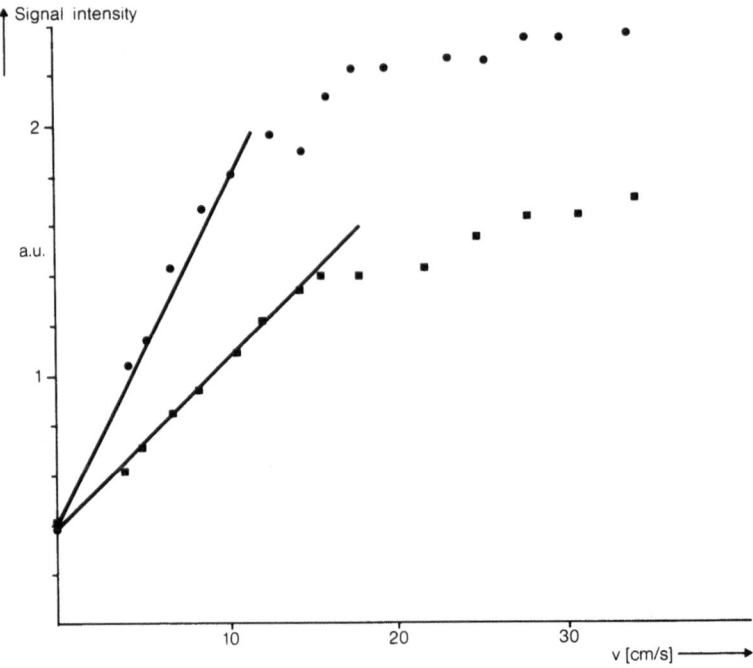

Fig. 12. Plot of received signal as a function of flow
velocity for a FLASH sequence. No outflow effects are seen. (● SL = 10 mm, ■ SL = 20 mm)

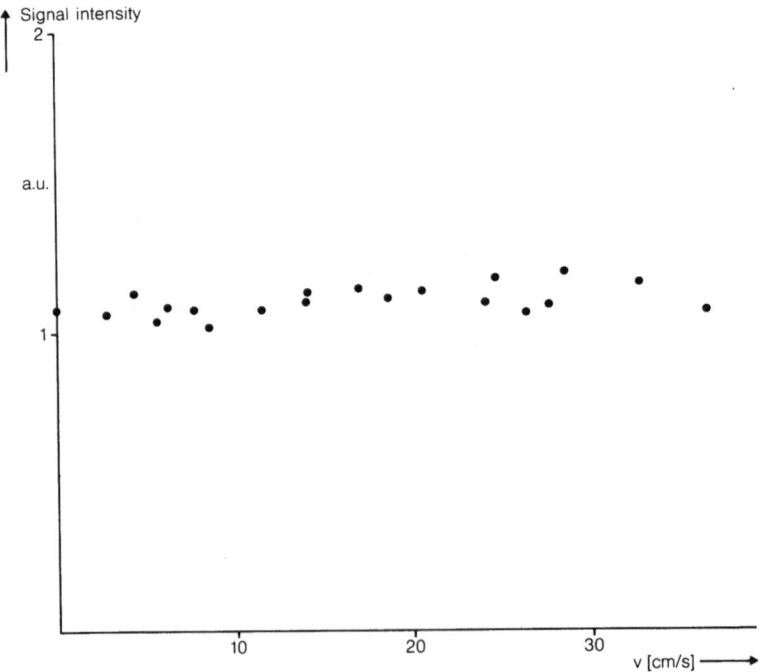

Fig. 13. Plot of received signal as a function of in
plane flow velocity. No inflow/outflow effects
are seen.

Figure 13 demonstrates the lack of transport effects when flow occurs within the image plane. This "Null"-effect can only be observed when the phase distribution (which is determined by the spatial velocity distribution within the pixel, see equation (9)) is flat. A flat phase distribution can be obtained by, for example, choosing short echo times in gradient echo sequences, using a plug flow phantom, or cutting thin slices in the center region of blood vessels where the spatial velocity distribution is normally flat.

b) Time of Flight

Laminar flow was quantitatively investigated in MR images using the series of Hahn spin echo experiments described in figure 14. The slice excited by a 90 degree pulse changes its geometry over time (before the application of a 180 degree pulse) because of the velocity distribution of spins within the initial slice (parabolic in this case).
In the case of laminar flow, and assuming rectangular slice profiles, we expect a signal behavior according to equation (5). The deviation between experimental and theoretical results at low d values is due to the overlap of the non-ideal excitation profiles. This effect can be more clearly studied in plug flow experiments. Figure 15 shows the relationship between the measured signal and the slice gap, d, at a constant plug flow velocity of 6.6 cm/s. Ideal rectangular slice profiles lead to the solid line. The dotted line represents a better approximation of the measured data and was obtained by calculating the autocorrelation function of the measured slice profile.
The experimental data thus show that it is possible to reconstruct unknown spatial velocity distributions in flowing fluids. This result has an interesting impact not only for medical applications but also for technical applications because the fluid flow properties are not influenced by the measurement technique (disturbing the nuclear spin system).

c) Phase Effects

Equation (7) was tested with the basic Hahn spin echo sequence applied under plug flow conditions. The results are plotted in figure 17 for velocities between 0 and 5 mm/s. The solid line corresponds to equation (9), with the flow direction parallel to the readout gradient of figure 1. This quantitative agreement demonstrates that flow quantification can be performed by phase analysis.
The influence of spatial flow distributions and pulsatile flow make in vivo quantification much more difficult with this technique. The phase shift due to constant flow can be considerably enhanced by applying additional pulsed magnetic gradient fields symmetrically about the 180 degree pulse of a Hahn spin echo experiment. This pulse sequence is well known from classical NMR diffusion experiments and was proposed by Stejskal and Tanner about 25 years ago (17).

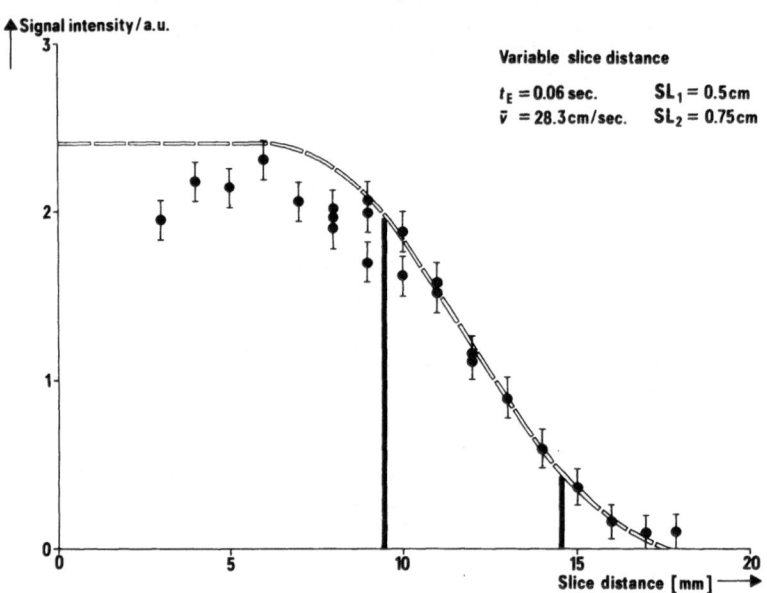

Fig. 14. Plot of received signal as a function of the
distance, d, between the application of the
90 degree pulse and the 180 degree pulse in a
Hahn spin echo sequence. Plot is fit using
equation (5).

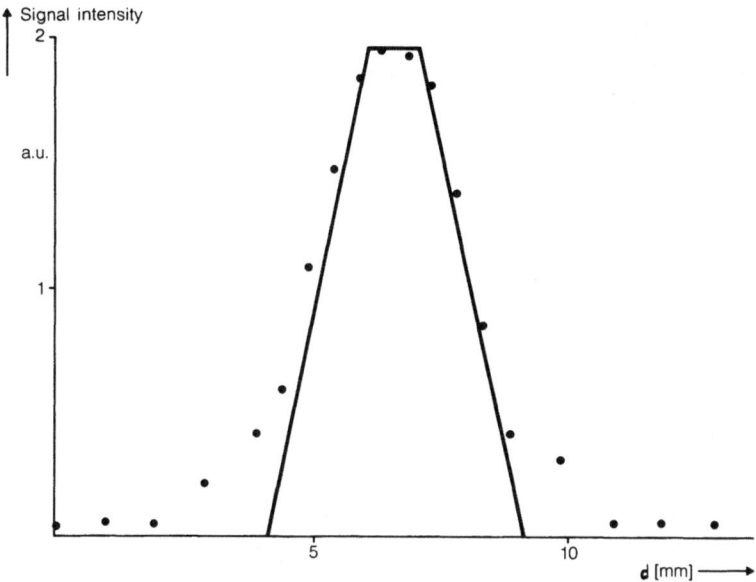

Fig. 15. Plot of received signal as a function of d.
The solid line shows a first order approximation of the signal response based on ideal
slice profiles.

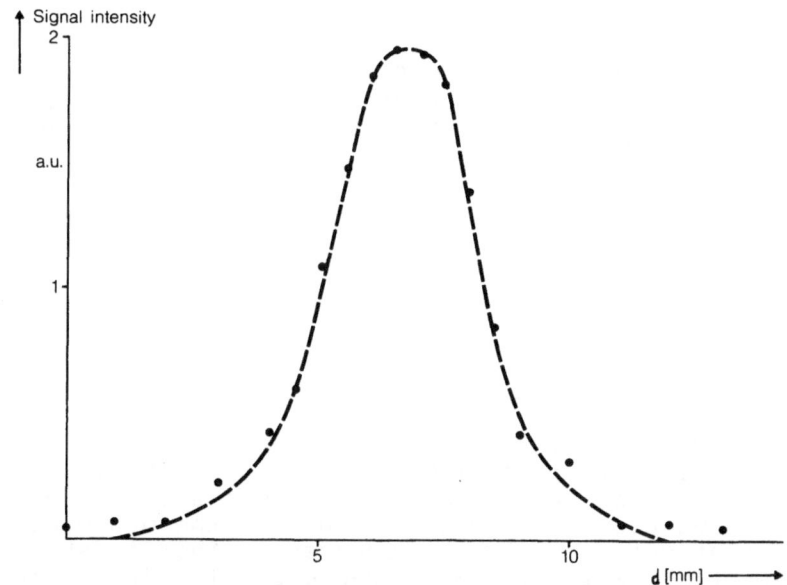

Fig. 16. Plot of received signal as a function of d.
The dashed line shows an approximation of
the signal response using the slice profile
autocorrelation function.

Figure 18 again shows the measured phase difference between the moving and stationary cylinders in the plug flow phantom versus the velocity of the moving cylinder. Note that the constant in equation (9) is now larger because the additional gradient fields lead to a more sensitive measurement of slow flow.

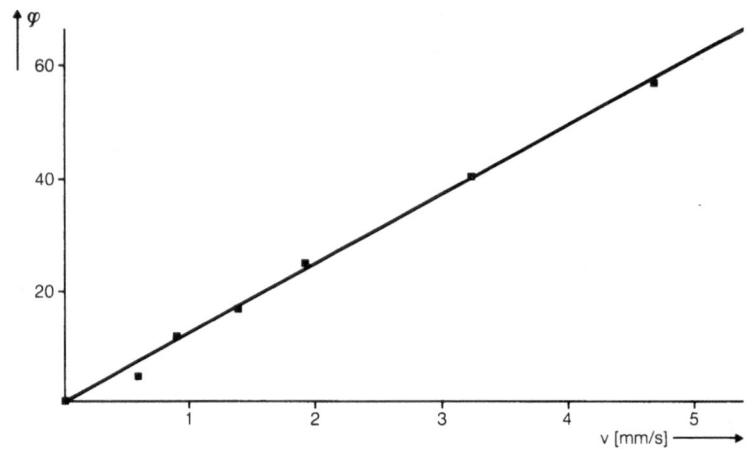

Fig. 17. Plot of phase difference between moving and stationary cylinders in a plug flow phantom as a function of flow velocity. The data are fit using equation (9).

Figure 19 demonstrates what can be achieved using a pulse sequence which was proposed by Packer (2). The lowest velocity in this plot corresponds to 50 μm/s and the measured phase can be clearly discriminated from the measured phase in the stationary cylinder. Compared to the Stejskal/Tanner experiment, Packer's pulse sequence results in a strong reduction of diffusion effects and therefore allows a separate measurement of slow coherent flow. In contrast, the signal in the classical Stejskal/Tanner method is influenced by both diffusion and coherent motion.

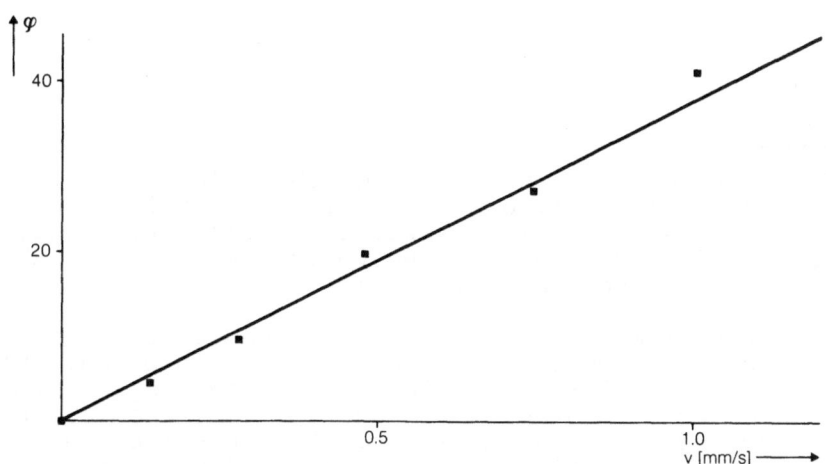

Fig. 18. Plot of phase difference between moving and
stationary cylinders in plug flow phantom as a
function of flow velocity. Flow velocities are
lower than those in fig. 17 and the data are
again fit using equation (9)

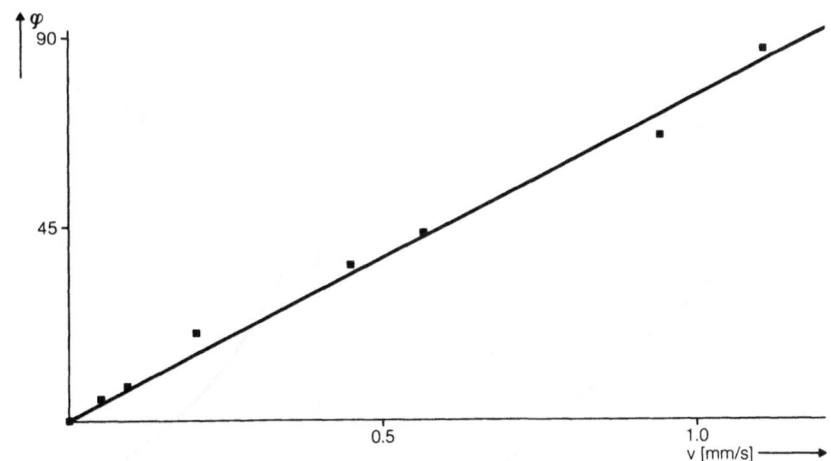

Fig. 19. Plot of phase difference between moving and
stationary cylinders in plug flow phantom
as a function of flow velocity. Data acquired using the Packer sequence and are
fit using equation (9)

These results lead one to strongly consider the possibility of measuring microcirculation of blood with these techniques. However, as we will discuss later in section 4, various other technical problems make this measurement impossible to perform at this time.

d) Blood

In previous sections, we have seen that flow quantification is possible with MR imaging systems at least in flow phantoms containing simple Newtonian fluids. With this foundation firmly in place, one would like to apply those same techniques to human blood flow.

A basic requirement for some of the techniques previously discussed is a simple relaxation behavior in blood. This situation is simplified considerably in MR imaging, where we look at water protons which undergo fast chemical exchange during the MR examination. The concept of fast chemical exchange of spin phases (figure 20) was introduced by Zimmermann and Brittin (18) and is explained in detail in a previous chapter (fast exchange means that the residence time within one phase is short compared to the relaxation times T1 and T2).

SLOW EXCHANGE

$$M(t) = \sum_{i=1}^{n} P_i \, e^{-t/Z_i}$$

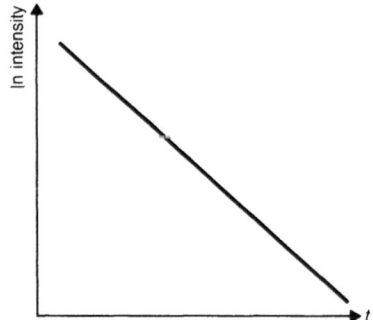

FAST EXCHANGE

$$M(t) = \exp\left[-t/\left(\sum_{i=l}^{n} P_i/T_i\right)\right]$$

$$1/T = \sum_{i=l}^{n} P_i/T_i$$

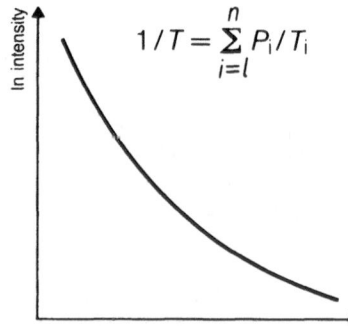

Fig. 20. Graph of the natural log of received signal intensity as a function of time, t. The graph and equation on the left represent the results which might be obtained from a material undergoing slow exchange. The graph and equation on the right represent the results which might be obtained from a material undergoing fast exchange.

Blood can be treated roughly as an aqueous protein solution (19). In this solution, proton spins within a hydration shell around protein molecules relax faster than spins tumbling in the bulk water. When fast exchange between different spin phases (e.g., intracellular water, extracellular water tightly bound in a hydration shell, and free bulk water) occurs, we observe a monoexponential relaxation behavior.

Figures 21 and 22 show that, based on its transverse and longitudinal relaxation times and at least in large vessels, blood can be treated as a fast exchange material for flow studies. Figure 21 shows the semi-logarithmic plot of signal intensity versus echo time in CPMG experiments at 42 MHz and 270 MHz. Note that the transverse relaxation time gets shorter with higher field strength as first mentioned by Thulborn, et al. in 1982 (21).

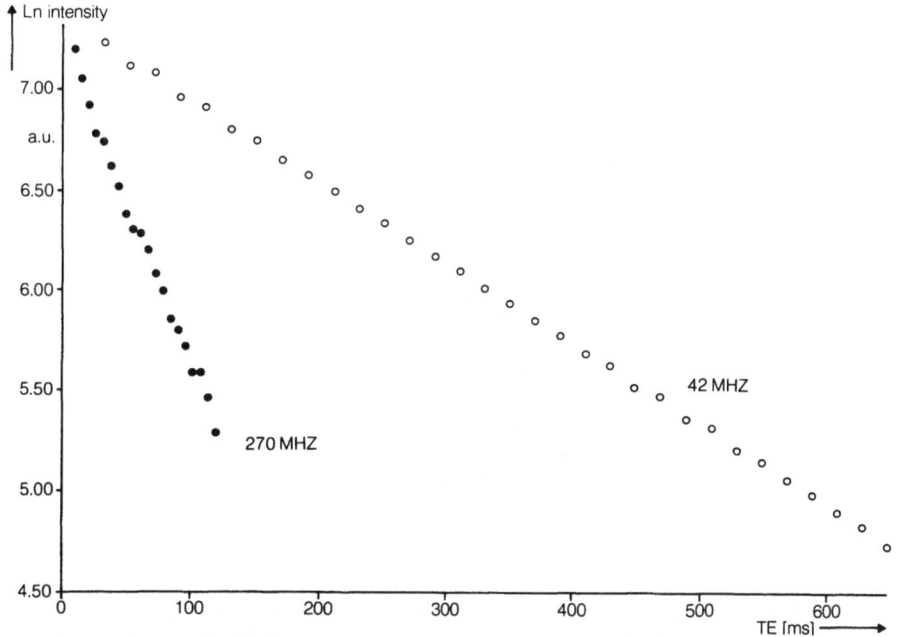

Fig. 21. Natual log of received signal versus echo time in CPMG experiments at 42 MHz and 270 MHz. Data demonstrate a linear behavior in this type of plot, characteristic of a material undergoing fast chemical exchange.

Figure 22 shows the results of a saturation recovery experiment performed with a 1.0 Tesla MRI unit (42 MHz). The solid line corresponds to a monoexponential T1 of 1.15 sec. In large vessels, blood can be treated as a homogeneous Newtonian fluid. One must be careful, however, with the interpretation of measured data from regions where the vessel diameter increases in size. The relaxation times and phase information of flowing blood can be altered in some way by the increasing influence of surface effects, and the fact that blood can no longer be treated as a homogeneous fluid.

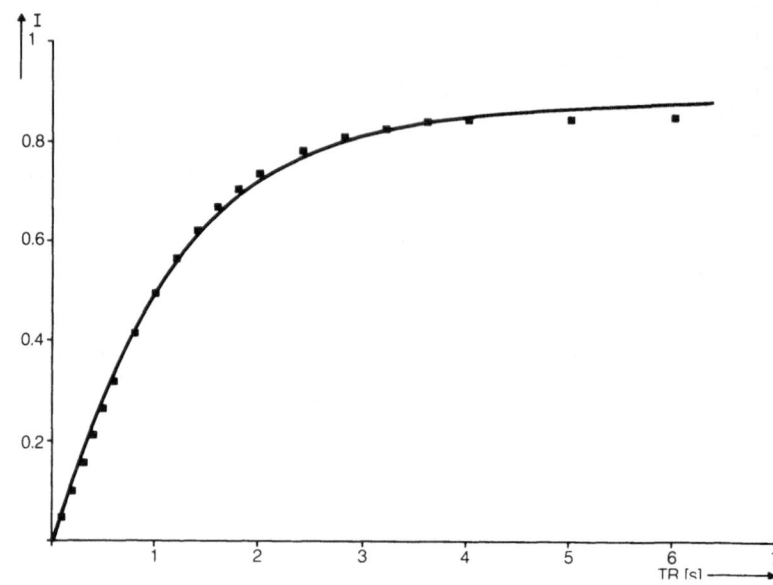

Fig. 22. Plot of signal intensity as a function of
repetition time. Data are fit using a
monoexponential curve with a T1 of 1.15 sec

3. APPLICATIONS TO MAJOR VESSELS

There are some fundamental differences between the phantom
experiments discussed in the previous section and in vivo
applications. Pulsatile blood flow through elastic vessels
cannot be treated in an analytical manner, even in normal
patients. Furthermore, many diseases lead to an even more
complicated (spatial and temporal) blood flow behavior in
the major vessels. One possible strategy to test different
MRI methods in patients is to compare these methods to currently accepted gold standards in vascular studies. One
example might be Doppler ultrasound studies in the carotid
arteries.

Despite the complex underlying physics of human blood flow
and MRI, a reasonable agreement between MRI methods and
Doppler ultrasound studies can be achieved. Although both
time of flight and phase information can also be used with
MRI, we chose the outflow analysis mentioned above as the
method to use in our attempts to expand the application of
flow techniques from phantom studies to in vivo studies.

We used ECG gated multi-echo sequences applied to transverse slices through the vessel (12). The advantage of the
multi-echo sequence compared to the single echo experiment
is that one obtains the time resolved pulsatile blood flow
within one measurement. The signal intensity over time
(taken from images at increasing echo times) will be modulated in some way if the velocity of the fluid flow changes
during the application of the CPMG echo train. If this
change in signal can be described by equation (4), velocity
information can be extracted by simply analyzing the slope
of the signal intensity versus echo time data on a semi-logarithmic plot (figure 23). If one also knows the cross

sectional area, A, of the vessel, the volumetric flow in ml/min can be calculated by multiplying the integral of velocities over one R-R interval by A (equation 10).

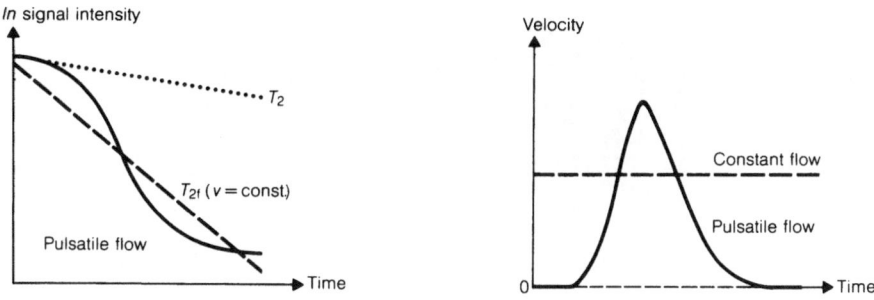

$$\text{Flow} = A \cdot \int_{R\text{-wave}}^{\text{next R-wave}} v(t).\,dt \qquad (10)$$

Fig. 23. Semilogarithmic T2 graphs (left) for samples under varying conditions. The dotted line represents the behavior of a stationary fluid and demonstrates the normal T2 behavior. The dashed line represents the behavior of a fluid flowing with a constant velocity. This graph demonstrates a slightly different T2 behavior based on the apparent T2, T2f. The solid line represents the modulated T2 behavior of a fluid undergoing pulsatile flow. The graph on the right shows the velocity information which was extracted from the T2 curves on the left.

Figures 24 and 25 demonstrate an example of the measurement of arterial blood flow in the descending aorta. The transverse slice can be first defined by a coronal scout view and is approximately 5 cm thick in this example. An ECG gated CPMG sequence is applied to the chosen slice. A region of interest can be chosen in any of the images obtained with the CPMG experiment (figure 25) and a computer program used to analyze the signal decay in this ROI in all spin echos. The result, based on the assumption that the simple equation (4) of the previous section is correct, is plotted on the MR image. The exact procedure is described elsewhere (12).

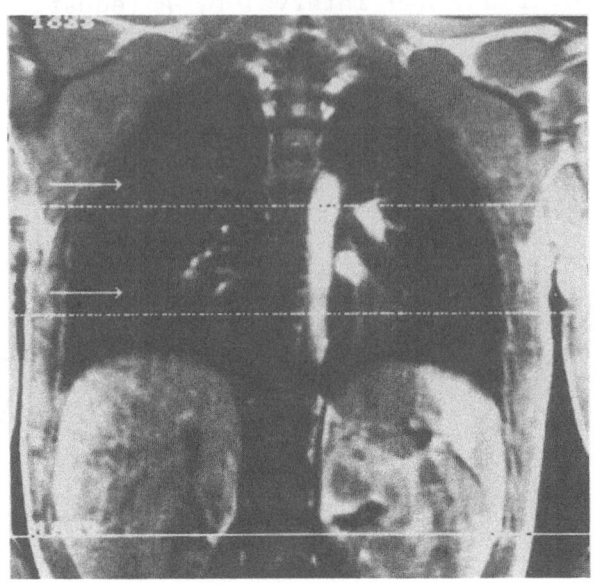

Fig. 24. Coronal scout view used to define the place-
ment of the slice used for evaluation purposes

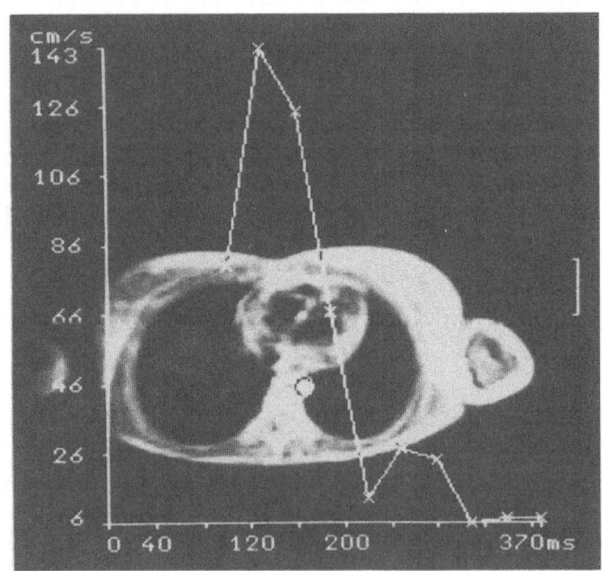

Fig. 25. Transverse slice through the descending
aorta. The superimposed graph shows the flow
velocity as a function of time within the
roughly circular ROI in the image

A comparison of the peak flow velocities obtained in the center region of 30 carotid arteries was performed between the MRI technique previously described and pulsed Doppler ultrasound (22). The result is plotted in figure 26.

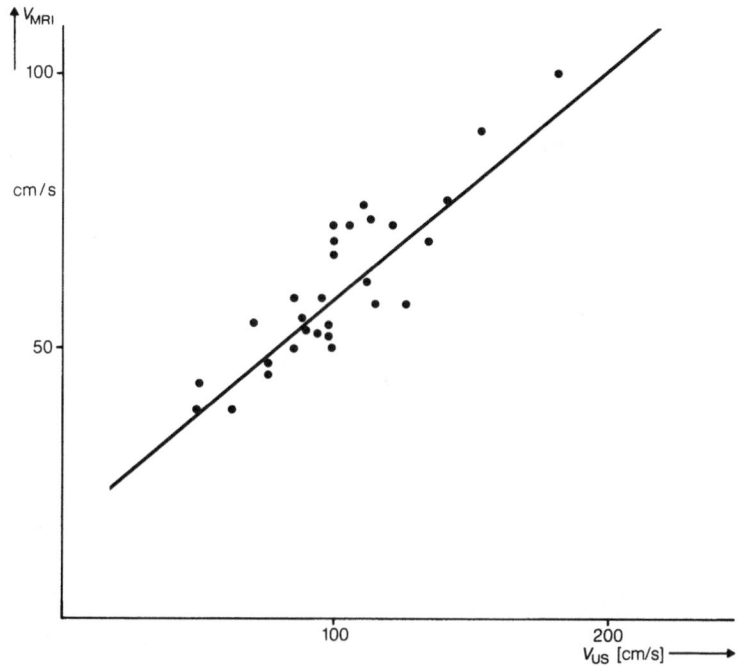

Fig. 26. Correlation of the flow velocity measured with MRI techniques (vertical axis) and with Doppler ultrasound (horizontal axis). A correlation factor of 0.84 is obtained for these data

The correlation factor of 0.84 is quite high considering that blood flow velocity was measured distal to stenosis in 15 vessels. This correlation demonstrates that, in principle, MRI is capable of noninvasively quantifying human blood flow in major vessels. A clinical example is shown in figure 27.

In this example, aortic backflow, demonstrated by negative velocity values in the plot, is observed in early diastole. The examination required 4 minutes and the MR result corresponds to backflow seen during cardiac catheterization of this patient suffering from an aortic insufficiency (23).

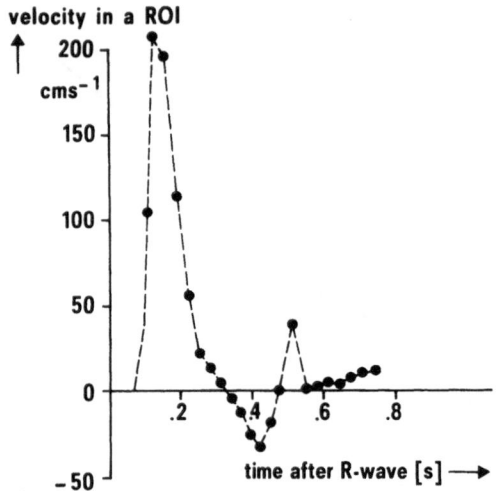

Fig. 27: Plot of flow velocity as a function of time after the peak of the R wave. The negative velocity, indicated by the shaded region, is indicative of backflow in the region of interest

An additional example is shown in figure 28, where the pulsatile blood flow of a patient with cardiomyopathy is compared to the same measurement in a normal young adult (12). The peak velocity in the patient is significantly lower than in the normal subject. For comparison, the nearly constant flow in the vena cava of the patient can also be calculated from the same MR data set.

238

In another study, blood flow through the descending aorta was measured in 8 patients with MRI and compared to the cardiac output estimated by thermodilution measurements during cardiac catheterizations (figure 29). As expected, because of blood flowing out to various branches before reaching the descending aorta, the MR values obtained are consistently lower than the values obtained by thermodilution. The correlation coefficient r=0.79 demonstrates

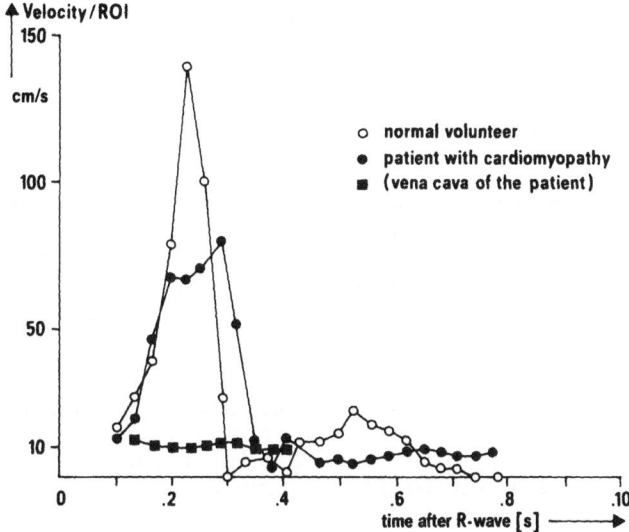

Fig. 28. Plot of flow veloc.   as a function of time after the R wave. Tr velocity profile of the subject with cardiomy :athy (black circles) demonstrates a significantly reduced peak velocity behavior in comparison to the normal subject (white circles). Almost constant flow through the vena cava can also be seen in the subject with cardiomyopathy (black squares)

that the accuracy of MRI in quantifying flow is quite poor at the moment. One major problem in this field is outlining the vessel wall. We estimate an error of at least 20 % in defining the cross sectional area of the vessel. Nevertheless, one should keep in mind that MRI provides the possibility of noninvasively estimating blood flow in large vessels.

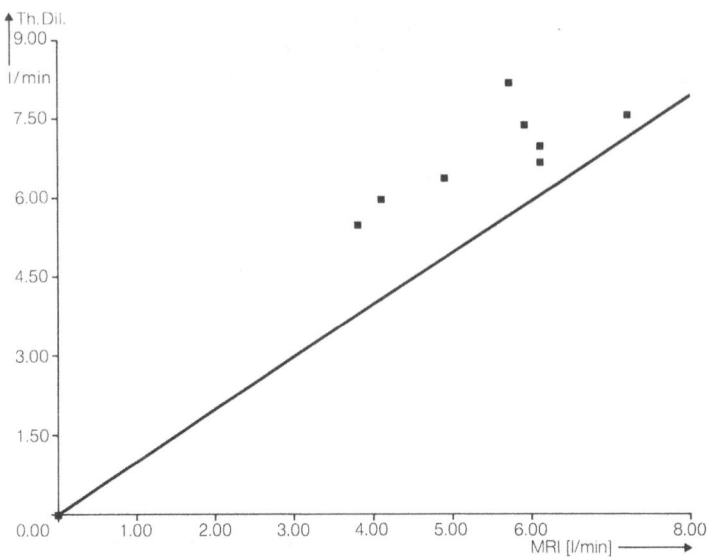

Fig. 29. Correlation of MRI flow data with thermodilution data obtained in the descending aorta of
8 subjects. The straight line represents ideal
correlation between the two types of data

Currently, efforts in MR are primarily directed toward MR
angiography. Two examples of MR arteriograms are shown in
figures 30 and 31 in which the gradient motion refocussing
technique (6) was used to image the vessels.

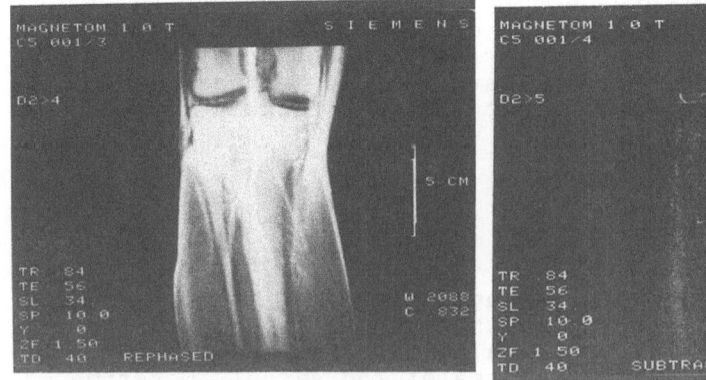

Fig. 30. Hahn spin echo sequence using bipolar gradient
rephasing. The image on the left clearly shows
signal enhancement in the femoral artery. The
image on the right was obtained by subtracting
a non-rephased image from the image on the
left so than, in principle, only the enhanced
flow signal remained

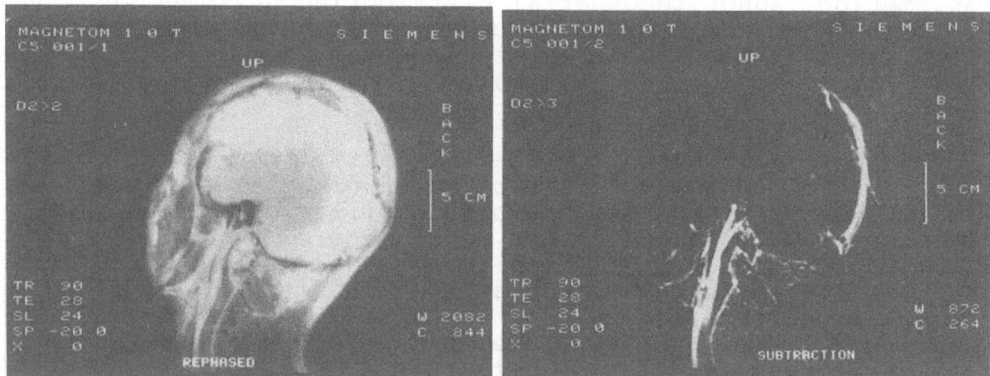

Fig. 31. Hahn spin echo sequence using bipolar gradient rephasing. Once again, the image on the left clearly shows signal enhancement in the carotid and sinus arteries. The image on the right was obtained by subtracting a non-rephased image from the image on the left so than, in principle, only the enhanced flow signal remained.

4. CEREBRAL BLOOD FLOW

We have seen in previous sections that methods well known from classical NMR studies can be modified to suit the special properties of MR imagers, and can now be applied to clinical studies to evaluate the potential of measuring blood flow in MR imagers. There has been some success in studying larger vessels, but to date there have been no conclusive data supporting monitoring of tissue blood flow in vivo. There are several reasons for this. First, one must consider that the flow velocity in these small vessels is approximately 3 orders of magnitude slower than in the large vessels studied up until now. Therefore, one has to be able to accurately measure velocities in the sub-mm/s range. In addition, one must separate the influence of this slow motion from any other motion, such as movement of organs due to respiration, or pulsation of the brain due to pulsatile flow.

Perhaps the technical demands of precise respiratory gating can be fulfilled in the future, but other physical problems also remain to be solved. For example, phase sensitive techniques like the Packer sequence mentioned above have been used to measure plug flow of 0.1 mm/s in one direction (see section 2). If one assumes that two vessels have anti-parallel flow of the same magnitude (v, -v), then when both vessels are small and are located within the same pixel, the phase information would suggest "no flow". The spatial resolution which can be achieved with commercial MR imagers is an order of magnitude worse than the diameter of arterioles and venules, and more than an order of magnitude worse than the diameter of capillaries. Because of this, assumptions about the spatial distribution of small vessels must be made to allow a calculation of average flow velocities within a pixel containing many small vessels. Even with this information, however, no indication of the volumetric flow rate can be obtained.

Furthermore, the effect on the transverse (and longitudinal) relaxation times caused by the enormously increased surface area/bulk volume ratio in capillaries must be determined. Blood can no longer be considered a homogeneous fluid. Additional local field inhomogeneities caused by erythrocytes which have changed shape and travel single file through capillaries must be taken into account. Inflow/outflow methods as well as time of flight techniques suffer from the demands of very thin slices. Thin slices are necessary to enhance changes in signal intensity caused by slow motion. Because of slice profile effects, flow velocities lower than 1 mm/s are difficult to measure, even in phantoms.

In closing, two optimistic remarks should be mentioned. The first point is that an indirect estimation of cerebral blood flow may be performed by measuring the blood flow through the large vessels which supply the tissue in the region of interest. The second point is that technological and methodological developments in MRI will continue for at least a few more years, and new promising models are now being described.

Acknowledgement

The author gratefully acknowledges Dr. P. Ruggieri of Case Western Reserve University in Cleveland, USA, J. Hofler of SMS Iselin, USA, and Mrs. E. Rose for their help in producing this paper.

We also wish to thank Dr. A. Weikl of the Cardiology Department of the University of Erlangen and Dr. M. Seiderer of the Klinikum Grosshadern in Munich for doing the patient studies discussed.

LITERATURE

1) Stejskal, E.O., J. Chem. Phys. 43; 3579 (1965)
2) Packer, K.J., J. Mag. Res. 17; 355 (1969)
3) Carr, H., Purcell, E., Phys. Rev. 94, 630 (1954)
4) Laude, D.A., Lee, W.K., and Wilkins, C.L., J. Mag. Res. 60; 453 (1984)
5) Singer, J.R., Science 130; 1652 (1959)
6) Moran, P.R., Moran, R. A.; Magn. Res. Imag. 1, 197 (1982)
7) Dijk, P. van, J. Comp. Ass. Tom. 8, 429 (1984)
8) Wehrli, F.W., Shimakawa, A., McFall, J.R., Axel, L. et al., J. Comp. Ass. Tomgr. 5; 537 (1985)
9) Bryant, D.J., Payne, J.A., Firmin, D.N., Longmore, D.B., J. Comp. Ass. Tomgr. 8; 588 (1984)
10) Category 8: Flow Measurements 4th SMRM London, Book of Abstracts (1985)

031 01 13

11) Crooks, L.E., Mills, C.M., Davis, P.D., et al.,
   Radiology 144, 843 (1982)
12) Müller, E., Deimling, M., Reinhardt, E.R.,
   Magn. Res. Med. 3, 331 (1986)
13) Meiboom, S., Gill, D., Rev. Sci. Instr. 29, 688 (1985)
14) Graumann, R., Oppelt, A., Stetter, E.,
   Magn. Res. Med. 3, 707 (1986)
15) Prototype of Bosch Corp.
16) Deimling, M., Müller, E., Lenz, G. et al.,
   Diagn. Imag. Clin. Med. 55, 37 (1986)
17) Hasse, A., Frahm, J., Matthaei, D., Haeinke, W., and
   Merboldt, K.D., 4th SMRM, London 980 (1985)
18) Stejskal, E.O. and Tanner, J.E., J. Chem. Phys. 42; 288
   (1965)
19) Zimmermann, J.R. and Brittin, W.E., J. Chem Phys. 42,
   288 (1965)
20) Brooks, R.A., Battocletti, J.H., Sances, A. et al.,
   IEEE Trans. Biomed. Eng. BME-22; 12 (1975)
21) Thulborn, K.R., Waterton, J.C. et al.,
   Biochem. Biophy. Acta 714, 265 (1982)
22) Seiderer, M., Spengel, F., Müller, E. et al., SMRM
   1986, Montreal, 1103 (1986)
23) Weikl, A., Müller, E. and Reinhardt, E.R., SMRM 1986,
   Montreal, 366 (1986)

PERFUSION STUDIES AND FAST IMAGING

R. Turner

University of Nottingham
University Park
Nottingham, NG7   2RD

Distinctions between flow, diffusion and perfusion

Flow, diffusion, and perfusion are different terms to describe the motion of molecules, in this context usually water molecules, from one point to another. Flow and diffusion are the fondamental physical processes, which can be measured using NMR by well-established techniques. Perfusion is a related physiological quantity, which can be measured fairly accurately by surgical or radiological techniques (as described elsewhere in this volume). The problem at hand is to specify the precise relationship between these phenomena.

We speak of flow when the motions of large numbers of particles of a fluid are very similar; the velocities of all the particles within a particular volume are highly correlated. The average flow velocity vector $\underline{v}$ may be unambiguously specified. Diffusion, by contrast, describes uncorrelated, random motion. Each particle performs a "random walk", in which its average position remains the same, while the probability of finding it at some other position increases with time. Since diffusion is a consequence of the random thermal vibrations of particles, it always accompanies flow, but in many cases its effect is too small to be important. A straightforward classical argument (see, for instance Reif,1965:486-488) demonstrates that the self-diffusion constant D is related to the average distance which a particle of the fluid moves before scattering, l, and its root mean square velocity v, by the formula

$$D=lv/3 \tag{1}$$

Since the argument is entirely general, we can define an effective diffusion coefficient in any sistem of particles where motion is random and isotropic, given appropriate estimates for l and v.

Perfusion, one the other hand, is a measure of the volume of fluid F which passes through a given mass of biological tissue per unit time. It is an empirical quantity which takes no account of the detailed paths followed by elements of the fluid, but gives a good representation of the degree to which the tissue is irrigated. In what follows the perfusion is taken to be equivalent to the volumetric flow rate.

In order to understand how the transport processes in question are related to each other in a complex inhomogeneous substance such as tissue, it is essential to formulate a model which is simple but represents all the important physical features. To begin with, we distinguish two types of fluid molecular environment: blood vessels and cellular tissue. At this stage we classify all extracellular space in which flow is extremely slow, such as lymphatic vessels or cerebral ventricles, together with intracellular fluid. We also specify a characteristic volume of tissue which contributes to the NMR amplitude of a single picture element, or voxel. We may take this to be a cube with an edge of 5mm.

Consider the tissue contents of such a cube (Figure 1), assuming that it is inside a larger organ of interest. In general on this scale we will find cells of various types, capillaries, arterioles and venules, and perhaps portions of terminal branches of arteries and veins. Now let us examine the fluid motions in such a volume. In most organs, unless the volume includes a larger blood vessel, the pattern of flow is likely to be essentially isotropic, as a consequence of the random branching of the capillary network. Where highly organised capillary beds exist, as for instance in skeletal muscle tissue (Middleman 1972:117), this will not be the case, but such configurations are easily dealt with and will be discussed below. In more detail, flow takes place along the capillaries, typically with a velocity of 0.5 mm/s (Mcdonald 1974:41), and through the capillary walls into and out of the cells, with a velocity typically a factor of 1000 smaller (Middleman 1972:166). At the same time, water molecules diffuse within the blood and intracellular fluid, with a diffusion constant not very different from that of pure water (Mansfield and Morris 1982:31).

Collectively these motions cause the perfusion of the tissue volume. Since the flow velocity within the cells is very much smaller than that in the vessels, we need consider only the inflow and outflow of blood along those vessels intersected by the faces of the cube. We note that there is very little pulsatile flow in capillaries. Let the average entering flow velocity be v, the total cross-sectional area of intersection of entering vessels with the cube surfaces be A, and the total volume of tissue be V. If the tissue density is p, the perfusion can now be written as F, the volumetric flow rate:

$$F=Av/pV \qquad (2)$$

If the cube has a side of length a, we may write

$$A=6\alpha a^2 \qquad (3)$$

where α is the fraction of the area of one side of the cube which is occupied by vessels bringing blood into the volume. Since $V=a^3$,

$$F=6\alpha v/pa. \qquad (4)$$

It should be noted that perfusion will normally be scale-independent within an organ. Rearranging (4) we obtain

$$\alpha v=paF/6 \qquad (5)$$

In order to estimate the effective diffusion constant D in the volume

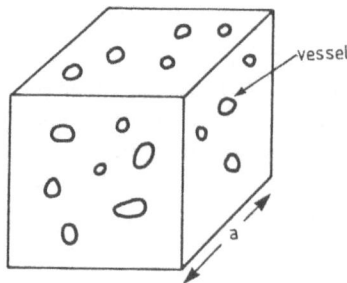

Fig. 1.   Diagram of a cubical volume element of tissue, or voxel, showing sections of blood vessels.

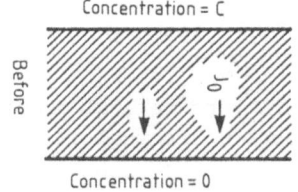

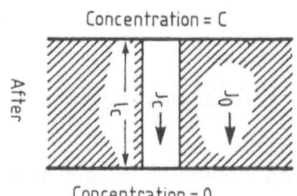

Fig. 2.   Diagram illustrating how a capillary provides a parallel path for flow af water through tissue.

element concerned we observe that D is strictly defined as the ratio of a current density to a concentration gradient. Consider the flow before, and after, a capillary is opened through the tissue (Figure 2). If Jo and J_c are the current densities in the tissue and in the capillary, and J is the resultant average current density, we may write

$$J=(1-\alpha)J_O+\alpha J_c \qquad (6)$$

On the scale of the 5 mm cube with which we are dealing, the capillaries are randomly oriented and the flow of blood is essentially isotropic. We may thus treat a section of capillary between branches as analogous to the free path before scattering of a diffusing particle. It is easy to see that the capillary acts as a shunt, in parallel with the transport caused by diffusion on a molecular scale, and hence

$$D_{eff} =(1-\alpha)D_O+\alpha D_c \qquad (7)$$

where Do is the diffusion constant in the absence of capillary flow. D_c arises from the flow in the capillaries and may now be estimated usind (1):

$$D_c =v_c l_c /3 \qquad (8)$$

where l is a typical capillary length, about 0.3 mm (Middleman, 1972:133).

For typical values of $v_c =0.5$ mm/s, $=4 \times 10^{-2}$, we find that $\alpha D_c =2 \times 10^{-9}$ m^2/s, compared with $D=2 \times 10^{-9}$ m^2/s for pure water. Since NMR can accurately measure D for water it is clear that the effect of perfusion should be easily observable. To relate the observed change in diffusion constant directly to the perfusion F we may write, using (5) and considering a cube of side l_c ,

$$v_c =p l_c F \qquad (9)$$

to obtain

$$D_{eff} =(1-\alpha)D_O+pF l_c^2 /18 \qquad (10)$$

The ratio α is much less than 1, and several drastic approximations have been made in deriving this result, so that it is best to write simply

$$D_{eff} \cong D_O+cF \qquad (11)$$

where c is a constant of order 10^{-6} kg/m, depending only on structural details of the microcirculation. This implies that the perfusion F can be inferred from measured values of D, as

$$F \cong (D-D_O)/c. \qquad (12)$$

Similar arguments have recently been put forward by Le Bihan et al (1986), and Ahn et al (1986).

In pratice the calibration constant c and the intrinsic diffusion constant Do may be obtained using animal models and biopsy tissue, since it would present ethical problems to vary the perfusion of a vital organ (such as the brain) of a live human subject sufficiently for these parameters to be measured directly.

We return to the case of skeletal muscle mentioned earlier, where the capillary bed is ordered and anisotropic. Typically the bed consists of nearly straight and parallel sections, in which the blood flows mostly in one direction (Figure 3). Here is possible to dispense with the concept of an effective diffusion constant. We need only specify an average flow velocity u along the capillary direction, which is given by the relation

$$u=paF \qquad (13)$$

Normal NMR imaging methods for measuring flow are adequate to measure u, from which F can be derived as

$$F=u/pa \qquad (14)$$

Typically, u is about 5mm/s, requiring a fairly sensitive method of measuring flow.

Use of diffusion gradients

As we have seen in previous chapters, the key to the measurement of flow and diffusion using NMR is the application of a magnetic field gradient (Taylor and Bushell, 1985). This causes the nuclear spins at different sites in the sample to precess at different rates. Reversal of the gradient, (or alternatively the spin polarity using a 180° pulse), causes the spins to reverse their precession and thus to re-focus. A simple translational flow of the spins with a component parallel to the gradient direction results in a net phase shift of the resulting radio-frequency signal, while isotropic diffusion of the spins away from their original sites causes an overall loss of coherence, manifested as a reduction in signal amplitude:

$$M(2t)=M(0)exp(-2t/T_2)exp(i\gamma \underline{G}.\underline{v}t^2)exp(-2\gamma^2 t^3 G^2 D/3) \qquad (15)$$

Here the gradient G has been kept on for a time t, then reversed and maintained for a further time t. M(t) represents the net transverse magnetization of the sample, and γ is the magnetogyric ratio.

Assuming that M can be mapped pixel by pixel by some imaging technique, all that is necessary to produce diffusion images is to divide the data taken for some specific value of G by the data for G=0, for constant delay 2t. This eliminates variations in proton density, flow (where this is uniform from one experiment to the next) and T. What remains is a map of the spatial variation of $exp(-2\gamma^2 t^3 G^2 D/3)$. Figure 4 shows the characteristic variation with 2t of the moduli of the exponents of the three final terms in (15), for realistic values of T_2 (100) ms), G (10 mT/m), D ($2x10^{-9}$ m^2/s) and v (0.5 mm/s). It is clear that once 2t is larger than 100 ms the effects of diffusion become very important compared with transverse relaxtion. In voxels without a major blood vessel the average flow velocity is zero and here only the relaxation and diffusion terms contribute.

The problem of motional artifact

A major problem arises, for most NMR imaging techniques, from the fact that the data from a large number of rf pulses, spaced over several minutes, must be used to reconstruct a flow or diffusion image. The application of a diffusion gradient gives the imaging experiment an exquisite sensitivity to motion of any kind whatever. This means that small involuntary motions of the part of the body being imaged during the sequence of up to 256 rf pulses cause large variations in the signal, which are unlikely to be coherent from pulse to pulse. Even when (for instance) the head is clamped, motions of the eyes, tongue and soft sinuses during breathing can

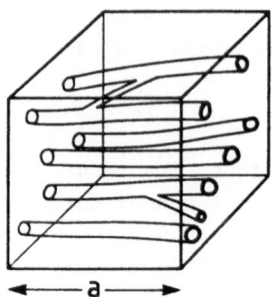

Fig. 3. Diagram of a voxel of tissue in skeletal muscle, showing the order
ed capillary bed.

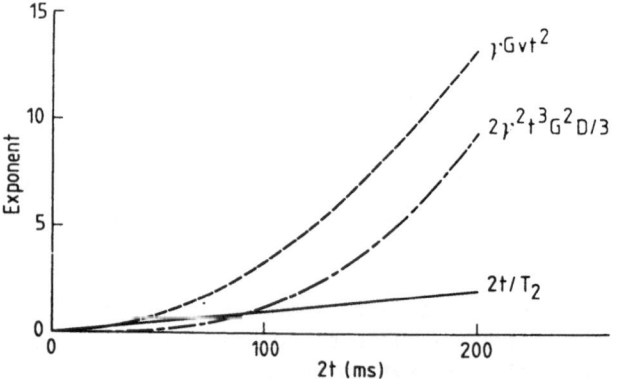

Fig. 4. Comparison of the time dependence of the exponents in the three
terms in equation (15) which express the variation of the magnetization:
--- relaxation term, - - velocity term, - · - diffusion term.

give spurious signals. The result is extremely severe motion artifact whenever sufficiently large gradients are applied to make small flow velocities
observable.

The corollary to this observation is that for successful diffusion
imaging in vivo the fastest possible imaging techniques must be employed.
For cerebral perfusion studies this has an additional advantage, as we shall
see in the final section. There are two methods presently available for
very rapid image acquisition: FLASH imaging, first described in detail by
Frahm et al (1985), but already foreshadowed by Mansfield (1982:75-76); and
Echo-Planar Imaging (EPI), described by Mansfield (1977).

Signal : noise considerations

It should be pointed out that the NMR signal remaining after the application of the diffusion gradients arises almost entirely from extracellular
water. Transverse relaxation of other tissue protons generally occurs in times much less than 100 ms. Furthermore, when imaging at high static magnetic
fields (greater than 0.5 T) the shortness of T_2 even for extracellular water
could make it difficult to observe any effect of diffusion.

A reasonable estimate of the effect of perfusion on the size of the observed NMR signal can be obtained by reference to equation (8). When perfusion is present the effective diffusion constant may be increased by 100%.
Using the typical values given above the NMR signal will thus be reduced by
a factor of exp $(-2\gamma^2 t^3 G^2 D/3)$, about 3.2. Provided that after a time delay
of 100 ms prior to data acquisition, without applied diffusion gradients,
the signal: noise ratio of the reconstructed image pixel is still as high
as 10, there should be no difficulty in observing the presence of perfusion.

Fast imaging techniques

The phase-space trajectory approach to NMR imaging

To make easy and meaningful comparisons between different methods of
producing NMR images, it is helpful to introduce the phase-space trajectory
concept of Ljunggren (1983). We start with the Larmor equation

$$\omega = \gamma B \qquad (16)$$

where γ is the magnetogyric ratio and ω is the precessional frequency of
the spins. We now decompose B into a steady part B_0 and a variable part
arising from a magnetic field gradient $\underline{G}$, and move into a frame of reference rotating at a frequency $\omega_0 = \gamma B_0$. The residual precession of the spins
at a point $\underline{r}$ in the sample may be described as follows:

$$d\,\theta_i/dt = \gamma \underline{G} \cdot \underline{r_i} \qquad (17)$$

where θ_i is the total angle (in the rotating frame) through which the spins
at $\underline{r_i}$ have precessed at time t.

Now the quantity measured in an NMR experiment is essentially the projection of the total transverse spin magnetization in a given direction (the
x ' direction in the rotating frame). Let the original transverse magnetization at time t=0 be $M(0)=cf(\underline{r})$, is the density of nuclear spins and c
is a constant. Then at time t we will measure a magnetization (neglecting
relaxation)

$$M(t) = \sum_i M(0) \cos\,\theta_i = c \int f(\underline{r}) \cos\theta' \underline{dr} \qquad (18)$$

This can be straightforwardly re-written

$$M(t)= \mathrm{Re} \left\{ c \int f(\underline{r}) \exp \ (i\theta) \ \underline{dr} \right\} \tag{19}$$

where Re indicates the real part of what follows. Integrating (16) we obtain

$$\theta = \gamma \int_{0}^{t} \underline{G} \cdot \underline{r} \ dt' \tag{20}$$

Substituting into (18), and writing

$$\underline{k}(t)= \gamma \int_{0}^{t} \underline{G}(t') dt' \tag{21}$$

we obtain finally

$$M(t)= \mathrm{Re} \left\{ c \int f(\underline{r}) \exp \ (i\underline{k} \cdot \underline{r}) \ \underline{dr} \right\} \tag{22}$$

This can be recognised at once as the Fourier transform of $f(\underline{r})$, the spin density, with respect to a variable $\underline{k}$, which is the time integral of the magnetic field gradient. We refer to the space in which the vector $\underline{k}$ is defined as phase space. Writing M (t)=M($\underline{k}$), corresponds to a sampling in this space of the Fourier transform of the spin density, along a trajectory defined by the time variation of $\underline{k}$.

For an NMR imaging technique to be adequate, the sampling must be sufficiently uniform and extensive for the inverse Fourier transform of M($\underline{k}$) back into real space to give a good representation of the original spin density. All established imaging techniques correspond to different trajectories in phase space; in some, the whole of phase space is sampled after one rf pulse, while in most a number of separate transects of phase space are necessary, corresponding to a sequence of rf pulses associated with varying gradients.The disadvantage of such latter methods is that in general it is necessary to wait some time between successive rf pulses, in order to al low the nuclear magnetization to relax back into the the z direction so that it can be re-excited. This time is usually at least 200 ms, allowing considerable opportunity for motional artifact to arise from voluntary or involuntary body movements, ranging from peristalsis to scratching an itchy nose.

Fast Low-Angle Shot imaging (FLASH)

In the most commonly employed method of imaging, known as 2-D FT, the nuclear magnetization is flipped entirely into the x' - y' plane by each consecutive 90° rf pulse. x and y gradients are applied; firstly the x gradient, which moves $\underline{k}$ out to a predetermined position k , and then the y gradient, maintained during data acquisition, which sweeps the position of $\underline{k}$ along a vertical line in phase-space. The magnetization is then allowed to recover, and the process repeated. Another transect of phase space is sampled by means of a new value for the x gradient.

To reduce the waiting time between rf pulses, during which no further useful data can be acquired, the FLASH technique uses rf pulses which are considerably less than 90°. The signal amplitude is of course smaller, because only a fraction sinß is available of the magnetization M(0) (where ß is the flip angle used), but this is more than compensated for by the quick ness with which pulses may be repeated. The rf pulse and gradient sequences can be found elsewhere and the sampling trajectory is shown in Figure

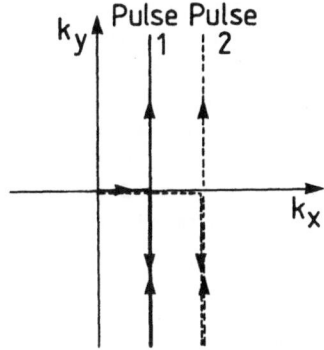

Fig. 5.   Sampling trajectory in phase space for the FLASH imaging sequence.

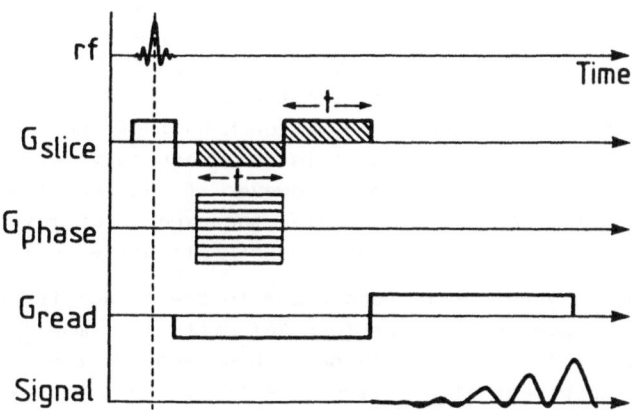

Fig. 6.   FLASH imaging sequence modified to enable diffusion or perfusion
imaging.

5. The data for a complete 64x128 image may be acquired in 1.15s, considerably reducing the danger og motional artifact.

In order to adapt FLASH for diffusion imaging, it is necessary to introduce a diffusion gradient (which is reversed in direction half way through) between the rf pulse and the period of data acquisition. This could be applied as a continuation of the slice-select gradient during the period of phase encoding (Figure 6), if sufficiently high gradient strengths were available, but normally it would follow phase encoding, thus lengthening the total data acquisition time. Even using gradients as large as 5 mT/m, for typical values of D (see equation 15) an evolution lasting about 100 ms is necessary for the effect of diffusion to be readily observable. This would largely negate the effectiveness of FLASH as fast imaging technique.

Echo-Planar Imaging (EPI)

Instead of scanning phase space in several passes, as with the imaging technique just discussed, it is possible, if the gradients can be switched sufficiently rapidly, to obtain complete coverage in one traverse. Following the rf pulse and slice selection, one of the transverse gradients is held constant at a relatively low value, while the other, much larger, gradient is oscillated back and forth as a square wave (Figure 7). Data is acquired during the application of the transverse gradients. The trajectory in phase space followed in one half of the experiment is shown in Figure 8. Here the continuous small gradient causes a steady drift of $\underline{k}$ upwards, while the oscillating gradient sweeps $\underline{k}$ from side to side, the overall effect being a triangular wave approximating to a raster scan. Since the resulting signal appears as a series of echoes produced by the successive partial refocussing of spins in the selected plane, the technique is called Echo-Planar Imaging (EPI).

To ensure equal resolution in both directions, it is necessary to employ pairs of rf pulses, after the second of which the oscillating gradient starts with the opposite polarity to the first. By making the first pulse a flip of 45°, and the second 90°, it is possible to start the second part of the experiment immediately after the first, with no time delay. Thus a complete image may be acquired in 64 ms, each half of the experiment taking 32 ms. To allow the spins to relax it is then necessary to wait at least 200 ms before repeating the sequence. However it has recently been demonstrated that by using smaller flip angles the experiment may be repeated up to 10 times per second, at the espense of signal: noise ratio.

If a diffusion gradient is used prior to the 'read' transverse gradients, it becomes possible to perform EPI diffusion imaging. This gradient must be applied as previously described, and identically in each half of the imaging experiment after slice selection. The total experimental time is markedly increased, increasing the possibility of motion artifact. If the subject's motion during the experiment is nonuniform it becomes difficult to collate the data from each half so that they are properly matched. Provided that this is correctly done, however, diffusion images as described earlier may be obtained by normalizing an image acquired when a diffusion gradient has been applied by division with an image obtained using the same timing but with the diffusion gradient switched off.

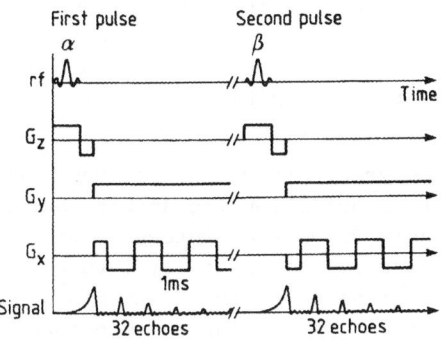

Fig. 7.   EPI (Echo-Planar Imaging) sequence.

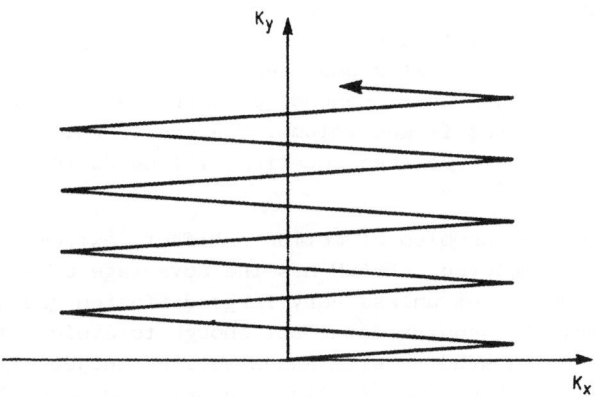

Fig. 8.   Sempling trajectory in phase space for echo-planar imaging.

Blipped Echo-planar Single-pulse Technique (BEST)

A new variant of EPI (Ljunggren 1983; Doyle et al 1986: Chapman et al 1986), which uses a single rf pulse in acquiring data for a complete image, will make diffusion imaging considerably easier. The gradient sequence for this method is shown in Figure 9. The basic idea is that the two transverse gradients, now of equal strength, are on alternately; the first for long square pulses, alternating in sign, which sweep k in successive horizontal transects, and the second for shorter pulses, which move $\underline{k}$ up a short distance in phase space to the position for a new transect. By using low flip-angle rf pulses it is possible to acquire images continuously at a rate of 20 frames per second.

Adding diffusion gradients to this sequence (Figure 10) of course makes such rapid data acquisition impossible, but since all the acquisition is done within a single 32 ms time interval, while diffusion gradient is off, the technique is clearly relatively invulnerable to motion artifact. It is worth noting that although the gradients required to scan phase space during image acquisition are similar in magnitude to the diffusion gradient, they are applied for a much shorter time, so that their interaction with diffusing protons can be neglected.

Cerebral Perfusion

The monitoring of blood flow in the brain is a major activity in neclear medicine studies of tissue perfusion. The NMR imaging techniques which have just been described are likely to contribute further to our understanding of CBF. MRI is a genuinely non-invasive imaging modality, and with fast imaging it is possible to make quick examinations of subjects who may be nervous or even critically ill. The link between the observed diffusion constant and the volumetric flow rate of blood in tissue appears to be sound and reliable, similar values for D arising from a number of fairly realistic tissue models. It seems clear, however, that the precise numerical relationship between D and F depends on characteristic features of the micro circulation inside the imaged volume. The way to which the parameters c and D_O vary with tissue type (see equation (12) needs to be determined experimentally.

Because of the problem of motion artifact, extremely fast imaging methods must be employed. FLASH has the advantage of a relatively simple gradient sequence, but unless very large diffusion gradients are provided it is impossible to acquire data fast enough to avoid artifact (except possibly with brain studies with anaesthetized subjects). EPI in its original form still requires two rf pulses to build up a complete image, and since the diffusion gradient must be applied after each of these, the possibility of artifact again arises, though less severely. The new technique of BEST should ultimately prove of greatest value in the proposed studies. The total time of 130 ms during which subject movement could cause artifact is short enough to capture most body motions, except that of the heart.

Because of the rapidity of data acquisition, the signal: noise ratio of 'single shot' echo-planar or BEST images is no more than about 10 for static magnetic fields of 0.1 T. This is marginally adequate for observation of perfusion but not really good enough to observe the changes in perfusion caused by changing functional activity which have been studied

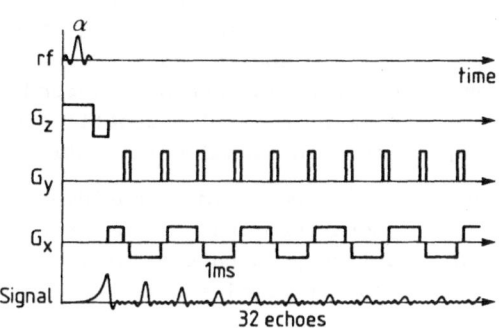

fig. 9. BEST (Blipped Echo-planar Single-pulse Technique) imaging sequence

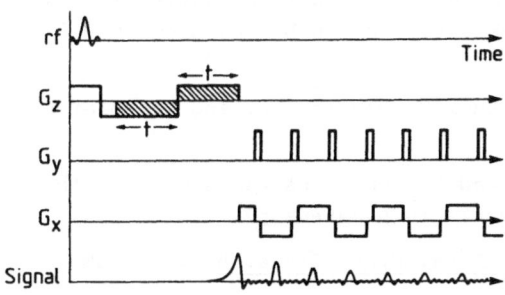

Fig. 10. BEST sequence modified to enable diffusion of perfusion imaging.

by Lassen and co-workers using radioactive tracer methods. However, since even with the delays caused by the application of diffusion gradients it is possible to produce three echo-planar images per second, a simple running average over, say, 100 images, (which should give the signal noise ratio of 100 desirable for such investigations) would not take unduly long. Note that such an average, in real space, does not introduce the intractable artifacts inevitably associated with the averaging in phase space entailed by conventional imaging methods. A fringe benefit in this context of using a running average in the averaging-out of cardiac cycle effects. Such a procedure permits a time resolution of about 30 seconds, which makes feasible experimental studies analogous to those performed with tracers mentioned above. Using BEST makes it possible to image faster, at seven frames per second, albeit with a poorer single-shot signal: noise ratio, but with an improved cumulative signal: noise ratio when a 30-second running average is employed. Finally, an increase in static magnetic field to 0.5 T should result in a five-fold improvement in signal:noise ratio.

In summary, the prospects for real-time, non-invasive MRI monitoring of abnormalities and fairly short term changes in cerebral blood flow have never been brighter. This is likely to have major implications for medical and psychological research, and for clinical diagnosis and follow-up.

Bibliography

Ahn CB, Lee, SY, Nalcioglu O and Cho ZH (1986) Abstracts of the 5th Annual Meeting of the SMRM, Montreal, Vol 1, p102

Chapman B, Turner R, Ordidge RJ, Doyle M, Cawley M, Coxon R, Glover P, Coupland RE, Morris GK, Worthington BS and Mansfield P (1986) Magn. Res. Medicine., in press

Doyle M, Chapman B, Turner R, Ordidge RJ, Cawley M, Coxon R, Glover P. Coupland RE, Morris GK, Worthington BS and Mansfield P (1986) The Lancet, in press

Haase A, Frahm J, Matthai D, Haenicke W, and Merboldt K-D (1986) J. Magn. Res. 67,258

Le Bihan D, Breton E and Gueron M Abstracts of the 5th Annual Meeting of the SMRM, Montreal, works in progress p7

Mansfield P (1977) J. Phys. C:Sol.State.Phys. 10, L55

Mansfield P and Morris PG (1982) NMR Imaging in Biomedicine (Academic Press, London)

McDonald DA (1974) Blood Flow in Arteries (Edwards Arnold, London)

Middleman S (1972) Transport Phenomena in the Cardiovascular System (Whiley-Interscience, New York)

Reif F (1965) Foundamentals of Statistical and Thermal Physics (McGraw-Hill, New York)

Taylor, DG and Bushell MC (1985) Abstracts of the 4th Annual Meeting of the SMRM, London, p612

CONTRIBUTORS

A.M. Baldassarri
NMR Laboratory
IRCCS San Raffaele
20132 Milano, Italy

G.W. Bennett
Brokhaven National Laboratory
Upton, New York 11973

C.A. Boicelli
NMR Laboratory
IRCCS San Raffaele
20132 Milano, Italy

F. De Carli
Institute Neurophysiopathology
University of Genoa
Genoa, Italy

J. Feeney
MRC Biomedical NMR Centre
National Institute for Medical
  Research, Mill Hill
London NW7 1AA, United Kingdom

K.B. Gerald
Department of Statistics
University of Kansas Medical Center
39th & Rainbow
Kansas City, Kansas

Maria Carla Gilardi
Consiglio Nazionale delle Ricerche
Centro Studi Fisiologia
  del Lavoro Muscolare
Via Olgettina 60
20132 Milano, Italy

M. Giomini
Istituto di Chimica Generale
  dell'Universita'
00185 Roma, Italy

A.M. Guliani
ITSE-CNR
00010 Montelibretti (Roma), Italy

Laurance D. Hall
Laboratory for Medicinal Chemistry
Addenbrooke's Hospital
Cambridge CB2 2QQ, United Kingdom

Giovanni Lucignani
Consiglio Nazionale delle Ricerche
Centro Studi Fisiologia del Lavoro
  Muscolare
Via Olgettina 60
20132 Milano, Italy

S. Marenco
Institute of Neurophysiopathology
University of genoa
Genoa, Italy

J.H. Matis
Department of Statistics
Texas A&M University
College Station, Texas

G. Mies
Max-Planck-Institut für neurologische
  Forschung
Abteilung für experimentelle
  Neurologie
Ostmerheimer Strasse 200
D-5000 Cologne 91, West Germany

Edgar Müller
Siemens AG, UB Med.
Henkestrasse 127
Erlangen, West Germany

G. Novellone
Institute of Neurophysiopathology
University of Genoa
Genoa, Italy

Aldo Rescigno
Institute of Experimental and
  Clinical Medicine
University of Ancona
Ancona, Italy

G. Rodriguez
Institute of Neurophysiopathology
University of Genoa
Genoa, Italy

G. Rosadini
Institute of Neurophysiopathology
University of genoa
Genoa, Italy

L. Sokoloff
National Institute of Mental Health
Laboratory of cerebral Metabolism
Bldg. 36, R. 1A-05
Bethesda, Maryland

A. Todd-Pokropek
Department of Medical Physics
University College
London, United Kingdom

R. Turner
University of Nottingham
University Park
Nottingham NG7 2RD, United Kingdom

Steven C.R. Williams
Laboratory for Medicinal Chemistry
Addenbrooke's Hospital
Cambridge CB2 2QQ, United Kingdom

INDEX